Author's impression of a gravitationally propelled spacecraft hovering in a lunar crater.

UFOS AND ANTI-GRAVITY

Leonard G. Cramp

Adventures Unlimited Press

Other books by Leonard G. Cramp:

The Cosmic Matrix
The A.T. Factor
Space, Gravity & the Flying Saucer

Other books in the Lost Science Series:

The Anti-Gravity Handbook
Anti-Gravity & the World Grid
Anti-Gravity & the Unified Field
The Fantastic Inventions of Nikola Tesla
The Free Energy Handbook
The Tesla Papers
The Time Travel Handbook
The Energy Grid
The Bridge to Infinity
The Harmonic Conquest of Space
Tapping the Zero-Point Energy
Quest for Zero-Point Energy

UFOS AND ANTI-GRAVITY

Leonard G. Cramp

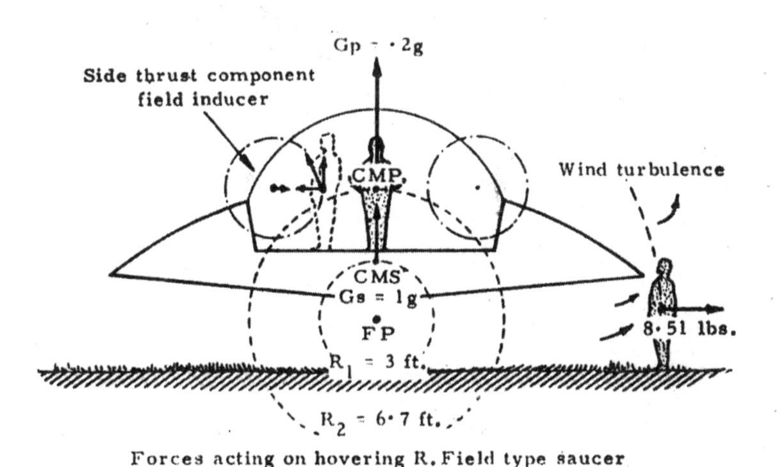

Forces acting on hovering R. Field type saucer

UFOs & Anti-Gravity
New Expanded Edition

ISBN 978-1-939149-56-5

Published by Adventures Unlimited Press
One Adventure Place
Kempton, Illinois 60946 USA

www.adventuresunlimitedpress.com

This edition includes material from
Space, Gravity & the Flying Saucer

Printed in the
United States of America

To my wife Irene, without whose untold patience this work would never have been possible.

To little Susan and Jane. I am grateful for the opportunity not to leave them out.

To Dad and Alf, and a dear friend from another place . . . .

Acknowledgements

If my attempt to record some of the technical evidence for the existence of visiting space craft has been accompanied by a sense of inadequacy, then so also is my wish now to thank those who have so willingly assisted me.

It is with much gratitude that I offer my thanks to Mr G. Bainbridge who virtually made this book possible, may his good faith be in some measure rewarded when he reads it.

Fred Smith, Sam, John, Dave and the members of the Isle of Wight UFO Investigation Society, who are doing a wonderful job and who have listened so patiently to my theorising.

I wish to thank Mr and Mrs Rampton (sorry about the custard), Yvonne, without whose help I could not have met the publishing date, the various Peters (too many to quote) and the other good friends who kindly assisted me.

Phil Fricker for the cold trips to photograph my test rigs and his quiet patience (and amusement) when they frequently did not work.

Antony Parker for the re-use of the abridged version of his wonderful 'Unity of Creation' theory which will always be an inspiration to me.

Charles Bowen for his kind help and co-operation for the use of material from 'The Flying Saucer Review', and the directors of NICAP for quotations from the 'UFO Investigator', and Antoni W. Szachnowski, Chairman, Anglo-Polish UFO Research Club, for quotation in Chapter 8.

Mr R. Mockford and the staff of Sir Joseph Causton & Sons for achieving next to the impossible in getting the book completed in such a short time.

I would also extend my grateful thanks to The Bodley Head for quotations from my previous book. The directors of the magazine 'Aeronautics' for quotations from some of my articles and those of the late A. R. Weyl, to whom I am deeply indebted, and Vauxhall Motors Ltd. for their kind co-operation.

In addition I would extend my thanks and acknowledgements to the following concerns for their prompt encouragement and use of photographic plates.

Plate

1 North American Aviation Inc.
2, 3 } Flight International.
4 Howard Levy.
6a Associated Press.
6b, 8 } Interavia.
9 Crown Copyright, Science Museum, London.
10 Howard Levy.
15 General Dynamics, Convair Division.
21 Mrs Birch.
32 NICAP.

CONTENTS

List of Plates

Foreword

AT the time of going to press, it is twelve years ago since Stephen Darbishire photographed the Coniston saucer. Twelve years since the writer first examined and correlated the Darbishire and Adamski photographs by orthographic projection in *Space, Gravity and the Flying Saucer*.

At that time the scientific world was clamouring for 'just one tiny shred of scientific evidence' to justify the existence of flying saucers. 'Show us just one' they said, 'and we might begin to think there is something in it'.

The analysis of these two photographs was in the strictest sense scientific, and the resulting conclusions were fair and unbiased. Yet although these offered something a little better than the 'tiniest shred of scientific evidence', for the only alternative amounted to a world-wide conspiracy, the conspicuous silence which followed, both in the national dailies and the scientific press, left no doubt as to their interest. Perhaps it was simply a case of the lay public press not being able to understand, despite my attempts to portray the claim simply. Maybe it was out of sheer scientific aloofness that the technical press chose to ignore it. But the fact remains—the analysis did not fail to impress all those who read it. And further, the claim is just as valid today as it was then; it still stands up to sensible consideration.

Stephen Darbishire was nearly fourteen years old, when, accompanied by his eight year old cousin Adrian, he photographed a saucer hovering near Lake Coniston.

Then, over a decade later, the scene was almost exactly duplicated in the Sheffield area. Although on this occasion the players have been changed, the circumstances remain strangely the same. This time, another small boy, fourteen year old Alex Birch and his friends were chosen for the principal parts. The circumstances of the event which bear an almost uncanny similarity to the Coniston sighting, have already been published elsewhere.

It is interesting to note that, although it was Stephen Darbishire who had the camera and subsequently photographed the saucer, it had been his small cousin Adrian who had first spotted the object and had drawn Stephen's attention to it.

So with their more recent counterparts near Sheffield. It was Alex Birch who had the camera and took the picture, but his young friend

Stuart Dixon had first seen the saucers and excitedly pointed them out to Alex and his other friend David Brownlow. Neither does the pattern end there, for once again there has emerged far more than a 'tiny shred of scientific evidence', analysed in these pages, which is difficult to refute or explain away as mere coincidence.

Once again the writer has been called upon to play a small part in this up-to-date play, and time has forged our metal a little sharper. Indeed, so much so, that the technically corroborative evidence for the flying saucer is set out in the following pages in the form of an open challenge to the scientific sceptic.

Over the last seventeen years or so, I have become increasingly convinced that flying saucers, among other things, are extra-terrestrial space ships powered by a form of gravitational manipulation (g field) the fundamental concept of which was set out at some length in *Space, Gravity and the Flying Saucer.*

The dual purpose of this subsequent book is to reconsider the 'G field theory' in terms of more recent sightings and to offer evidence of a mechanical nature for the consideration of both the layman and the technician alike. To this I would hasten to add, that those who might hope to find the know-how of 'anti-gravity' will not find it in these pages, for obviously a scientific break-through of such magnitude could hardly find its way into a book such as this. But I assure you, herein you will find many, many clues, while allowing for such a technique to be realisable, the reader will find accompanying engineering problems which dramatically supports many flying saucer witnesses' claims. It is accepted of course that some of these facts will be more acceptable sometime in the future than they are now.

Right from the beginning I would like to make one thing quite clear. After many years of study, I do not know where flying saucers are coming from, or why they are coming, though I have a suspicion why officialdom chooses to keep the public ignorant of this truth. But I am certain they are coming, as I am certain any unprejudiced person will be if he studies the facts. Naturally we can all be prolific with theories and the ideas expressed in the following pages are also theory and I would add that although mechanistic in conception, I think may be the correct ones.

To offer some of the technical evidence for the existence of flying saucers to the lay public is no mean task, for, however much we try to simplify, it still remains technical. Yet I honestly believe, even the most untrained person will be able to identify the pattern herein outlined. A pattern which is there for the finding. I am merely pointing the way, the theory itself requiring little more than an understanding of the inverse square law. Because of the valuable evidence frequently left after landings, much of the information herein was taken from the French records of that epic year of 1956/7. But elsewhere the reader will find on independent investigation similar evidence in abundance to endorse this testimony.

This approach to the subject is deliberately dual in nature and for a dual purpose.

Dual Nature

1. To make a brief appraisal of current developments and general state of the art of aero-astronautics and to draw conclusions on where this is taking mankind.
2. To review some technically corroborative evidence among flying saucer sightings and incidents taken from a hard core of reliable cases, to draw conclusions from this and compare with the previous conclusions.

Dual Purpose

1. To offer information to flying saucer enthusiasts which may help to designate what such craft are not, and what it may all imply.
2. To offer generally to the public evidence of a kind which has not been heretofore presented, to verify the existence of visiting space ships called Flying Saucers.

In the eighteen years or so since Unidentified Flying Objects were first brought to the attention of the general public, much has been said about Flying Saucers. Books have been written, groups have been formed, and an ever increasing number of ordinary people are becoming convinced that 'there is something going on', and despite talk to the contrary, reports of sightings are just as numerous now as ever they were, no matter how valid they may or may not be. In fact, since the publication of *Space, Gravity and the Flying Saucer* in 1954, several important factors have emerged which have caused a great deal of controversy among sceptics and the followers of 'UFOlogy' alike. Furthermore in presenting the evidence, I am acutely aware of two radically opposed points of view which has prompted me to be pertinent to volunteer as referee.

There have been books on the origin of Flying Saucers, books debunking Flying Saucers, books on contacts with visitors from Flying Saucers and a whole host of conflicting and often fascinating literature. But paramount among this debris of confusion stands one solitary and concrete fact which I have ventured to put forth in all humility and tolerant goodwill. It is this. It is becoming all too apparent that there exists on both sides of the 'camp' a great deal of intolerance and prejudice which is never worthy of any kind of scientific investigation. We are—by now—most of us acquainted with the ostrich technique of the interplanetary flying saucer sceptic, and unfortunately, the indifference of the so-called scientific world at large. But equally there are students of UFOlogy who will not, or cannot, appreciate that mankind has to crawl before he can walk. The modern aircraft and the rocket may by comparison seem clumsy and 'brute force-ish', but in their own right they represent near miracles of engineering achievement which should be given their due nevertheless. It must be stressed therefore that the inclusion of the first three chapters is intended primarily to establish a comparison of the present 'states of the arts' without which much confusion is apt to arise.

It is from this standpoint then that I would ask indulgence as we review and attempt to correlate, step by step, some of the relevant issues of this most important enigma of our times. But by all means let us keep to the facts, then maybe we shall have earned the right to further romantic speculation.

In the following pages evidence is re-examined which indicates all too clearly, that if there is the remotest possible chance of a better way for mankind to identify himself with the infinitude of the cosmos, then we should seek it. It is the author's sincere belief that the accumulation of this evidence illustrates a pattern which clearly exhibits unchallengeable proof that there is a better way. It is for us to be, as it were, unprejudicial judges at a scientific hearing, while endeavouring to remember that if this is a degree we cannot attain, then by all the truth in the meaning of the word, let us gracefully leave the court room, for true scientific pursuit of *any* portal, has no place for scientific snobbery of any kind.

Leonard G. Cramp.

Isle of Wight.

PART ONE

The beginning
of
a journey

1

Crossroads of Aerodynamics

AT the outset the inclusion of the subject matter of this chapter may seem to have little to do with flying saucers or unidentified flying objects, dealing as it does with the classical development in aero-astronautics, at best it may appear to be only a brief appraisal of man's present attempts to fly faster and higher in the gaseous bubble he calls the atmosphere. But it is set out here so that we might envisage a limit to feasible aerial travel as we know it today, at what point development may finally stop and into what exciting avenues it may lead designers of the future. At this point I would only ask indulgence towards flying saucers while we examine for ourselves where modern development is taking us. Moreover it is hoped this and the following two chapters may help to give a little insight to the laymen who otherwise might often and quite understandably make erroneous deductions concerning some aerial phenomena.

If by considering this information the reader arrives at the gravitational threshold and the gravitational space ship it suggests, then the objective of this book will have been partly realised. But if in addition you are guided to an acceptance of visiting space ships to this planet, then it will have been fully justified.

In all scientific pursuits, as in nature, there are signposts available for the guidance of the individual who takes upon himself the task of exploration and in the science of aero-astronautics he will find no exception to the rule.

Therefore in order to lend a little colour to this story which has been set out in the form of an enquiry, may I suggest that from now on we imagine ourselves as travellers on a scientific exploration into unknown country, bearing in mind that the journey before us may have been trodden by others, long, long ago, and we have to look for clues left by them, or natural signposts to guide us on the way. In the course of this book we shall see plenty of them, they might be regarded as markers down the exciting and unknown avenues which may one day lead mankind on his ultimate journey, a fantastic journey through time and space to the distant stars.

We begin our story, not in the days of Wilbur Wright and his brother, but in the present, with an up-to-date aeroplane, for we are here to visualise the end of an era rather than the start of one.

There was nothing very special about the morning of 17 July, 1962. People slept, people dressed and prepared themselves for another day at the office, the factory, the shoe shop, and the hundred and one things ordinary people do on an ordinary day. Yet in a way it was a very special day, for an earthman was about to qualify as an astronaut—flying an aeroplane.

On that particular day at Wendover Air Force Base, California, Major Robert White of the U.S.A.F. was strapping himself into the confined and profusely instrumented space called the cockpit of the world's fastest, highest flying machine, the X-15. As far as the Major was concerned it was to be a day like any other, a little more exciting perhaps for he was to push the little craft 'over the top'. But it did not work out quite so uneventful—there was something else.

Before we re-acquaint ourselves with this, by now well-known occasion in the field of aero-astronautics, we shall benefit later on in these pages if we take a brief look into the background of what has been called the most thoroughly tested man-machine system for astronautical research, the North American X-15.

It is an aircraft only in the strict sense of the word, for with a fuselage length of some 50ft, and a wing spread of only 22ft, certainly the X-15 resembles a streamlined dart rather than the hypersonic aeroplane she really is. Primarily designed as a research vehicle for high speed missions of more than 3,600 m.p.h. and altitudes above 100 miles, the X-15 is a rocket powered, sleek looking little craft of immense power.

Prodigious fuel tanks serve to form the major part of her belly and her stubby wings have been formed with solid leading edges, machined from a special alloy, known as Inconel X, which resists the intensified aerodynamic heating through friction the little aircraft suffers when re-entering the denser regions of the earth's atmosphere.

The X-15, the machine which was designed to meet the challenge of launching a human being from earth into space with a glide return journey, is the result of a national effort managed jointly by N.A.S.A. (the United States National Aeronautical and Space Administration), Air Force and Navy. In the Spring of 1952, the National Advisory Committee for Aeronautics gave their orders to its laboratories, 'to study the problems likely to be encountered in flight beyond the atmosphere and recommended methods to explore these problems'.

The result of these labours was a decision in favour of an aeroplane and in December 1955 a contract was placed with the Los Angeles division of North American Aviation Inc., to carry out basic research, development and manufacture of three aircraft.

October 1958 saw the completion of the first, and in March 1959 the machine had its first captive flight. In appearance it differs from most other conventional high speed aircraft chiefly by the extraordinary small wings and the fin which has a surprisingly thick trailing edge, Plate 1.

On re-entering the earth's outer tenuous layers of atmosphere, the aircraft could become unstable, therefore the wedge shaped fin has been designed to act in a similar way to a small parachute—it drags the machine back a little and maintains its weathercock stability.

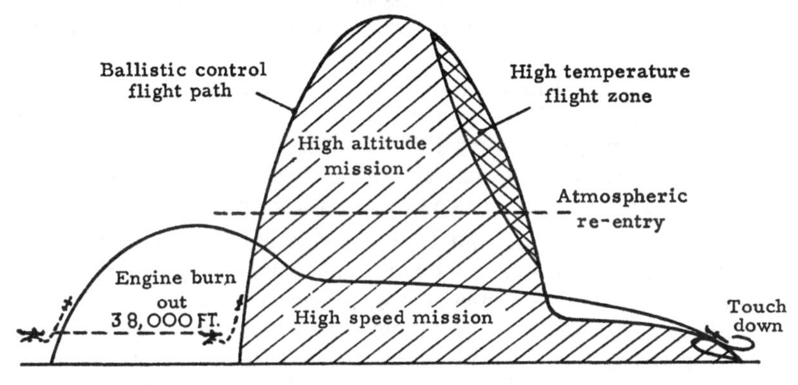

Fig 1. *Basic flight programme of the* X-15.

The aircraft has two basic flight programmes planned well ahead, Fig 1. Research into high altitude ballistic flight and pure high velocity flight and associated problems such as aerodynamic heating of the machine as it plummets back through the atmosphere, and physical and psychological effects on the pilot. We shall hear more about these effects when dealing with the Flying Saucer later on.

The North American Co., planned to carry out a total of twenty test flights before handing the aircraft over to the U.S.A.F. and N.A.S.A. Repeated use was made of the large room sized flight simulator which the manufacturers had developed to simulate and check all the control functions in the space aircraft. More than 2,000 such 'flights' were made by Scott Crossfield, North American Test Pilot and others on this electronic monster in order to familiarise themselves with the X-15's whims before they finally took her into the air.

The test pilot wears a fully pressurised space suit and special means are provided to keep the cockpit temperatures normal. In the event of an emergency in space, he will try to stay with the aircraft through the re-entry stage, then using his ejector seat, he will eject himself from the machine. Should the velocity of the X-15 still be very high, the ejection seat has been fitted with a small plate which creates a shock wave in front of the pilot, thereby offering him some protection against the supersonic blast he will be subjected to. The ejection seat has also been fitted with small fin-like stabilisers which will keep the seat and pilot on a straight flight path, until the parachute deploys.

The X-15 is the first research aircraft to be fitted with three different control systems; an ordinary conventional control column between the pilot's knees, a second aerodynamic system fitted with a miniature control column for wrist operation when high accelerations are encountered, and finally a ballistic control system, also fitted with a short control column which operates jet control nozzles fitted to fuselage and wing tips. These are used during the ballistic phase of the flights, where at extreme altitude the outer layers of the atmosphere are so tenuous as to render ordinary control surfaces inoperable. In order to control or re-orientate the machine, the pilot operates the small column which controls the jets so that the wings are virtually pushed up or down as he wishes. Note, we shall observe similar stabilising techniques elsewhere as we progress, meanwhile the reader is asked to retain it in mind.

As the speed of descent from extreme altitude increases, the plummeting X-15 begins to get hot through air friction and the pilot must guide her back into denser atmosphere straight and true, otherwise the machine may suffer an extreme buffeting which may cause it to break up. At a predetermined height, the pilot begins to pull the machine out of its screaming dive, and but for the 'anti G' suit he is wearing he would black out in the process. Even so, the majority of untrained people could not tolerate pressures of the magnitude he has to withstand.

But we must now return to that day of 17 July when Major White piloting the X-15, was at the top of his climb. The height was 58 miles, making him the first man to qualify as an astronaut by flying a winged craft, and the fifth man eligible to wear the United States Space Wings. But at that moment such thoughts were far from his mind, for Major White dramatically reported over his radio, 'There is a thing out here, there absolutely is!' He said later, 'I have no idea what it could be. It was greyish in colour and about 30 to 40ft away'.

In this respect it is perhaps appropriate therefore to begin our story with the X-15, for it will be UFO reports like this, given by men like these, which will ultimately help to smash the official silence about extra-terrestrial visitors. An isolated case perhaps? Hardly, for another pilot of the X-15 reported a sighting which very much upset the flying saucer applecart. This time it was Joe Walker, but he had a film to prove it. When showing his film to the Second National Conference on the peaceful uses of Space Research in Seattle, Washington, U.S.A., he said 'I don't feel like speculating about them. All I know is what appeared on the film which was developed after the flight'. The altitude had been 50 miles above the earth.

The objects numbering about five or six, appeared to be cylindrical or discoid in shape and Walker admitted that this was the second occasion on which he had filmed UFOs in flight. The camera was mounted in the rear of the X-15 and the objects appeared as he reached the arc of his flight and began heading back for earth.

A later communique from N.A.S.A. stated that the objects had since been identified as ice flakes breaking off the aircraft. We shall hear more about chunks of falling ice as the story unfolds.

As this book goes to press the latest information on the X-15 rocket aeroplane says that in addition to heights of 67 miles and speeds of equal to Mach 6.3 (about 4,100 miles per hour), the little craft had made important contributions to the subject of man's ability to exercise his powers in space and to the study of manned flight generally.

Now her designers are taking steps to develop a ramjet to fit under the rear part of the fuselage. With this she will be capable of speeds up to Mach 8!

With the X-15 man has met the challenge of pricking the gaseous bubble which surrounds the globe. It is a great pity that such an achievement has not received the publicity the rocket shots have received, for the dart-like aeroplane will ultimately do all that a capsule can do, and more, for it can be brought back to base fully controlled.

But space in this chapter is running out, so we must leave the X-15 feeling a little satisfied we have given her some of the public attention she and her team deserve, and return now to developments in nearer to earth aircraft. What for instance of the huge transporters, the service and civil aircraft?

Well, in a matter of ten years or so, flying speeds of civil and service aircraft have increased by no less than 50 per cent and their speeds are still going up. But one of the associated problems confronting the aircraft designer of today is keeping the landing and take off speeds correspondingly down. This is usually achieved only by a compromise in the design requirements and to some extent, cruising speed suffers most.

In the case of the military aircraft of course the compromise is not so rigidly adhered to, therefore as cruising speeds increase, so landing speeds are apt to get frighteningly higher and higher. A present day combat aircraft for instance will touch down at something like 150-200 knots, while the supporting wings have been reduced in area and thickness, so drastically, that it is often difficult, if not impossible for designers to find sufficient space to accommodate the undercarriage, as has been the practice in the past.

In order to match these increased touch down speeds, runways have become prohibitively longer, while a great deal of research work has been done into the problem of checking the speed of an aircraft after touch down. In the case of the older propeller driven aircraft, this was achieved with the aid of the reversible pitch propeller such as those employed in the Britannia, Vanguard and the Lockheed Electra.

The introduction of the pure jet brought its own particular problems and one solution was to install baffle plates or deflectors, which in effect reversed the thrust of the jet efflux, thereby reducing the landing run of the larger aircraft. But this offers only a partial solution to the problem and at best will offset the main difficulty a little longer.

So we arrive at the first signpost in modern aeronautical science, for there must be a limit to runway length as speeds get even higher. Therefore quite clearly aircraft must be designed to take off and land in shorter distances and ultimately, vertically.

In this respect the helicopter type aircraft wins hands down, but on the debit side of the comparison, it is drastically handicapped by forward speed limitations, which restricts its employment to specialised fields of application. Therefore designers are being encouraged to seek solutions in other aerodynamic avenues.

This has given birth to aircraft with swivelling wings, swivelling propellors, ducted fans and downward facing jets. All of which are fundamental approaches to the same problem, that of getting an aircraft off the ground vertically. The advent of this approach to the problem has introduced new terms into the aircraft industry, VTOL (Vertical take off and landing) and V/STOL (Vertical/short take off and landing).

The Doak experimental aircraft built for the army by the Doak Aircraft Company and powered by an 850 h.p. Lycoming T53 gas turbine in the centre of the fuselage, was a typical example of this type of aircraft, Plate 4. For take off, the ducted fans are swivelled about the longitudinal axis of the short fixed wing, so that the thrust is absorbed in a downward direction. Once into the air, the fans are tilted a little forward so that they give a slight forward thrust component to the aircraft.

As the machine moves slowly forward a small amount of lift is generated on the wings which gradually builds up as the fans are lowered further. Lowering the fans still more produces a greater forward speed, until finally all the thrust from the propellers is absorbed in a longitudinal direction and the aircraft now moving horizontally at normal speed is fully supported by wing lift.

Most of the aircraft being built to investigate the problems of vertical take off and landing have a common design difficulty, the transition from the hovering condition to the normal forward flight pattern. That point where the wings are at too steep an angle to offer useful lift from forward motion and the point where the aircraft might lose lift from the fans when they are being lowered down. This is the transition period from fan lift to wing lift which is presenting designers with many new and difficult problems.

As propellers and ducted fans are not suitable forms of propulsion for high speed aircraft, downward facing and swivelling jets are also being used for vertical lift. This was the principle employed with the British Rolls Royce 'flying bedstead' and then later, the Short S.C.1. Plate 2. This aircraft was successfully flown both as a VTOL aircraft and as a conventional airplane off the runway. It was also the first VTOL aircraft to be put through the transition manoeuvre successfully. In fact her test pilot Tom Brooke-Smith of Short Bros., claimed that there was no change at all in the behaviour of the machine. He said that it was as easy as getting into a car and driving away. The S.C.1. rose vertically, gained speed and transferred from jet lift to wing lift so smoothly that the uninformed passenger would have been quite unaware of the rather wonderful feat of aeronautical engineering taking place.

Experimental aircraft like the S.C.1. were the forerunners of the high performance aircraft now being developed. For example the Hawker P.1127, VTOL fighter aircraft which employs a fixed turbo-jet, fitted with turbo fan and swivelling jets for take off and forward flight, Plate 3. Hovering flights on this machine commenced on 21 October, 1960 and were successfully completed. The Griffiths supersonic VTOL 'dart' air. liner designed by Dr A. A. Griffiths of Rolls Royce, Plate 5, was to employ a multi-banked lift engine system for take off and landing, and forward facing turbine ramjets for transitional flight. In this configuration the fuselage of the aircraft virtually forms the lifting wing itself.

The vehicle would be capable of speeds around Mach 2-3. With the exception of this and a few similar designs, many VTO projects are somewhat penalised in the vertical take off rôle by the fact that they have to lift the weight of their wings.

This may yet prove to be one of the final radical developments in the history of the aeroplane as such, for already plans are being laid for a type of aerial vehicle which can hardly be termed aircraft at all. In fact one can seldom browse through an aeronautical magazine nowadays without coming across several of these unusual futuristic configurations.

Topping the ever growing list is the work being done on a new concept called the 'Aerodyne' by Dr Alexander M. Lippisch of Cedar Rapids, Iowa. Dr Lippisch, designer of the famous war time Me.163 'tailless fighter', suggests that the modern aircraft is after all only a powered glider and that the drag penalty suffered by most modern winged aircraft is largely a needless one. He points out that a wing is a means of deflecting a relatively large mass of air downwards, which in turn gives an upward component of reaction. If the propulsive efflux—be this a high velocity stream from a thin jet, or a low velocity stream from a large propeller—is partially directed downwards through a duct, then the nett result can be comparable to a wing and the ordinary wing becomes obsolescent, Plate 6(a) and (b). This is of course the principle expressed in simple terms, it becomes far more complicated in actuality.

The aircraft, as we know it, might well disappear therefore, and its counterpart of the future may look something like a streamlined shell which houses the payload and power plants, with jet streams facing rearward and downwards, to both propel and help support it in mid air.

Of course, with less drag, such a vehicle can attain far higher speeds, in fact its efficiency increases with forward speed. This gain can either be used to transport payloads cheaper, or faster, or both. But if past experience is any guide, it will almost certainly be ultimately employed 'to get there faster'.

Other prominent designers visualise the day when aircraft speeds permit the use of pure ballistic flight in the rarefied atmosphere at high altitudes, in which centrifugal force generated by the trajectory of such a craft, will virtually replace aerodynamic lift for the major part of the journey, then the combined effort of extendable wings and jet lift will bring it safely down at its destination.

If in this particular rôle centrifugal force might offer some assistance, it can be a positive disadvantage in another, for because of it, manoeuvrability of high speed combat aircraft is seriously restricted. An idea of the stress magnitudes generated by centrifugal force can be gained by remembering that a modern fighter aircraft flying around Mach 2, must turn in a circle no tighter than *three miles radius* in order to comply with the regulation 6g limits imposed, and in order to withstand this, the pilot must be wearing an 'anti-g suit' or he would black out.

As there is a limit to the altitude in which an air breathing engine will function and as hypersonic aerodynamic heating can become a formidable problem which may inhibit further increases in speed at lower altitudes, we might be justified to pause a while and ask the question 'whither now?' Will we in fact be able to fly aerodynamic passenger aircraft faster, and higher, or are there other factors which will set even greater limits? Indeed, at present, it rather looks as if the latter will prove to be the case, for there are two other major problems which we have not yet considered—the sonic boom and cosmic radiation.

Although taken for granted by many, the sonic boom may still yet prove to be one of the greatest insoluble limitations of the aeroplane. For instance, as a measure of its severity, a 180 ton Mach 2 or 3 airliner flying at 70,000ft, perhaps carrying 100 passengers or so, would sweep across the Earth's surface with a thunder-like noise along the entire supersonic flight path, often rattling and breaking windows in its wake and awaking people from their sleep within a band disturbance of some 70 miles wide! More recently it has been estimated that the Concord airliner flying overhead will probably sound like thunder at five miles range.

On this possible future nuisance, no nation has yet decided its policy, but they must very soon.

One leading expert has been quoted as saying, 'Never before in history would so many have been disturbed so much, by so few'.

We might ask the question, 'is it justifiable that millions of people the world over in populous areas, should have disturbed sleep, so that a relative handful of passengers might gain an hour or two in some other place on our globe?' Clearly the answer is no. For there could be no deterrent, no moving away from the intolerable nuisance as one might from a noisy railway shunting yard. We should have the unbelievable task of having to learn to live with it. We shall see later that other kinds of research may help to solve this problem in a revolutionary kind of way, down another road.

Perhaps Air Marshal Sir Victor Goddard was strolling along with us for a while, when in the 'Shell Aviation News 1959', he said, 'Is there no royal road to excellence in all the essentials of economic long-range aviation? Can progress towards an ever receding summit of economic excellence only be gained by those who, following well-trodden paths as far as they go, reserve their courage and energy for the later part of the ascent where they aspire to demonstrate their own particular contribution to progress? Must we continue for ever to beat the air as though it were the enemy of flight? Or, is it possible that all paths have hitherto been leading to a barrier that cannot possibly be economically scaled, but may be avoided by abandoning not the ideal itself but the latterly-accepted paths? All history shows that when the irresistible force of the spirit of man meets an immovable body of technical opposition, the spirit of man eventually overcomes his pride and the dilemma by going round the obstacle by a new route'.

We may now ask, what is this new route? An aircraft is essentially a device which obtains support by air displacement. The bulk of this air mass displacement and the velocity at which it is displaced are the limiting factors. So we go on improving design, making aircraft less susceptible to aerodynamic heating and drag as velocities get higher, but somewhere along the line we can see the end. Somewhere along the road where designers contemplate such limits as protecting aircraft surfaces by magnetic shields to repel the ionised searing gas which, split seconds ago was the earth's atmosphere. A typical example is seen in Plate 7. Problems like this are not so very far off—we are beginning to see the turning point.

The possibility of finding a solution to the problem in some major advance in aerodynamics is remote indeed, for while the very nature of our airborne vehicles demands that they rend and push their thunderous way through the air, so there must be a reaction and that reaction is experienced by human beings as noise.

If the sonic boom caused by future airliners is likely to prove a major problem for the earth's slumbering inhabitants, the passengers of such vehicles may themselves be subject to an even greater hazard, that of cosmic radiation. For it is now well known that at 70,000ft altitude —at which such aircraft will probably operate—the so-called total ionisation due to cosmic rays can reach a maximum. The intensity in this region is often 200 to 300 times stronger than at the earth's surface, and the primary cosmic rays (chiefly extremely energetic protons) can penetrate to this level.

During solar storms, such radiation can intensify greatly, accompanied with an abundant production of neutrons where the rays encounter solid matter like an aircraft. On the possible harm caused to passengers who might constantly fly at such altitudes, we can only be guided by the experts. But even the experts are in disagreement. Some are inclined to believe that there will be no danger at all, while others are concerned at the possible cancer risk—and with the younger passengers—long standing genetic effects of radiation. Certainly these are the limitations which must be tackled first and in tackling these we may well find fresh and more hopeful avenues to explore. We might find down these avenues signposts which clearly indicate alternative and exciting approaches to our problems. We may find that harmful radiation, air drag and sonic phenomenon can all be by-passed quietly and without fuss, down another road.

To the Flying Saucer sceptic, may I repeat once more the inspiring words of Sir Victor Goddard, 'Must we continue for ever to beat the air as though it were the enemy of flight?'. We might find that there is an alternative, and what is more, the technically corroborative sightings of UFO behaviour which we shall examine later on, suggests that others may have achieved this very alternative and are employing the technique to visit us now!

2

Aerodynamic Saucers

THE material in this chapter is offered to both sceptics of flying saucers and believers alike, for its primary intention is to show some of the things that saucers are definitely not. For despite some of the excellent aerodynamic qualities of disc type aircraft, it must be understood such designs are subject to all the aerodynamic limitations discussed in the previous chapter.

It is of course natural when attempting to diagnose certain phenomena, to interpret the unknown in terms of the known, but caution must be applied diligently to avoid unnecessary misinterpretation. In fact one might say that one of the natural stages in acceptance of extra-terrestrial flying saucers is the aerodynamic interpretation, which, if there appears to be evidence to substantiate it initially, there is more to dismiss it later on. In order to present more fully some of the available information on disc aircraft, it is necessary to look back in time, but first a more up to date example.

Several years ago a small gathering of people stood watching as a test pilot climbed into the cockpit of a small prototype 'flying saucer'. Mechanics busied around the sleek looking craft for a few minutes preparing her for a test flight. Barely audible, from within the internals of the vehicle, a soft whine murmured, gradually rising in intensity and pitch. Around the perimeter of the craft a slight shimmering haze was seen. Then slowly the dark shadow underneath became deeper and deeper, until a line of daylight separated it from the circular vehicle, testifying that it was now airborne. As the shining craft lifted, it turned slowly on its axis, reflecting shafts of light from the early morning sun.

Watching intently, the onlookers saw the vehicle lift higher and higher, then gracefully move off to circle the airfield. Suspended beneath could be seen the three dark blobs which was her undercarriage and as the pilot manoeuvred the craft, a semi plan form view indicated nothing in the way of wings, fins, or control surfaces. A dark space in the centre and a peripheral slot were the only distinguishing markings. Apart from that the craft resembled a rather streamlined, perfectly circular doughnut of polished metal.

This particular account is not fictitious, neither is it an account of a futuristic film set. It is in fact a brief review of a test flight made on the AVRO Canadair aerodynamic flying saucer, originally designated 'Omega' and now more officially known as the AVRO 'Avrocar'. This particular aircraft—on which the author was allowed some pleasing speculation in *Space, Gravity and the Flying Saucer*—was originally designed as an experimental vehicle for the Canadian Government in 1953, then before the design work on the project was completed, it was quietly and mysteriously cancelled. Rumour had it at the time, that the real reason was, 'there had been a major break through in anti-gravitics', rendering the project redundant. On the validity of this report and others like it, there seems to be no substantiation. As it was, the American Government retrieved the drawings of 'Omega' from the aeronautical waste paper basket and work on the project was finally recommenced under contract for the American Defence Department. Even so, this machine although extremely advanced, was an aeroplane in the strict sense of the word, Plate 8.

No matter how sceptical about flying saucers some aeronautical engineers may be, continued reports of UFO sightings sooner or later stirs many a hushed little pipe dream into activity, and with pencil and slide rule doing overtime, the aerodynamic merits of the flying disc is once again born.

From the hardcore of reliable UFO sightings, the Topcliffe R.A.F. Station incident offers an admirable case in point.

A strange thing was seen by two R.A.F. officers and three aircrew while standing near Coastal Command Shackleton Squadron H.Q. at Topcliffe one day in November 1953.

They had just landed after a flight and were watching a Meteor coming in to land at the neighbouring Dishforth R.A.F. Station.

One of them, Flight Lieut. John W. Kilburn, 31, of Egremont, Cumberland, then spotted 'something different from anything I have ever seen in 3,700 flying hours in a variety of conditions'.

'It was 10.53 a.m. on Friday. The Meteor was coming down from about 5,000 feet. The sky was clear. There was sunshine and unlimited visibility.

'The Meteor was crossing from east to west when I noticed the white object in the sky. This object was silver and circular in shape, about 10,000 feet up, some five miles astern of the aircraft. It appeared to be travelling at a lower speed than the Meteor, but was on the same course.

'I said: 'What the hell's that?' and the chaps looked to where I was pointing. Somebody shouted that it might be the engine cowling of the Meteor falling out of the sky. Then we thought it might be a parachute. But as we watched the disc maintained a slow forward speed for a few seconds before starting to descend. While it was descending it was *swinging in a pendulum fashion* from left to right. We shall hear more about this pendulum motion later, meanwhile the reader is asked to retain it in mind.

'As the Meteor turned to start its landing run the object appeared to be following it. But after a few seconds it stopped its descent and hung in the air rotating as if on its own axis. Then it accelerated at an incredible speed to the west, turned south-east and then disappeared. It is difficult to estimate the object's speed. The incident happened within a matter of 15 to 20 seconds.

'During the few seconds that it was rotating we could see it flashing in the sunshine. It appeared to be about the size of a Vampire jet aircraft at a similar height.

'We are all convinced that it was some solid object. We realised very quickly that it could not be a broken cowling or a parachute.

'There was not the slightest possibility that the object we saw was a smoke ring or was caused by vapour trail from the Meteor or from any jet aircraft. We have, of course, seen this, and we are all quite certain that what we saw was not caused by vapour or by smoke.

'We are also quite certain that it was not a weather observation balloon. The speed at which it moved away discounts this altogether.

'It was not a small object which appeared bigger in the conditions of light. Our combined opinion is that it was about the size of a Vampire jet—and that it was something we had never seen before in a long experience of air observation.'

Flight Lieutenant Marian Cybulski, 34, who was in a Polish Squadron during the war and has flown 2,000 hours said:

'I agree with everything that Flight Lieutenant Kilburn says about this mysterious object. There may be flying saucers and there may not be. But this was something I have never seen before.'

Master Signaller Albert W. Thomson, 29, of Abbey Road, Barrow-in-Furness, who has been with the R.A.F. for 14½ years said, 'I saw just the same. It was there in the air, a round shape which hung for a few seconds. What it was I simply don't know.'

Sergt. Flight Engineer Thomas B. Deweys, 20, of Bedworth, Warwickshire, also saw the object.

L.A.C. George Grime, 22, of Salford, said, 'I saw a sort of halo shining in the centre of the object. It appeared to be going round and to shine as it turned. It was a solid object with no marks on it.'

A sixth flyer who saw the incident, Flight Lieutenant R. M. Paris of Brighton, was on a flying exercise the day before and could not give a personal account.

Now we shall see later that a bi-conic shaped disc will in fact oscillate from 'side to side' (or more technically, stall, from side to side) when descending, as in the Topcliffe incident, and witnessing this, many an aerodynamicist might naturally assume the craft to be aerodynamic in operation, but it would be wrong to assume from such behaviour that saucers are in fact totally aerodynamic.

The aerodynamic qualities of the disc are of course by no means unknown. In fact every complete work on aerodynamics of the aeroplane mentions at least one or two designs.

Of the earlier attempts, perhaps the following deserves the highest credit, for born as it was in the days when even the conventional fuselage-tail-wing system was still in the early experimental stage, a disc wing embodying VTOL capabilities was indeed years before its time.

It concerns the work of Capt. Alexander Sipowicz, who in 1927 responded to an official circular issued by the Tenth Department of the Polish Ministry of War, Air Force Command, inviting ideas for improvement in Service equipment. Sipowicz's contribution was a VTOL vehicle of annular disc plan form: he called it the Helipan.

Fig 2. Artist's impression of Sipowicz's VTOL 'Helipan'.

Fig 2 shows this to be of the ducted fan type aircraft in common use today. The annular wing (1) supported the 'cabin' (2), over the middle of the annulus on four struts. Beneath this and mounted within the annulus was a four bladed fan, powered by two engines (3). The bottom ring of the annulus was extended by attachment of the central duct (4) which also housed four sets of differentially operated rudders or control surfaces (5).

In operation, the fan was to draw air from above and below the wing, thereby producing a pressure differential plane which would augment the downward directed air jet. In addition Sipowicz claimed that the parachute type wing would reduce the rate of descent in the event of an engine failure.

Stability, manoeuvre and rotation about the longitudinal axis was to be effected by the use of the differentially operated rudders. To this end, Capt. Sipowicz conducted exhaustive tests with flying scale models

Plate 1. *Rocket powered North American X*-15 *high altitude hypersonic research aircraft.*

Plate 2. *Short S.C.*1, *VTOL research aircraft hovering.*

Plate 3. *Hawker P*1127 *VTOL aircraft hovering.*

Plate 4. VTOL sequenc of the experimenta DOAK aircraft wit ducted swivel rotors.

powered by 0.5 h.p. engines. From this he estimated that a 200 h.p. engine, or two 100 h.p. engines geared together, would provide sufficient power to lift a craft of some 1,764lb weight and that under these conditions a 13ft diameter propeller would create an airstream velocity of about 65.61ft per sec, enough for the internal control surfaces to provide adequate manoeuvrability.

The April 1929 issue of the Polish Aviation monthly, 'Lot Polski' carried an appeal by Capt. Sipowicz for practical assistance in completion of the Helipan, but owing to the loss of many records during the last war, it is not certain if in fact he received such help, or if the project was ever completed, but there can be little doubt, Sipowicz's VTOL machine was indeed years before its time.

Although Capt. Sipowicz's disc wing aircraft was essentially a VTOL project, down through the romantic years of early aviation there are records of earlier attempts to exploit the pure aerofoil quality of circular wings. One of the earliest successful attempts was the Lee-Richards annular wing monoplane, a model of which is exhibited in the Science Museum at Kensington, London.

In 1910, a Mr G. H. Kitchen did some original experimental work on the annular wing and accordingly took out patents which were later procured by Cedric Lee and Tilghman Richards.

After numerous attempts, a promising design was arrived at and a little monoplane took shape in the sheds at Shoreham Aerodrome. Gordon England, an experienced pilot of that time, having been flying since 1909, signed on as an independent test pilot, meanwhile the work continued behind locked doors and a shroud of great secrecy, to say nothing of the armed guards patrolling the establishment at night!

Then the great day arrived, and in the early hours of a bleak morning, the little craft was wheeled from the sheds. At 30 miles per hour, Gordon England had covered little more than 400ft of the first taxiing trials, when he found to his extreme surprise that he was airborne and climbing rapidly! But so responsive were the controls he decided to stay in the air.

Climbing with nose well up, the speed of the machine increased to about 85 miles per hour and very soon England was at the 2,000 feet level, where he carried out a series of gentle turns. Thus he continued for 30 minutes before descending to the aerodrome. Then at 700 feet the little Gnôme engine decided to cut out without warning and despite the pilot's efforts to regain control, up shot the nose until the little plane was completely inverted, whereupon it completed an exemplary loop.

Unfortunately, recovery from this unexpected manoeuvre brought it very close to the ground and after striking some telephone wires the little craft ended up very ungainly in a nearby ditch, close to the aerodrome sheds.

There was a breathless pause, then the somewhat bedraggled figure of Gordon England painfully extricated himself from the pitiful wreckage, stood looking at it for a moment, then somewhat shaken but none-the-less enthusiastic about the little annular wing monoplane's latent possibilities, he walked away.

A later examination of the wreck revealed that the fuel tank (although undamaged and the right way up) was completely empty, whereas it had originally contained sufficient fuel for some three and a half hours' flying. The cause of this mystery was never discovered.

Later the aircraft was redesigned and rebuilt and many successful flights were made until the official interest in it changed and the project faded into obscurity. Plate 9 shows how advanced the little craft was.

Then in 1913 a French Engineer in Dijon, M. Bourgoin, began experimenting with a similar annular-wing aeroplane. True the tests were unsatisfactory, but one interesting feature of the design allowed for the incidence of the annular-wing to be varied during flight.

In 1933 the German sculptor, Antes, created much local interest by successfully demonstrating his annular-wing type models.

Then more recently in 1937, N. H. Warren and Th. R. Young secured a patent for what they claimed was a non-stallable monoplane of 'rhomboidal shape', *i.e.* the forward wing curved backward and the trailing wing curved forward, so that the wing tips merged together. A conventional tail was provided at the stern of a long fuselage. In 1943, a model for a two-seater fighter of similar design was brought out, but now the tail had been omitted, special emphasis being laid on the triangular shape of each wing. Although this design was based on sound aerodynamics, nothing more was heard of it.

Later, in 1944, L. Peel brought out a further claim for the annular wing, in which two engines and their two airscrews were set in line and facing each other primarily as a means of offsetting torque.

It is interesting to note that while wind tunnel tests proved beyond doubt the admirable stalling properties of wings of very small aspect ratios, *i.e.* ratio of wing span to width or chord, this was never seriously utilised by subsequent designers. Yet even in the early days of the old 'box-kites', when the centre of gravity was often far too far back, the square shaped tailplane may have saved many a pilot's life, by refusing to stall even under extremely provoking conditions. But the science of aerodynamics was rushing ahead by leaps and bounds and one of the revelations was that a tailplane of 'good' aspect ratio was more efficient, so it was, but this in turn made the stall worse when the centre of gravity was moved somewhat aft.

It is none-the-less interesting to note that wings of circular or square plan form were tested in the early days at incidences of up to 90 degrees, while normal aerofoil tests were restricted to rather small incidences which generally excluded the range of stall. Despite the extraordinary capacity of circular wings to produce a very gradual stall even at very high incidences, it is strange that apart from the barest few, designers seemed to ignore the fact that such wings promised more safety in flight, though it was long ago established from practical experience that flying at the stall, known as the 'second regime', can be positively dangerous. On the other hand, spinning at that time was attributed to high wing incidences and no doubt many of the lightly loaded rectangular wings of that day made the stall comparatively harmless. Even so, the nose dive following accidental stalling was known to be the cause of most serious crashes.

Early researchers such as Eiffel, Riabouchinsky, Prandtl, Dines etc., conducted wind tunnel tests on aerofoils of low aspect ratio, and Eiffel's results showed clearly that while the ratio of the resulting forces was highest for wings of low aspect ratios and that slender wings gave greater drag at 90 degrees incidence, disc wings gave the least resistance of all. The work of Riabouchinsky established that disc wings attained their greatest lift at only 12 to 14 degrees incidence, and beyond their critical incidence they gave a *gradual* decrease of lift. Whereas it is well known that wings of normal aspect ratio give a very abrupt and unsteady decline.

But the real pioneer of the disc wing was Charles H. Zimmerman, an engineer of N.A.C.A., who in 1930 subjected the properties of disc wings to extensive wind tunnel investigation and it is interesting to note that not only does the report of this work still form the basis of some present research, but a good deal of it confirms qualitatively some of the experiments made 20 years before.

Zimmerman set himself the task of developing a really fool-proof aeroplane which anyone could fly, with particular regard to the stalling problem. Among other things, this work showed that very small variations in aspect ratio and wing tips produced marked differences, and that induced drag of circular or square wings is by no means as prohibitive as theory would indicate.

One of the chief advantages which Zimmerman's research revealed was the fact that disc wings gave less profile drag at small incidences, due to the relative reduction of the thickness of the aerofoil sections. Indeed this might be one of the chief advocates for the disc wing where high speed flight is concerned, for at twice the velocity of sound the drag is almost entirely dependent upon the thickness/chord ratio of the wing. As an aeroplane, the structural simplicity of the disc has much to offer. Concerning this and other attractive features, the author published an article entitled, *The Disc Type Aircraft*, in *Aeronautics*, December 1958. It is felt that an abridged version of it might be of interest here, even if to show the UFO student that the aerodynamic qualities of the disc have been given due consideration, though it is obvious from the beginning that in many respects saucers do not fit in with established aerodynamic technology. It is fair also to point out that the following similar conclusions were arrived at independently, the above material concerning the work of the pioneers not being available at the time.

The Disc Type Aircraft. Aeronautics, December 1958

The object of this article is to urge that the case for the disc type aircraft be given further investigation, for it is felt that it does in fact possess certain inherent advantages which have been largely overlooked and are here briefly outlined in this text. The following summary shows some of the advantages which are a natural consequence to an aircraft employing this type of wing.

1. It has been found that a wing so constructed offers a vastly superior strength-weight ratio to more conventional types.
2. The design inherently advocates the employment of vertical take off and landing.
3. Because of the extremely light structure plus VTOL characteristics, the aircraft offers its crew greater chances of survival than those now accepted.
4. It has now been established that the disc type aircraft offers the best compromise for re-entry into the earth's atmosphere.
5. In addition to the above, it is suggested that by rotating the disc wing, leading edge aerodynamic heating may be considerably delayed.
6. Due to the extremely light structure, the aircraft would have a greater payload capacity and/or greater range.
7. It is believed that this type of aircraft does not require a conventional fin, rudder and tailplane assembly, which again offers a considerable weight and drag saving.
8. As practically the whole aircraft contributes to lift in forward flight, most of the dead weight of a conventional fuselage is largely eliminated as in a normal flying wing.
9. Being a VTOL aircraft, a conventional type undercarriage is unnecessary—small castering type shock absorbers would suffice. This, together with a greatly reduced hydraulic system, also represents a considerable saving in weight.
10. Due to the total wing area being formed into a circle, obviating the need for fuselage and tailplane, this type of aircraft would greatly facilitate stowage in restricted spaces for instance, aircraft carrier hangar decks, Fig 3(a).

It is accepted that some of the above items could be true of any other 'flying wing' type of aircraft, but the prime reason why they would be especially applicable to this design is discussed below.

In the past, employment of a plano bi-convex wing would not have been considered, but with present day supersonic aircraft it can be shown to have certain advantages.

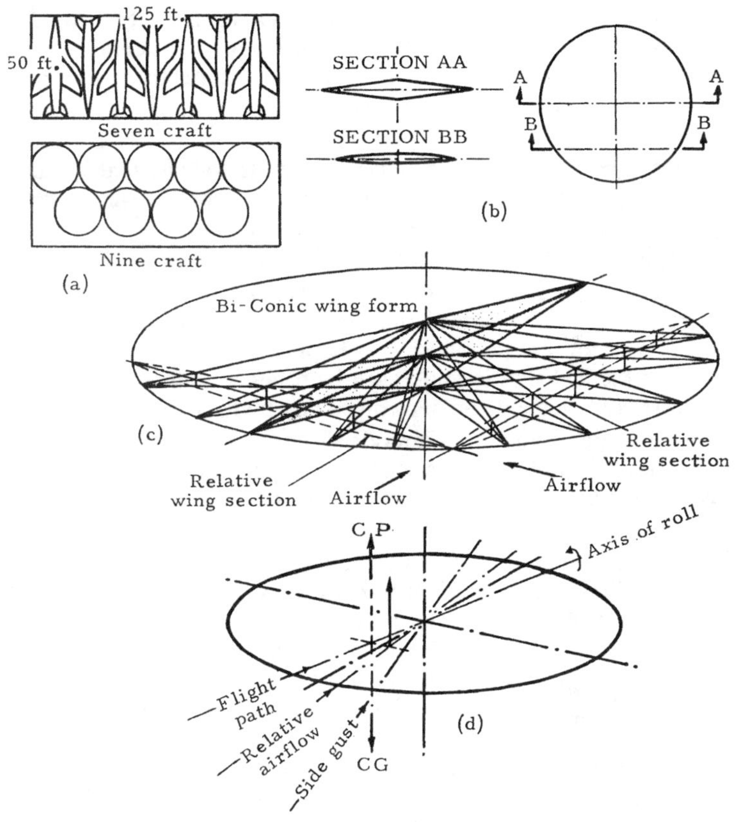

Fig 3. Two plain cones automatically provide aerofoil shapes for a very lightweight structure.

The basic proposal of this incorporated design suggests that a plane circular wing be constructed in much the same fashion as an ordinary bicycle wheel, which, because all interconnecting members are under tension loading, is immensely strong and resilient. It follows, to illustrate the principle simply, that a glider type aircraft could be built comprising little else than an outer rim, central disc or cupola, connected by a series of turnbuckle tensioned wires, the whole then being covered with doped fabric. This structure employs neither ribs nor spars, but in fact is the stronger. This has since been done in the Soviet Union. Reports say the pilot finds the aircraft will land itself and is practically crash proof. A typical part structural version of the author's is shown in Plate 12.

By way of example, a model aircraft with a wing span of 7ft and a chord of 1ft, having a wing area of 7sq.ft, has an all up weight of 6.5lb, giving a wing loading of some 14ozs/sq.ft. A 3ft diameter circular wing having approximately 7sq.ft of wing area was built on the wheel principle, which weighed some 15ozs, giving a wing loading of about 2ozs/sq.ft. This model when dropped from a considerable height on to the rim, sustained barely any damage. The weight saving in this case was no less than eighty-five per cent.

Fig 3(b) shows a cross section A-A through such a structure, represented by an upper and lower cone. It follows that section B-B is, in fact, part of a parabola and forms a bi-convex wing section. This holds good for approximately two-thirds the radius of the wing, gradually changing to section A-A. Fig 3(c) depicts this more graphically.

It must be stressed that this is a natural function of a plane conical shape, therefore no formed ribs are necessary; the bi-convex shape being relative to air stream, which can be in *any* direction. A feature of this wing is the fact that it could even be rotated as stated, but because of its uniformity it presents an aerofoil to any flight path.

Due to this, should the aircraft slip or receive a side gust, it meets the flow normal, and lift is maintained. There are certain rolling characteristics associated with this, but these can be compensated. Fig 3(d)

It is suggested that it would not be necessary to rotate the whole aircraft on change of course, reorientation of the centre portion may suffice. Should a disc of asymmetrical bi-convex section be rotated at speed, a vertical lift component due to the 'coanda' effect would be generated, Fig 4(a), the exact value of which can be determined by experiment. It is also suggested that a rotating disc may suffer less drag in a moving fluid than a stationary one, due to the local pressure rise at the 'trailing' edge, Fig 4(b).

A limited amount of work has been done by the author in this respect, without any satisfactory conclusions being reached. But the theory indicates that the combined effects of centrifugal and lineal flow induce a divergent and consequently decelerated air stream with an accompanying pressure rise. To some useful amount, this may offset normal drag. A 'magnus' or side lift effect from this might be offset by counter-rotation (later discussed in terms of UFOs).

We have seen that as aircraft flying speeds have increased, one of the accompanying disadvantages has been the correspondingly increased landing and take off speeds, which have in turn demanded longer runways. To this end, wing leading and trailing edge flaps on more conventional type aircraft have been successfully exploited. Fig 4(c) shows this to be in effect an attempt to change a high fineness wing section to one with high-lift characteristics, which, of course, has certain aerodynamic and mechanical limitations. It is suggested that a variant of the conic

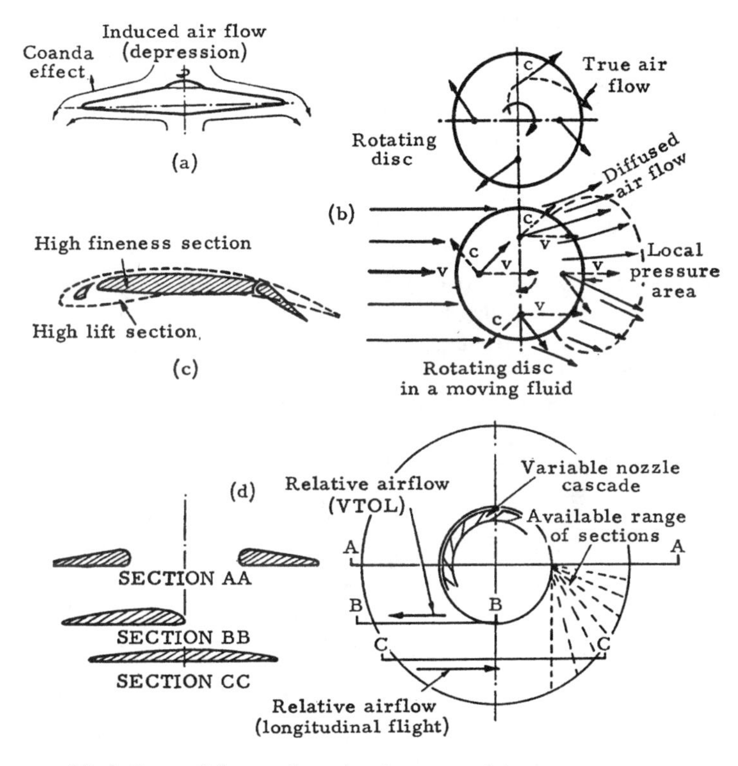

Fig 4. Some of the aerodynamic advantages of the Disc-type Wing.

wing aircraft however might contribute a useful solution which is briefly outlined as follows.

Fig 4(d) shows a modified bi-conic wing having an annulus in the centre which is formed by a radiused wall, section A-A. It follows that as with section B-B in Fig 3(b) a development at the section making a tangent with the inner annulus, reveals a naturally formed high lift type aerofoil section of usable shape. This would seem an advantage, as it has been arrived at without the employment of complex and weighty mechanisms. It follows therefore, that if an airstream is passed outwards over the wing, tracing a near tangential path to the annulus, a usable lift should be generated. Work has been done on this and the system works quite successfully, although complications are experienced due to the rather unusual divergent flow conditions over the wing, and the lack of data concerning it. It will be appreciated that between sections A-A and B-B, Fig 4(d) there are effective aerofoil sections of varying thickness/chord ratios which can be selected by a variable nozzle type cascade device such as that employed by the writer.

Unlike the comparatively thin trailing edge of the conventional wing, the disc wing leading-trailing edge is more rounded and to some extent lift is augmented by the downward deflection due to 'coanda' effect over this part of the wing as in Fig 4(a). It is possible that the effect might be even further enhanced by the provision of an annular flap situated at the periphery of the disc.

In the VTOL rôle the centre of lift would normally be situated at the centre of the wing, which would demand a similarly placed c.g. position. With the aid of the variable guide vanes however, an asymmetrically-placed centre of lift can be arranged, giving the aircraft a bias in any chosen direction. Work has been done in this respect, and the result suggests that at low speeds helicopter type manoeuvrability may be possible.

Rotating a portion of the wing would offer an attractive alternative means of stabilising the vehicle, particularly when operating in rarefied atmosphere at high altitudes, or hovering.

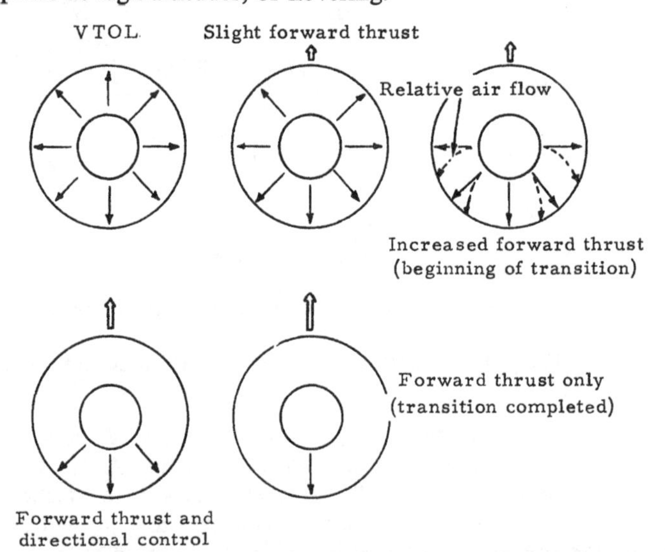

Fig 5. Changes of airflow during VTOL and Transitional flight.

Airflow conditions from radial to longitudinal, such as would be experienced in transitional flight, may not involve insurmountable difficulties as might be expected, but careful matching of the radial airstream velocity over the wing, and that due to the forward velocity of the aircraft will be necessary.

Fig 5 gives a diagrammatic idea of the general conception of hovering, transitional and directional control technique as at present visualised by the writer. While the sectional model and the test rig shown in Plate 12 illustrates the general idea further.

Only the simplest structure may be necessary in a high Mach number aircraft of this type, the two conical sections perhaps being fabricated from sheet steel, having a surrounding built-up rim leading-trailing edge. Such a structure would permit a wide range of internal layout without necessitating a comparable amount of structural alteration.

The above conclusions are the author's own, but I would like to take the opportunity here to say, that although it is natural for some to assume that one's work is influenced by that of others, even when this is not so, it is nevertheless sometimes irksome to be accused thus and even misquoted. The work the author has done on disc wings is entirely original and therefore I take full responsibility for any erroneous deductions I have made. For instance, when it occurred to me that aerodynamic heating might be *delayed* on atmospheric re-entry by employing a rotating leading edge on a disc wing, I was unaware that the well known aerodynamicist Dr W. F. Hilton of the Hawker-Siddeley group had advocated the use of the disc for similar purposes.

When my article in 'Aeronautics' was published however, some of my critics were quick to suggest that my idea was *based* on Dr Hilton's, and not content with that, they went on to say that it was quite erroneous for the author to claim that the leading edge of a disc wing could be *cooled* on re-entry by rotating it! This despite the fact that there were my printed words to verify that I said otherwise.

The fundamental idea behind this proposal was simple enough for even a child to understand, requiring no more than to imagine a copper disc pivoted at its centre. If the flame of a candle is held under the rim at one point, say for one minute, that part will be heated considerably, but if the disc is spun and the flame held at the periphery for an equal amount of time, then obviously the heat is distributed over a greater area. The disc will still get hot, but *local* heating will be *delayed.*

Lest the reader feels we have strayed a little out of context, I beg indulgence for this reiteration, but it may serve to illustrate a little the kind of barrier instinctively put up by some people when a new idea threatens long cherished pipe dreams of their own. I endorse this by asking the question, if such hasty and inaccurate judgement can be made about ideas which came within the framework of present day aeronautical technology, how much greater prejudice and misquotation can we expect when we dare to investigate the mysteries of gravitation, allied to those who are visiting us from outer space? The author has no idea what Dr Hilton's opinion may be concerning UFOs, only that the doctor was quick to use the term 'flying saucers' when describing his proposal for a re-entry vehicle at the conference held at Cranfield in August 1957, jointly organised by the College of Aeronautics, the British Interplanetary Society and the Royal Aeronautical Society.

He pointed out that if you approached the Earth in a 'flying saucer type craft' along a hyperbolic path at 7 miles a second (Mach 34), you must do something very quickly, for unless your hyperbolic path was speedily converted into a braking ellipse at approximately 5 miles a second, you would skim past the Earth, and head out into space again.

Therefore Dr Hilton advocated inverting the 'flying saucer' at an angle of attack of 30°—45° so that it developed negative lift, so as to cause it to 'hug the Earth's atmosphere' until drag slowed the vehicle to the required lower velocity. In this manner only the lower surface of the vehicle would get hot, whilst the pilot's canopy situated in the shadow of the shock wave would remain protected by the area of absolute vacuum where there would be no heat transfer. At the time of conference, various speakers felt that Dr Hilton was being over optimistic in 'expecting to remain cool in such a situation', but Dr Hilton replied with his characteristic puckishness:

'I'll put a girdle round the Earth in Forty minutes.'
(*Midsummer-Night's Dream*)

The French 'Aerodyne' designed and built by René Couzenet, is yet another example of the aerodynamic flying saucer. Almost 27 feet in diameter and powered by three 135 h.p. engines—which were said to lift the machine vertically, the craft was capable of transition and forward flight derived from a small turbo-jet engine suspended from the centre of the disc, Plate 11.

Hovercraft

This chapter would not be complete without mention of the nearer to earth counterpart of the aerodynamic flying saucer, namely the hovercraft, over which there has been a great deal of misrepresentation in the lay Press, and for this reason has been included here.

The author well remembers the memorable occasion when the Saunders-Roe N.1 hovercraft first publicly 'took to the air' over the waters of the Solent. An aged sightseer among the crowd remarked drily that 'something must have gone wrong, for the darned thing hadn't taken off yet'. No amount of argument could convince him that indeed the machine had been airborne by at least a foot for some minutes!

The aerodynamics of the hovercraft are of course somewhat different from the true flying disc wing, and indeed it is now well known that the plan form shape need not necessarily be circular. Further, unlike the wing type saucer, hovercraft, or more strictly correct, ground effect machines, as this name implies, demand the near proximity of the ground or supporting surfaces in order to function at all.

At the present time all over the world, an increasing interest is being shown towards the hovercraft vehicle. In Great Britain designers encouraged by the results of the first flown—SR. N1 built by Saunders-Roe, embarked on similar enterprises with the result that many exciting and strange shapes are emerging and growing in various workshops. While in the United States the widespread interest in ground-effect machines appears to have begun officially in April 1957, when the Lewis Laboratory of the then NACA issued a publication describing some ground-effect jet experiments. No doubt stimulated by this report, the Navy set out to establish a research programme at the David Taylor Model Basin.

Some three months later, in July 1957, this programme produced 'a comprehensive understanding of the fundamental annular-jet ground-cushion phenomena, complete with working formula'. However it has emerged since then that a surprising number of people, including companies, large and small, inventors, hobbyists and scientists, were already hard at work on ground-effect devices of one kind or another, even before the publication of the NACA report in 1957.

Probably one of the earliest experimenters in ground-cushion effects was Dr Andrew A. Kucher, now vice-president of engineering and research at the Ford Motor Co., for he was thinking in terms of 'sliding on air' way back in 1928.

Yet another pioneer Mr Toivo J. Kaario of Finland, experimented with a craft in 1935, first as a ram wing glider and later powered with a 16 h.p. Davidson motorcycle engine. The craft measured 6ft by 8ft, and was accredited with a speed of 12 knots over ice on its first trial. The next attempt was a machine measuring 8ft by 10ft, which took off from water and was capable of lifting four men. As with so many other projects, World War II interrupted this development.

In America alone some of the known projects include: Spacetronics, 'air-leakage' craft, to carry eight men, on order for the Marine Corps; National Research Associates, Pegasus 1; Ford, Levacar; Bertelsen, Aeromobile; Curtis-Wright, Air-Car; Gyrodyne Corporation of America, research craft (Bauer contract); and model work being carried out by the Navy's David Taylor Model Basin, NASA's Langley and Ames Research Centers and Princeton University. Aircraft concerns such as Convair, Grumman, Hiller, Lockheed, North American, Sikorsky and Boeing are also known to be actively interested in ground-effect craft.

Of all the attempts, and there are many, at approximations to a land-borne flying saucer, perhaps Princeton's X-3 comes the closest. Plate 10. Measuring 20ft in diameter, it is only 5ft high at the fin. Its construction is of aluminium tubing with fabric covering. It is interesting to note that aerodynamic lift is obtained over the 'wing' or aerofoil shaped body in forward flight, identical to the author's aircraft conception mentioned earlier.

The X-3 which made its first 'flight' on 18 October, 1959, has a 44 h.p. Nelson H-63B engine which drives a 4ft diameter propeller at 4,000 revolutions per minute, both are mounted in the central duct. Directional control is obtained by electrically operated vanes mounted in slots which ring the base of the machine.

A 5 h.p. Power Products engine turning a 26in diameter propeller is mounted in the fin; actuated by foot pedals, this can swing through 90 degrees to the right and left to provide yaw control. It can also be turned 'face on' to provide thrust for forward flight.

A bubble type canopy is situated in the forward part of the 'Saucer', while power packs, fuel tank, battery etc., are placed at the rear near the fin to help balance the machine. The little craft weighs some 850lb empty with a gross weight (including pilot) of 1,000lb.

Broadly then, the hovercraft type of vehicle offers to fulfil the need for transport of heavy payloads at fairly high speeds, up to 70-80 knots or so, and its use over unmade roads, swamps and lakes is obvious. But it will occupy a definite marginal bracket of its own, that is, somewhere between the pure land and sea vehicle, and the aircraft.

These then are but a few of the many attempts to exploit the aerodynamics of disc shaped wings. To find better and safer ways of getting aerial vehicles off the ground and back again. To fly faster and higher, to carry greater payloads more economically, they had to be included in this story, for in their own way they represent more signposts, no doubt in terms of gravitational control these attempts may seem fundamental, yet they serve to indicate all too clearly the trend of *wings* to come!

3

Limitations of Rocketry

IN an age when the rocket is very much in vogue, the title of this chapter may sound more than a trifle disparaging, but as sincere explorers we have to keep to the hard straight road of fact. Here we shall have another quick look at the scenery. In this case the scenery is in the form of all types of rocket motor. We are going to look at them and assess their merits soberly and justly without prejudice and to equally soberly judge their shortcomings. For only by doing so can we clearly see the way ahead, and in what direction this development is likely to take us. It may well be, that the signpost we shall discover in this chapter may have fascinating and far reaching implications of a method of space ship propulsion as yet not seriously contemplated. We must therefore ask patience of the rocket technician as we explore his ground, and beg indulgence when in the next few chapters we try to introduce evidence that there *is*, beyond much doubt, a more attractive method of space ship propulsion.

In order to attempt this we must stand to be bold and venture into the unknown realms of gravitation, but first let us despatch the rocket.

To begin with, the ordinary chemical or solid propellant rocket is one of the simplest engines known to science, and in our young days, and even now, most of us enjoy playing with the miniature version on special occasions. Yet the function of the rocket motor is still misunderstood by many people who are becoming space minded. Perhaps a brief word or two for their benefit may not be amiss. In the first place a rocket does not get thrust by the exhaust gases 'pressing' on the surrounding air, it functions according to Newton's Third Law of Motion, which states, 'To every action there is an equal and opposite reaction'.

One of the simplest ways of visualising this behaviour in a rocket motor, both in air and vacuum, is to imagine that the rocket motor consists of a hollow sphere within which there is contained an explosive mixture, Fig 6(a). Were the charge to be ignited, the sphere, or combustion chamber, would experience a unilateral pressure or 'push' within, and of course it would probably explode. But if it did not, then no movement would take place, for the 'push' is cancelled out in every direction. Now should the experiment be repeated, but this time a hole is made in one side of the sphere as in Fig 6(b), then the push from within is cancelled out everywhere, with the exception of the hole and that portion of the sphere diametrically opposite it. Therefore the sphere literally receives a push on one side and not the other, and consequently moves in that direction. It will be apparent that this would happen even if the sphere *is* surrounded by air, the only contribution made by the air would be to retard the movement, not help it. From this it will be obvious that the escaping gases or ejection mass issuing from the hole, will in no way impart a forward thrust by pushing on the surrounding air, as is often wrongly assumed, in fact, were the same thing to be repeated in vacuum, the sphere would move faster for two reasons.

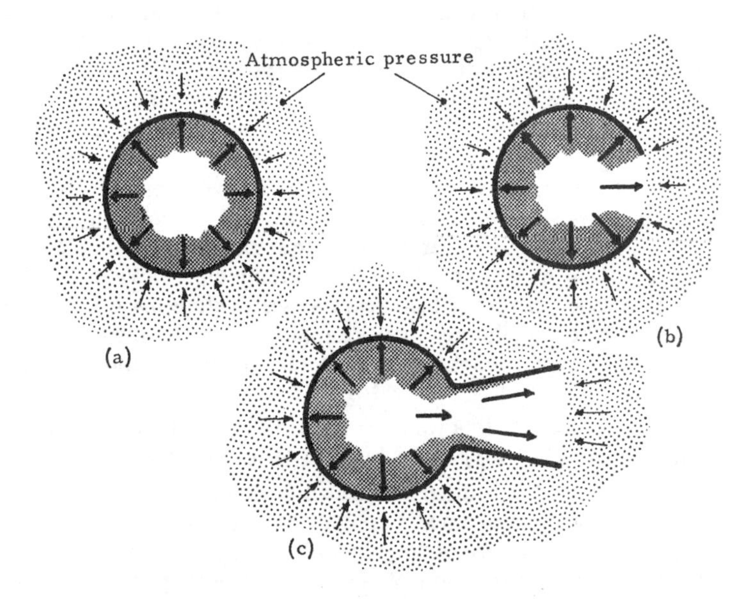

Fig 6. *Rocket motors derive thrust purely from the reaction of the expanding gases.*

(1) Because it would experience no friction or drag and (2) because the issuing gas jet from the hole moves faster, for it experiences no back pressure as would be caused by a surrounding atmosphere. Thus it will be seen there are only two masses to be considered, that of the sphere and that of the espanding charge. Therefore the rule is, the greater the mass of the escaping charge, or the higher its velocity, so the greater the thrust it produces on the chamber. The habit of smoothing out the edge of the escape hole and extending it into a nozzle as in Fig 6(c) merely improves the fluid flow or aerodynamic characteristics.

Broadly then this is basically the principle of the rocket, from the tiny penny firework to a huge and highly developed machine like the Atlas, there is no fundamental difference between the two. What difference exists at all is one of complexity, not of kind.

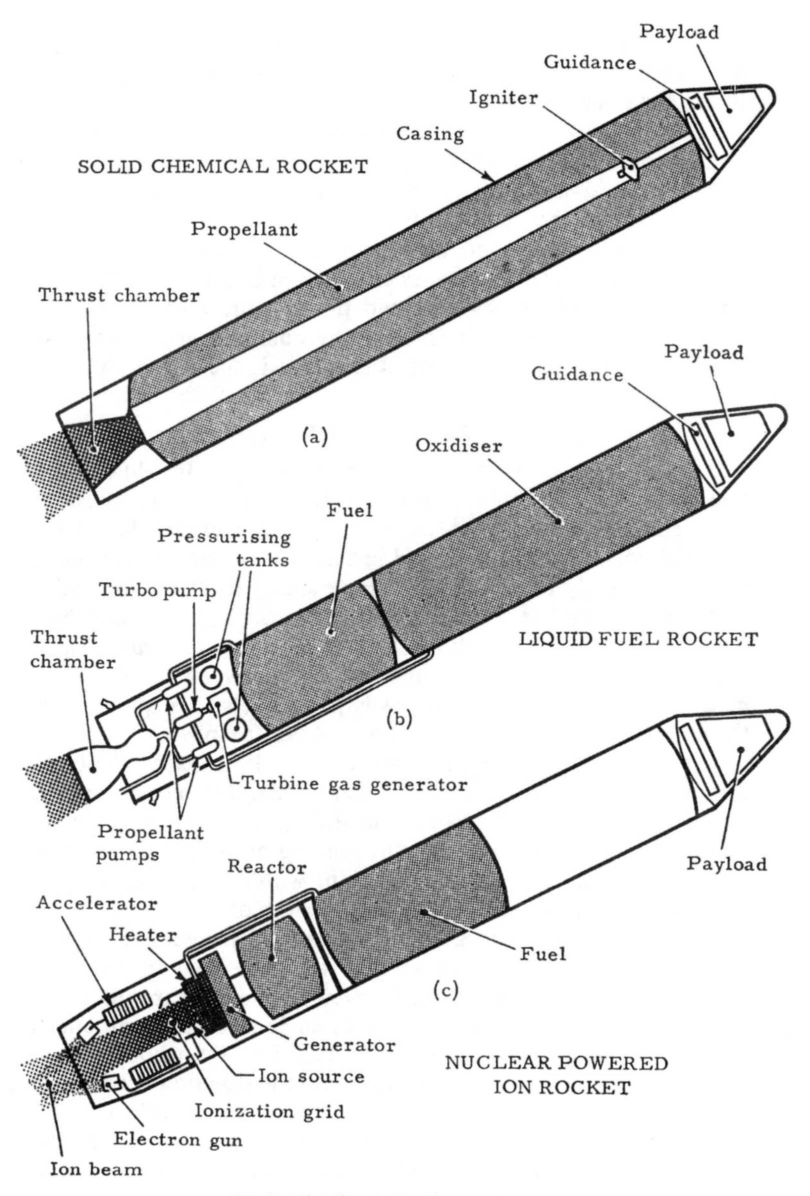

Fig 7. The three chief types of rocket.

The Chemical Fuel Rocket

Again the layman is often confused by the expressions chemical fuel rocket and liquid fuel rocket, for he is apt to interpret this as meaning two different types of motor. Again there is no fundamental difference in the method by which the rockets obtain their thrust, basically they both consist of a combustion chamber and an expansion nozzle, or thrust chamber.

The toy firework rocket is of course a chemical rocket and many larger motors are constructed in this way. In this case the fuel is prepared in solid chemical form, housed in a tubular container, which virtually forms the rocket tube itself. As thrust is a function of the available burning area of the charge, this is usually cast with a hollow centre. The charge terminates at the combustion chamber and when ignited, the resulting gases expand through the nozzle. As the charge burns along its length, it is as if the combustion chamber was itself stretching wider and wider, until the charge is spent, Fig 7(a).

The Liquid Fuel Rocket

On the other hand, the liquid fuel motor Fig 7(b) differs in that the fuel is contained in tanks which form the rocket casing. One tank containing liquid oxygen as an oxidizer, the other usually liquid hydrogen. The additional complexity found in this type of engine over the chemical rocket, is largely introduced by the necessity to pump the fuel into the combustion chamber in order to overcome the high pressure there. The pumps which do this work are turbine driven at extremely high revolutions and are in themselves engineering achievements of an extremely high order. For instance, in the larger rockets it is quite common for the pumps to deliver several tons of fuel per second.

The rocket then, is primarily suited to work in space, where having no retarding medium in the form of atmosphere, the motor can operate at maximum efficiency. Unlike air breathing engines such as the pure jet—which sucks in and ejects the surrounding air to obtain its thrust—the rocket has to take its ejection mass along with it. Obviously this imposes a weight penalty, which is impossible to overcome.

The situation is comparable to a steam locomotive which would normally have several stops to take on more coal to cover a certain journey, trying to do the same journey in one lap carrying all the coal. If we now imagine the journey to be so long that the amount of coal required would be several times the weight of the whole train, and the locomotive trying to pull this weight, then the situation approximates that of the space rocket. Thus the first limitation is established.

A rocket is basically a means of hurling a given payload into space, and like a thrown ball, the faster we throw it, the higher it will go, or in stricter terms, the greater will be its radius from the centre of the earth.

Orbital Techniques

At this stage it may be as well for us to get one point quite clear, for it is a pitfall which seems to trap so many. To the more technically minded it will be a fundamental, but the author knows from experience that even some people with a technical background cannot fully grasp the fact that there is no theoretical point in space where gravity does not exist. The belief that an artificial satellite stays in orbit because it has reached a point in space where gravity ceases to exist is quite erroneous, after all it is the earth's gravitational pull which maintains the moon in orbit over 240,000 miles away. The most remote star is affected by the earth's pull to an immeasurably small degree, but it is affected nevertheless. For the force of gravity, like all magnetic and electrical fields, gets less according to the inverse square law, the further out from the so-called point source we may go. At several hundred miles above our planet, the force of gravity is measurably subdued, that is all.

So with our thrown ball, the faster we throw it upwards, the further it will travel into these weaker fields, but we may have to impose other conditions to keep it from returning back to earth. In order to grasp this fully, let us imagine a rocket take off.

On ignition of the motor, the rocket vehicle receives an enormous thrust equal to more than its own weight, which imparts the initial velocity or momentum. Under the continual thrust the momentum increases, also seconds later, the vehicle is considerably lighter whilst its thrust remains constant; therefore the momentum is increased yet further and this continues until the fuel is expended at 'all burnt'.

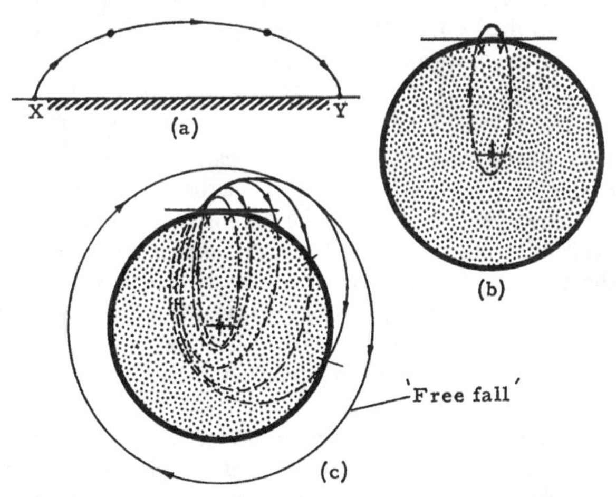

Fig 8. Even a thrown stone traverses part of an orbit.

At 'all burnt' the rocket has been inclined into a trajectory and has attained sufficient speed to carry it on under its own momentum, in the same manner as the thrown ball. It is not generally realised that even a thrown ball or firework rocket, if fired at an angle to the earth, is in fact moving in part of a very small orbit, and to see why, we should look at Fig 8(a). A ball has been thrown from X to Y, and to us it appears that point Y should have been the end of its flight. But this is not so. The natural and intended flight of the ball would be to follow the dotted path shown in Fig 8(b) or nearly so. It would in fact have been a partly elliptical orbit about the centre of the earth and back to its starting point X. That is, but for the fact that at Y its flight was interrupted by the presence of the earth itself.

In Fig 8(c) we can see that if the firing angle was kept the same, but the velocity increased, the missile would go higher and of course the distance it covered over the ground would be greater. The black dots indicate the position at which it would have fallen at different speeds, while the dotted lines show the path it would have followed back to X had it been allowed to do so. At the last (and highest) firing it will be seen that the missile never reaches the earth again, for its angle of 'descent' matches exactly the earth's curvature, hence the space flight term 'free fall' or 'fixed orbit'.

Of course there are added complications due to the retarding effect of the earth's atmosphere and a simple little experiment may help the reader to examine what effect this and varying velocities have on a satellite in fixed orbit. A small ball or suitable weight is secured to a length of elastic. The ball represents an earth satellite and the elastic represents the earth's gravitational pull, whilst the hand represents the centre of rotation, or earth. As the ball is whirled out into an 'orbit' a pull is exerted outward away from the hand. This is the pull due to so called centrifugal force and it is this same force which, acting against the force of gravity, keeps an artificial satellite in orbit above the earth.

It is a simple matter to observe just how these forces cancel one another out and by a little variation in the rotary speed, it will be seen what influence velocity has on the height or distance of a satellite from the earth. The faster the ball travels, the greater the distance it stays from the centre. In the case of the rocket, if this speed is *increased* to seven miles per second it will spiral outwards away from the earth. This speed is known as 'release velocity'. Also as in the case of our model, a *decrease* in velocity will allow the elastic *i.e.* Gravity—to pull it back in again.

The only inaccuracy in this space model lies in the fact that the further out the ball flies, the *greater* will be the tension on the elastic, whereas gravity works the other way round, for as we have seen the further out we go from the planet, the *less* is the gravitational pull.

It is planned that space ships of the future will be able to move to and fro between space stations in orbits about the earth by exactly the same principle as shown in Fig 9. An increase in speed will transfer a vehicle out to a further orbit (a) and the reduction of it will allow it to 'fall' back

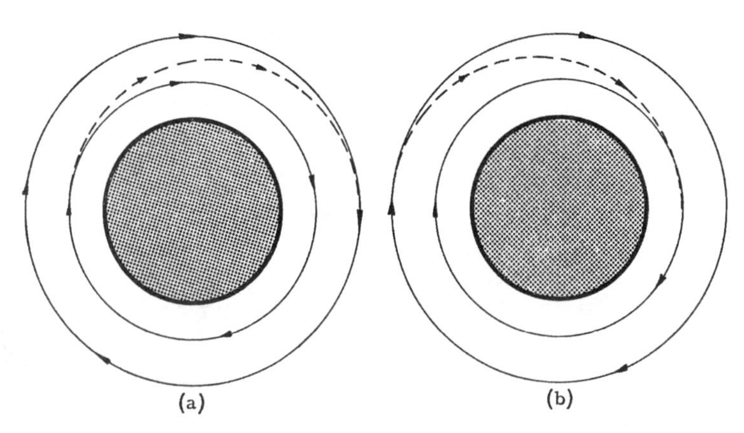

Fig 9. Transfer from an inner and outer orbit.

to a closer one (b). Indeed as this book goes to press, the American Gemini astronauts have done just this.

By the same technique, the space station itself will be placed in such a 'parking orbit'. Fabricated units, having first been assembled and tested on earth, will be dismantled and rocketed into space like pieces of a huge Meccano set. On each journey the ferrying rockets will join the established orbit, dump the sections into space and return to earth for another cargo. In exactly the same manner in which a dropped ball will take on the forward speed of a moving train, so will the dumped sections remain in orbit once tipped overboard by the ferry, where free from the corrosive effects of the earth's atmosphere, they will remain as the work continues. Gradually the large sections will be 'recaptured' and manoeuvred into place by space-suited work crews, who will appear to float in and out of the intricate framework, like strange and silent fish of the upper void.

Although the structural pieces will still have mass and therefore show resistance to being moved, the work will be greatly facilitated by virtue of the weightless condition.

It is from such a 'parking orbit' as this that future trips to the moon and further planets may be made, for although it is theoretically possible to build a manned circum-luna rocket, it will be more practical to achieve such journeys in stages, *i.e.* Stage 1, earth to parking orbit (refuel), Stage 2, orbit to 'soft' landing on moon, lift off and back to parking orbit, Stage 3, refuel and retro glide back to earth.

A trip to the moon and back will be feat enough, but imagine a future journey to a neighbouring planet, with the rocket crew confined to a restricted space comparable to a submarine craft, for perhaps a period of several years! Whilst the rocket ship coasts on motorless, weightless and silent through endless space. Clearly, should there be the remotest possibility of a better way, then we should seek it. Later we shall examine evidence which indicates all too clearly there may be a better way. But first let us acquaint ourselves with the rest of the problems thus far, for our signposts are not yet running out.

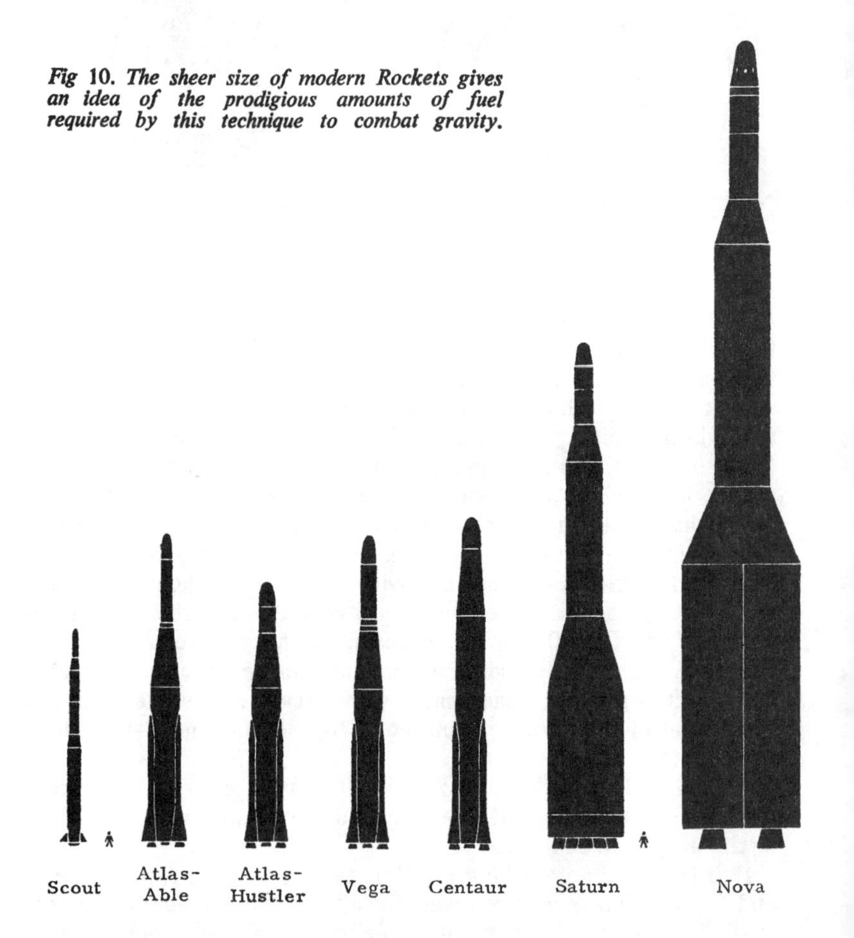

Fig 10. The sheer size of modern Rockets gives an idea of the prodigious amounts of fuel required by this technique to combat gravity.

Both the chemical and the liquid fuel rocket come within the category of heavy reaction mass motors. For although the exhaust velocity is extremely high (around 6,000 feet per sec) in both cases the mass of fuel burned and ejected is proportionally large. As already stated, in the larger rockets this can often be to the order of several tons per second. And again, as we have seen, this imposes a high weight penalty on the vehicle itself. Bearing in mind that 90% of the bulk of a modern rocket is accounted for by fuel tanks, their sheer size gives a good impression of the enormous amount of fuel required to place a vehicle in orbit about the earth or to send it to the moon, Fig 10.

In order to put an interplanetary probe vehicle into orbit there is no foreseeable improvement in rocket technique, but once that part of the journey has been completed, there are alternative solutions which rocket designers are finding attractive.

The Ion Rocket

One alternative scheme which may help to solve the large reaction mass problem involves the employment of a slightly different type of rocket motor, in fact it is an electric rocket. Again the operation is fundamentally simple, embracing as it does the same principle as the more common rocket motor, *i.e.* pure reaction.

We have seen how the thrust of a rocket depends on two factors, ejection mass and the velocity at which this is ejected. The higher the mass for a given velocity, the greater the resulting thrust, or the higher the exhaust velocity for a given ejection mass, the greater the thrust, Fig 11.

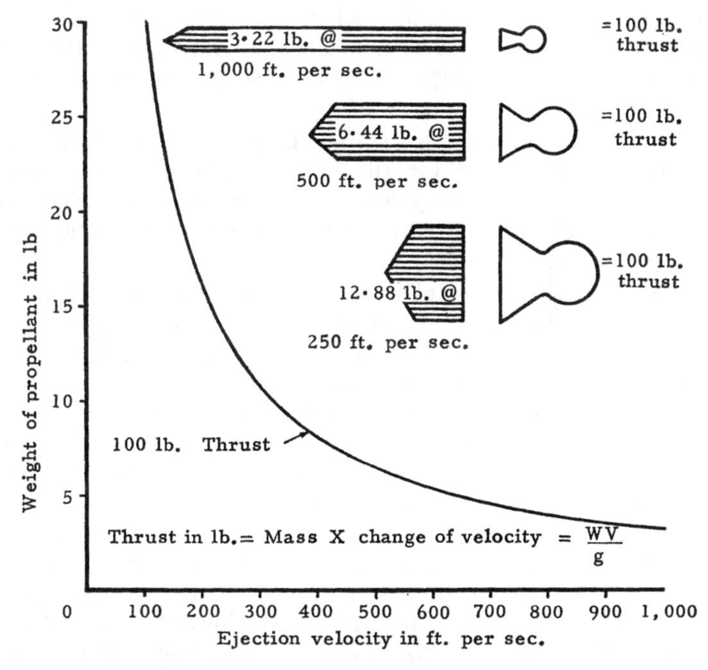

Fig 11. *Ejection mass required at different velocities to obtain a constant 100lb thrust.*

A simple way to visualise the effect, is to consider a small boy on roller skates throwing a brick away from himself at, say, six feet per second. He in turn is thrust away in the opposite direction. He might of course have thrown two such bricks at three feet per second, but the result would have been the same. Should he now succeed in throwing only one brick at twice the original speed, *i.e.* twelve feet per second, then he would be thrust back proportionally faster.

A measure of the relationship of thrust, mass flow of fuel and exhaust velocity, is usually expressed in terms of specific impulse, which is to the rocket engineer what miles per gallon is to the motorist, in effect it is a measure of the efficiency with which the propellants are generating thrust. For instance the V-2 rocket motor gave a specific impulse of 215 seconds, which means that it generated one pound of thrust for each pound of propellant over an operating period of 215 seconds. In the case of advanced chemical rocket engines, the specific impulse is in the order of 350 seconds. From this it will be seen that could we find a way to drastically increase exhaust velocities, with only a very small ejection mass we might obtain comparable thrust without the high weight penalty. Also, as the rocket vehicle would have to carry less weight in fuel, again it would move much faster. This relationship between the mass of the payload and the mass of the fuel is called the mass ratio and is a very important factor to the rocket engineer, as will be remembered from the steam locomotive analogy. Therefore the proposition of low fuel propellant consumption is a very attractive one.

A great deal of research has been done in this respect, but with heavy ejection mass rocket motors the limitation of exhaust velocities is set by the extreme temperature perimeters of the combustion chamber on the one hand and chemical reaction on the other, beyond which no increase in velocity can be attained.

The working temperature limits of some rocket materials lies in the neighbourhood of 1,500 degrees to 2,000 degrees C. In the course of time it may be possible to almost double this, up to about 3,000 degrees C. This might be obtained by employing magnesium oxide or something like that. Carbon for instance can reach temperatures around 3,200 degrees C. Above 3,500 degrees C however any material is in a gaseous state. This would appear to be the end of the line, the limit of the ordinary rocket.

But in this respect the electric rocket may offer a possible solution. The present state of the art involving complex studies into the new science of magneto-hydrodynamics, shows that small particles forming the ejection mass can be highly charged (ions) and further accelerated by powerful magnetic fields. From this standpoint, many UFO enthusiasts, in trying to find a solution to flying saucer phenomena have tried to explain them away as being some form of ion rocket, due to the electromagnetic and lighting effects often displayed. Deductions such as these are quite understandable but therein lies another of the pitfalls we are apt to make with insufficient knowledge of the facts. In discussing the ion rocket, it will be apparent that this is simply a technique for improving the efficiency of rocket exhaust velocities. The device remains a rocket and almost certainly would never have sufficient power-weight ratio to give the performance with which flying saucers are accredited.

The basic difference between the ion propulsion and chemical propulsion systems lies in the difference between an ion and an ordinary molecule. For an ion is an electrically charged particle, while an atom or molecule is normally neutral. If however an electron is added or removed, the resulting charged molecule is said to be an ion.

In a neutral gas, molecules are accelerated and acquire high velocities through collision with other neutral molecules, whereas an ion can be accelerated in an electric field. Moreover, in a gas composed of ordinary molecules, their directions of motion are completely random. Whereas a cloud of ions in the presence of an electric field will move uniformly in the direction of the field.

Just as the bombarding neutral molecules in a chemical rocket produce thrust on the exhaust chamber walls, so the ion exhaust in an ion rocket produces thrust against the electric or magnetic field.

This exhaust, now more commonly known as 'plasma' can be accelerated to enormous velocities, in fact much higher than that which can be attained by ordinary chemical reaction. For instance an ion motor will be able to accelerate charged particles of cæsium vapour at velocities of over 650,000 feet per second, compared to 6,000 feet per second of the conventional chemical rocket motor. Consequently for a given thrust, a greatly reduced exhaust mass is permissible.

Naturally designers are considering the ideal source of energy for the ion rocket would be a nuclear reactor and to this end a great deal of research is being conducted.

Basically the idea is to employ a nuclear reactor to produce heat in a circulating fluid. Energy will be absorbed from this via a turbine or suitable prime mover, which in turn will power an electric generator. The output from this will be employed to vaporise and ionise the propellant which is finally expelled from the unit as a jet plasma, which can be further accelerated by powerful electric or magnetic fields, Fig 7(c).

One of the most promising designs now under consideration is an ion rocket lightheartedly called 'Project Snooper' by the preliminary design section of North American Rocketdyne Division at Canoga Park, California.

The source of Snooper's power would be a fast or intermediate nuclear reactor, which it is claimed, could operate for a year or more before the slow accumulation of poisonous isotopes choked off the reaction. The reactor would have a total thermal output of about 1,000 kilowatts, sufficient to produce about 147 kilowatts of electrical power.

Sodium would be used as a reactor coolant, giving up its heat to mercury through a heat exchanger. The mercury would drive a turbo generator, then flow through a second condenser-heat exchanger where it would give up its heat to a sodium loop which would circulate through huge bat-wing radiators necessary to reject unwanted heat from the cycle. These radiators would be folded around the vehicle during the boosted portion of the flight, and unfold when the ion system commenced operation.

Cæsium was chosen as the propellant for a variety of reasons. It has the highest ionisation potential of all the alkalis, the lowest melting, boiling and vaporisation points of the alkali family, and the highest density.

In the projected vehicle, the cæsium would be contained in a tank heated by an electric blanket to about 100 degrees F, to hold it in a molten state. An automatically controlled metal expellant bag within the tank would pressurise the vessel and force the cæsium into an atomiser which would vaporise the metal at a temperature of approximately 1,500 degrees F.

The vapour would then flow through a sintered steel distributor, then impinge upon incandescent tungsten surfaces formed into a series of ionising grids. After a brief stay of only a few microseconds on the grid, essentially all the cæsium atoms would lose an electron and become ions.

At this stage the ions would be accelerated across a three centimetre gap to a velocity of 657,000f.p.s. by a direct current of 27,500 volts. The cathode grid would consist of a honeycomb cross-section to assure a uniform electrostatic field in a radial direction. The electrons are simultaneously ejected along with the ions by means of an emitting system located downstream of the cathode grid. The electrons thus mix with the ions at a point which does not interfere with the efficiency of the reaction and no excessive negative potential is allowed to accumulate in the vehicle which would otherwise tend to attract the ejected plasma back to itself, thereby cancelling out the thrust effect.

As the all up weight of such a system might prove to be prohibitively high, its employment may well be restricted to true space drive, where the vehicle having been put into parking orbit by conventional rocket booster, the ion motors will be turned on imparting a small amount of thrust. Snooper for instance will give only a total thrust of one-third of a pound, or about 0.0001 gravity. This would give it an acceleration of about 0.04 inch per second.

But even a small thrust continually applied over a long period of time is capable of moving a vehicle at enormous speeds. Therefore once orbital velocity has been obtained by conventional means, the nuclear or solar powered ion rocket may come into its own. Naturally, if it takes such a vehicle several months to gain a certain speed, then with reversed firing motors it must take an equal time to lose this speed, so that forward thrust must be strictly limited to the first half of any space journey. Soft landing on a planet will of course only be achieved finally by the use of the conventional high thrust motor stage.

In the constant search for higher propellant velocities, physicists are exploring every possible avenue with emphasis on the electric rocket motor. One novel application proposed by Electro-Optical Systems Inc. of Pasadena, California, is to obtain thrust by exploding wires as a propellant.

The company claim that such a propulsion method may provide more thrust for the amount of power consumed than many currently proposed electrical propulsion systems.

Temperatures up to 100,000 degrees C and pressures in the megabar range have been obtained. Specific impulses of 1,000 seconds have already been achieved by exploding several wires at once and a 1,000-6,000 seconds range appears possible.

Materials used in some experiments include:—aluminium, iron, copper, gold, silver, nickel, tungsten, tin, titanium, zinc, cadmium and bismuth. Some test wires measure one millimetre in diameter and a quarter inch in length.

The technique is simple. A capacitor of from 0.002 to 0.02 microfarads charged to 10-20 kilovolts is suddenly discharged into the wires for about seven millimicroseconds, reducing the wire to an explosive vapour. It is claimed the technique has enabled scientists to place many times the materials vaporisation energy into the wire. Such a vapour or plasma could be accelerated up to 20-30 kilometres per second, a speed far in excess of anything at present in laboratory use.

The Photon Rocket

But we are not yet finished with the electric rocket, for there is one more turning along our exploratory preamble. One more possibility before we come to the end of the pure reaction principle.

According to relativity, the greatest velocity in the universe is the velocity of light, 186,000 miles per second. A beam of light is made up of photons which have a rest mass of zero and therefore it is argued, a beam of photons now having mass according to relativity, should produce a measurable thrust. In fact it has been shown that an ordinary household electric torch, if deposited in outer space, would attain a velocity of something like one foot per hour, that is if the battery lasted long enough.

Its big brother, the photon rocket, has often been likened to just that, 'an electric torch of colossal candle power'.

In terms of thrust weight ratio it comes off worse than all the electric rockets. In fact such a device would have to be of enormous proportions to yield only one pound of thrust. But if it were solar or nuclear powered, it could function almost indefinitely and as with the ion rocket, even a meagre pound thrust constantly applied over months can produce very high speeds. Therefore, as an interplanetary vehicle, the photon rocket has exciting possibilities, but as with all low power/weight ratio electric rockets, it must first be ferried out to orbit piecemeal by conventional chemical boosters, then assembled in space.

Naturally not all the various methods of electric propulsion are included here, but the following table indicates the relative advantages offered by the different techniques.

Rocket Motor	*Specific Impulse*
V-2.	225 seconds
Advanced chemical.	300— 350 „
Nuclear with chemical working fluid.	300— 800 „
Colloid (electrically charged smokes).	500— 2,500 „
Exploding wire plasma.	1,600— 6,000 „
Ion.	5,000—10,000 „
Photon.	1,000,000 „

Although these various investigations will no doubt bring accompanying discoveries which might well lead to a major breakthrough in science, as a pure lifting device the electric rocket principle is at present useless, the amount of thrust per pound of weight being infinitesimal in all cases, and even in the rôle of an interplanetary vehicle, any of these motors must initially depend on the prodigious chemical booster to blast them into orbit at some 14,000 miles per hour!

Re-entry

On this note we come to the final limitation of the rocket, for this enormous speed, this fantastic amount of kinetic energy must be lost, retro-thrusted or dissipated in heat by friction as the plummeting vehicle makes the return journey to earth. This problem of ballistic re-entry into the earth's atmosphere has presented engineers with one of the greatest technical difficulties of modern space flight. Indeed, at one time, the dissipation of the intense heat generated by a returning capsule seemed impossible. The term 'heat barrier' took its aeronautical place with the 'sound barrier' and just as surely as its predecessor, it has been met and overcome. The final success of the Mercury capsule only came after radical changes in basic ideas. The departure from the streamlined ballistic missile to a blunt aeroform, was a slow painstaking process of often disappointing research, crowned finally with success.

Aerodynamic shape is one of the major factors determining the performance of a re-entry body. And right from the beginning it was known that a blunt shape, because of its greater resistance, slows down more quickly after entering the atmosphere, than a streamlined body. Therefore it loses its speed at a higher altitude and consequently in thinner air, where it will be subjected to less heating. But the angle of descent is steeper, which in turn introduces other problems. The mercury capsule has been designed to meet such aerodynamic requirements and many lesser informed UFOlogists have been quick to interpret some shapes of UFO as being similarly designed. Whereas in actual fact as we have seen, the pure bi-convex or plano-convex aeroforms are the only really practical shapes of UFO which would lend themselves aerodynamically. With perhaps the exception of the cigar type, none of the others with the bulbous conning tower projections would receive a second look in a modern high speed wind tunnel.

The Mercury capsule on the other hand was designed to re-enter the atmosphere with its *blunt* ended fibreglass heat shield foremost, while its afterbody was covered with corrugated cobalt-alloy shingles to dissipate the heat by radiation.

Nowadays of course even the man in the street accepts the fact that such systems seem to work quite effectively, but of course there are accompanying disadvantages. One involves problems of radio interference as the capsule plummets back to earth, due to ionisation of the adjacent air particles as they become heated through friction. This can cause quite a radio blackout for part of the return journey as in the case of Colonel John Glenn's Friendship 7 Mercury capsule shot in 1962.

Now this introduces another interesting fact as far as some UFO are concerned. As we shall see later, such an ionised belt may be a permanent and unavoidable complication for a device which may employ a field drive. Could it be that ionisation blackout problems of our own returning rockets will prepare scientists for another type of communication, which in the long run might render our present radios obsolete, perhaps answering many a sceptic's query regarding the apparent lack of radio noise from saucers?

Be it so or otherwise, this side effect of re-entry may serve to illustrate the enormous price in terms of patience as well as money that must be paid before we progress just one faltering stage further. Perhaps the cascading blinding inferno of yet another returning space venture, screaming its tortured way back to mother earth, brings dramatically with it a breathless message . . . there *must* be a better way!

4

Gravity and Magnetism

YOU have survived the last three chapters and have earned the right to a little respite. This was the stiffer, rougher part of our preamble, now the land is getting flatter, the view a little clearer. We can take a look backward over the track and perhaps recognise a pattern there. Observe, the last three chapters had one thing in common. Each in its own way tells a little of the story of man's continual battle to get at grips with the most baffling secret of nature, gravitation.

You will see that with the conventional aeroplane, no matter how sophisticated it may become, man is combating gravity with the aid of aeronautical stilts, no more, no less. He rises from the bottom of the aerial sea he calls his atmosphere, much as do the fish in the oceans, neither they nor he are one jot free of gravity.

You will have seen how the rocket is not much different. The principle is one of refinement, not of kind.

The aeroplane derives lift as a reaction to the downward displacement of a mass of air. The rocket on the other hand derives lift, or thrust, as a reaction to a downward ejection of a mass of gas. Fundamentally the principle is the same. The chief difference lies in the fact that the rocket is capable of taking some of this gas, or working fluid, outside the atmospheric envelope, where it can go on functioning in free space. We have seen how it opposes gravity by piling on an enormous velocity, which will either be fast enough to send the vehicle coasting outwards against gravity or if directed parallel to the earth's surface, create sufficient centrifugal force to exactly match the gravitational pull, and then it is said, the vehicle is in orbit. In any event, gravity is being fought, with man as the duellist, having not the slightest idea just what it is he is fighting. True we have discussed some of the limitations of the rocket, but are there any other grounds at all for considering that pure reaction machines may not be utilising their energy in the most efficient manner? Well, from the thermodynamic point of view, the efficiency of the modern rocket is pretty good and in this respect the thermodynamicist has done incredibly well. And it is obvious that as far as the available kinetic energy stored in the thrust jet is concerned, the engines can hardly get any more power out of this. But suppose an argument could be offered which although mechanically unrealisable, would nevertheless indicate that every reaction motor is theoretically capable of producing exactly twice its power, what then? We might be more than justified in saying to the rocket engineer, 'You're doing a pretty wonderful job, but are you sure you are using the available energy in the most profitable way?' Then naturally he would ask for your proof, and having offered it to him, he might be offended by the simple nature of the argument, that is, if he accepted the analogy literally. If on the other hand he was a wise man, he would not be offended, instead he would see the point and might begin looking elsewhere.

The author has two such arguments to offer, but I would stress to the lay reader that they cannot be interpreted literally, anymore than one could dream of building a bridge higher than the tallest mountain and circumnavigating the earth where it would remain suspended at its centre of gravity without any physical support, theoretically possible, but quite unrealisable.

The first analogy in Fig 12(a) is self explanatory, in which a rocket is just supported by its jet efflux, weight being equal to thrust. But the exhaust gases are conducted along a shaft running through the centre of the earth, and allowing for no losses in the kinetic energy of the exhaust stream, this is theoretically capable of supporting another rocket of equal weight situated diametrically opposite on the other side of the globe, the reaction on the base plate on this second rocket being equal and opposite to the reaction of the first. So that two Saturn rockets are theoretically capable of being supported by the blast of one, neglecting the loss of weight due to consumed fuel.

In actual fact this very situation exists on every rocket lift-off, for the deflected exhaust is in fact 'pushing' the earth in the opposite direction, but due to the vast mass of the earth such movement is undetectable, according to the law of reaction, Force is equal to Mass $\times$ Change of Velocity.

My second analogy is equally annoying, for while true and serving to illustrate the point, it can never be realised in actuality; it concerns centrifugal force. In Fig 12(b) we consider again the fundamental law which keeps a satellite in orbit, centrifugal acceleration of lg (being a function of the vehicle's forward velocity V) balances the acceleration due to gravity, *i.e.* lg. It will be seen from the formula $CF = \frac{MV^2}{R}$ that in order to maintain a constant centrifugal acceleration of lg, any reduction in the value of R must be accompanied by a proportionate reduction in the value of V. But obviously orbiting capsules are restricted to R_1 in Fig 12(b) due to the size of the earth. If however we could reduce R_1 to the relatively very small radius of R_2 in Fig 12(c) and a revolving mass of only a few pounds be permitted to enter the same shaft running through the centre of the earth, and revolve in the duplicate system at position Y, then two such identical 'earth satellites' of equal

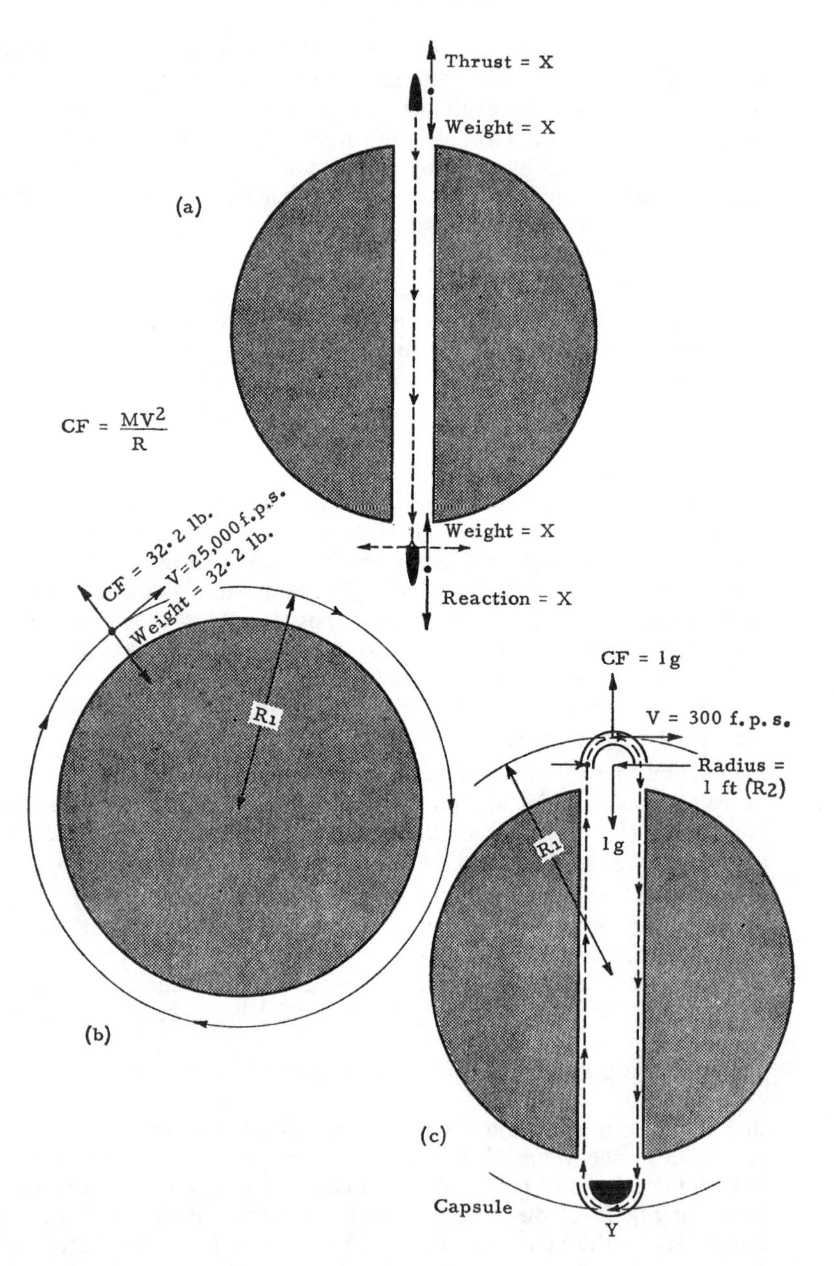

Fig 12. Two of the author's 'impossible' analogies derived by natural laws which nevertheless illustrate that there must be a more effective way of combating gravity.

mass to that in case (b) could be suspended above the surface of the earth for a fantastically reduced amount of kinetic energy, that is, if we neglect the time taken in transit of the mass from one side of the globe to the other. Just for the fun of it, this means, in the hypothetical case of an 'orbiting' weight of 32.2lb revolving round a 1ft radius arc and travelling at, say 300ft per sec, would yield a force of:

$$CF = \frac{MV^2}{R} = \frac{WV^2}{gR} = \frac{32.2 \times 300 \times 300}{32.2 \times 1} = 90{,}000\text{lb or a little over 40}$$

tons! In other words the above 'engine' under these conditions is theoretically capable of *supporting two* 40 ton space ships and allowing for no frictional losses in the system, once the revolving mass had been accelerated up to 300 feet per sec, it would go on for all time, just exactly the same as its earth orbiting brother.

But if it takes colossal boosters yielding millions of horse power to accelerate orbital capsules of several tons weight up to velocities in the order of 25,000 feet per sec, then there is no necessity for further recourse to calculation in order to show that the acceleration of a thirty-two pound weight up to a velocity of a mere 300 feet per sec, as in the later example, would take a correspondingly *microscopic* amount of energy to accomplish several times the work. If only we could dig that big hole! Briefly then, there would appear to be justifiable reasons enough to look into other techniques of space ship propulsion—the first step must surely be a closer look at gravitation.

Long before, and certainly ever since Galileo first demonstrated that all bodies, irrespective of their size or mass, fall towards the earth with the same acceleration, men have dreamed and pondered on the mystical nature of gravity. What is this strange force which permeates all matter?

Our space between these covers is limited and beyond doubt this and much more besides could be filled with many theories and pipe dreams on gravitation. We must resign ourselves to just mention a few. But how often, and if at all may we wonder, have some of these inspirations bordered on the truth, or does the secret of gravity reside so deep within some incomprehensible geometry of spacetime, that it will forever be hidden from the inquisitive gaze of man? I think it will not.

There can be little doubt that the late Sir Arthur Eddington was equally addressing the present when in 1933 he wrote, 'We have turned a corner in the path of progress and our ignorance stands before us, appalling and insistent. There is something wrong with the present fundamental conception of physics and we do not know how to set it right'.

The following observations of the author are by way of an honest appraisal, rather than a rebuff, but let us face it, almost daily the enormous strides made in physics take scientists deeper and deeper into an

ever darkening wood and somewhere within that wood lies the answer to gravity. Only a few years ago, physicists spoke of the electron as a particle, but then the thing behaved irrationally as a wave and clearly it couldn't be both . . . or could it? And the 'waveicle' was born. Perhaps in no clearer way can the situation be expressed than by quoting Hoffman's* chicken and the egg analogy:
Little boy: 'Daddy, what came first, the electron or the wave?'
Daddy: ' . . . eh yes'.

The late Sir James Jeans once made several pointed remarks with which such men as de Broglie, Max Planck, Einstein and Schroedinger associated themselves. They may be thus epitomised:—
'Thirty years ago we thought that we were heading towards an ultimate reality of a mechanical kind. Today there is a wide measure of agreement, which on the physical side of science, approached to unanimity, that the stream of knowledge is heading towards a non-mechanical reality. The universe begins to look more like a great thought than a great machine. Matter is derived from consciousness, not consciousness from matter. We ought to hail mind as the creator and governor of the realms of matter'.

On gravity, Sir Isaac Newton in his 'Principia' said:
'That there is some subtle spirit by the force and action of which, all movement of matter are determined', and again in his third letter to Bentley he says:—
'It is inconceivable that inanimate brute matter should without the mediation of something else which is not material, operate upon and affect other matter, without mutual contact, as it must do if gravitation in the sense of Epicurus be essential and inherent in it. That gravity should be innate, inherent and essential to matter so that one body may act upon another at a distance, through a vacuum without the mediation of anything else by or through which their action may be conveyed from one to another, is to me so great an absurdity that I believe no man, who has in philosophical matters a competent faculty of thinking, can fall into it. Gravity must be caused by some agent acting constantly according to certain laws, but whether this agent be material or immaterial I have left to the consideration of my readers'. We shall see at the end of this chapter that Sir Isaac may not be so wrong as many modern thinkers would have us believe.

On gravity, Michael Faraday said:—
'As the coil is to the magnet, so I believe the condenser may be to gravity', and again, 'I have long held an opinion, almost amounting to conviction, in common, I believe with many other lovers of natural knowledge, that the various forms under which the forces of matter are made manifest have one common origin; or in other words, are so

* Author of 'The Birth of the Quantum'.

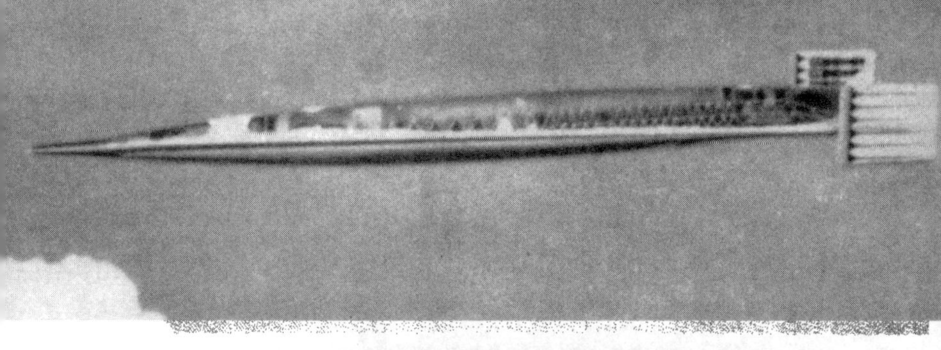

Plate 5. *Griffith supersonic VTOL aircraft.*

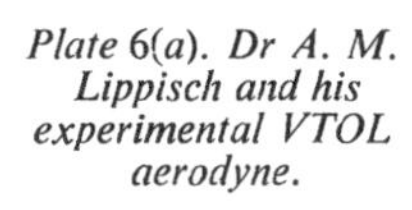

Plate 6(*a*). *Dr A. M. Lippisch and his experimental VTOL aerodyne.*

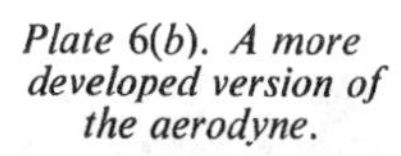

Plate 6(*b*). *A more developed version of the aerodyne.*

Plate 7. Stainless steel model subjecte to air blast conditions likely to b met at five times the speed of sounc The temperature is about 3,000F.

Plate 8. Canadian Avrocar uses an air cushion for VTOL. On reaching a certain translational velocity it functions as a conventional aircraft supported by aerodynamic forces.

directly related and mutually dependent, that they are convertible as it were, into one another, and possess equivalents of power in their action. In modern times, the proofs of their convertibility have been accumulated to a very considerable extent, and a commencement made of the determination of their equivalent forces.

'This strong persuasion extended to the powers of light, and led, on a former occasion, to many exertions, having for their object the discovery of the direct relation of light and electricity, and their mutual action on bodies subject jointly to their power; but results to this time have been negative.

'These ineffectual exertions, and many others which were never published could not remove my strong persuasion derived from philosophical considerations; and, therefore, I recently resumed the inquiry by experiment in a most strict and searching manner, and have at last succeeded in magnetising and electrifying a ray of light, and in illuminating a line of magnetic force'.

In this, Faraday may have been inspired by the observations of Newton, from whom were derived the Newtonian Philosophical rules of nature, the first of which in effect says: 'As nature herself demonstrates abundant economies, then in trying to understand her processes, man's reasoning must also be economical. Therefore the rule: whenever a complicated hypothesis can be embraced by a much simpler one, then the latter must predominate'. This is plain commonsense, yet it is amazing how often physicists are tempted to stray from the rule, when new conceptions clash, despite the long and tedious experience over the last few decades which has demonstrated otherwise. Bearing in mind the fact that practically every major advance in physics is usually the result of a 'hunch' in the very first place, it might now and again pay off to consider one or two more outrageous ones, examples, there have been many, which only today are being substantiated. One such 'strange coincidence' occurs to the author, and at the risk of straying a little out of context, I have recorded it here.

It concerns the work of John Worrel Keely, who among other things nearly brought Wall St down in 1890, due to his controversial claims of the Keely Motor. He also successfully demonstrated a gravitational device to leading scientists of that time. Incidentally, contrary to popular belief, Keely was never really exposed as a fraud; the whole affair still remains a mystery.

The basis of Keely's work was vibration, and let us remember, his sceptics loudly declared this was nothing but a front for his fraudulent claims. In performing one experiment employing vibrations of an extremely high order, Keely writes, 'The highest range of vibration I ever induced was in the one experiment that I made in liberating ozone by molecular percussion, which induced luminosity, and registered a

percussive molecular force of 110,000lb per square inch, as registered on a lever constructed for the purpose. The vibrations induced by this experiment reached over 700,000,000 per second, unshipping the apparatus, thus making it insecure for a repetition of the experiments. The decarbonised steel compressors of said apparatus moved as if composed of putty.'
Note: This effect, now known as sonoluminescence, is a comparatively new phenomenon, where electrons are encouraged to change their energy levels, due to ultra sonic vibration, thereby emitting quanta of light.

This was either a case of chance favouring a fraud, or Keely's claims were genuine. The point is, his work also was based solely on an original 'hunch'. To any of my readers who doubt the authenticity of Keely's claims, I say, investigate this man's work *fully* before you arrive at hasty conclusions.

Sometimes it is a good practice to go right back to square one and begin our thinking all over again, and in the case of gravity we have no alternative, for we know nothing about it at all. All we know is that matter seems to have the property of wanting to get together. For the want of better understanding, men have called this 'attraction'. Modern physicists call it a warp in space time—same thing—we still don't understand it. To some this may sound impudent, but let us face it, it happens to be true.

Now while we are resting up here on our little plateau, let us muse a little and go back to square one. We observe that all matter appears to be 'attractive' in the strictly material sense. An ordinary bar magnet has this property to a marked degree, over a piece of steel or iron. We note that if the mass of the magnet and the mass of the iron are the same, despite the 'origin' of the force residing in the magnet, in free space, the two would move together over equal distances and at uniform rates of acceleration. In other words, the motion is evenly shared. Now on the other hand, should we substitute for the iron another magnet of equal mass to the first and orient them with unlike poles facing, then they will also move together and the motion will again be evenly shared. But in this case, due to the greater field strength, there will be a correspondingly greater acceleration. All this is in accordance with Newton's laws of motion. But note, the increase of attraction in the latter case has been nothing to do with mass, for there has been no change in this respect.

Similarly an electro-statically charged body will exert an attraction on an uncharged body, and if two such bodies of the same mass were in free space they will move together at equal rates of motion exactly as the magnet and soft iron. Again should the two bodies be charged with opposite signs, negative and positive, then the acceleration will be increased as with the two magnets of opposite polarity. Again there has been no increase in mass.

Now by giving the magnets like polarity, north to north or the charged bodies a like sign, negative to negative, they will experience the same rate of acceleration as before, but this time away from one another. We do not really understand this behaviour, neither does modern physics explain it. Men the world over have spent their lives theorising and whole volumes have been written on it, but let us have no doubt about it, we are standing on very firm ground when we say the world is still waiting for an explanation for these phenomena. Of course there are theories in abundance. Paramount among them is the molecular theory where each molecule is likened to a small magnet, and the more popular current idea, the atomic theory, where each atom behaves again as a bar magnet. In effect, this says, if an electric current consisting of billions of electrons, flowing round a coil of wire, produces a magnetic field within the coil almost identical to a bar magnet, then similarly an electron revolving in orbit around the nucleus of the atom will form a miniature magnet.

A bar of steel is said to be magnetised when the poles of the atoms are lined up one with the other. Apparently the only difference with other materials is, this phenomenon of lining up does not occur quite so readily as it does with iron or steel atoms. Therefore it is difficult, though not impossible, to align a magnetic field in say, a piece of wood.

On the other hand, an electrified body exerts an attraction on a neutral body because it is said to have a surplus or a deficiency of electrons. The materials in a substance are said to be electrically neutral when the negative and positive charges within the atom exactly cancel each other out. In the simplest case, the Hydrogen atom, the electron having a negative charge, cancels out the proton's positive charge and the atom is said to be electrically neutral, and the same is true of more complex atomic structures.

When electrons are removed from the atoms of a substance, the balanced state is upset and the material is said to be positively charged. Similarly when the atoms of a body take on a surplus of electrons, the body is said to be negatively charged. We are told that such a charged body is attracted towards an uncharged body, or another body of positive sign, because the surplus electrons *want* to dissipate themselves, or the electron deficient positively charged body *wants* to borrow some of the surplus electrons.

Again within the structure of the atom itself, we find the same phenomena of attraction and repulsion, particle for particle and wherever there is a moving electron, there too is a magnetic field.

Summing up then, in all matter, from the microcosm to the macrocosm, the same basic pattern is there. Whether it be gravitational, electro-static, or magnetic, there is this moving to and from tendency inherent in all matter, constantly obeying the inverse square law. But gravity is *always* there, be matter electrically or magnetically neutral, it is always subservient to the force of gravity. Two magnets may repel each other, but the force of gravity is not one bit subdued by the magnetic effect. The same is true of two similarly charged bodies, gravitational attraction is still operative between them.

Now here we are at square one; shall we for the moment then, leave the well trodden paths of orthodox scientific exploration, and remember Newton's Philosophical rule of economy? By applying this rule, we would have to suppose that magnetic, electrostatic *and* gravitational phenomena were one and the same thing. But now we are still left with two other alternatives, one of which is claimed superfluous by the economy rule, which is it to be, attraction *or* repulsion, for the rule must still be applied and in any case, what if that which you and I observe as 'attraction' and 'repulsion' was incidental to matter, and all we were observing was relative motion, nothing more, what then? Naturally we would have to talk in terms of attraction and repulsion in order to designate the motion, but due to the economical rule, our hypothesis has been drastically simplified. We are left with matter and motion. This is all very breathtaking up here on the plateau, but be patient, we still have a long way to go. 'But', you might protest, 'this chapter was headed gravitation; we seem to have strayed into magnetics a little.' Then I shall answer, 'Again have patience, for if Newton's economy rule is to be of any use at all, surely it is far more convenient to examine magnetism as a kind of selective gravitation than it is for us not to be able to examine gravity at all.'

In *Space, Gravity and the Flying Saucer*, the author constructed some simple analogies in order to illustrate more abstract ideas to the layman. At the time I felt this usage might have offended some of my more technically capable readers, but this proved not to be so, in fact encouraged by favourable comments from many highly qualified people, I have taken advantage to use some analogies again.

The following example is merely an attempt to illustrate more graphically an imaginary relationship of the various field phenomena mentioned above. The important thing to be remembered of course is, as with all analogies, the experiment is not to be interpreted literally. In this it will be seen all the varying degrees of attraction and repulsion have their origin in a common source, while the condition which imposes the apparent difference is one of relative motion, that is all.

Consider the test rig shown in Plate 13, it comprises a base stand on to which are mounted two electric motors, having bar magnets secured to the shafts. The rig is self explanatory in layout, now for a word on its function.

Let us assume at the start of the experiment, that both motors are running at such speed that we, the observers, are aware only of the blur of the revolving discs. Up to this point the upper swinging motor has been held aloft, now it is allowed to descend. The speed of the motors is variable so that any particular phase or polar relationship of the bar magnets may be operative, and a little thought will show that if the speed of one motor is slowly adjusted, different settings will be found where the conditions of attraction, neutral and repulsion are manifest. In other words, when both magnets are revolving at, say 1,000 revs per minute, with opposite poles lined up, *i.e.* north to south, then they will be attracting. But if one of the motors is speeded up, then the lining up of the poles becomes alternate, north-south, north-north, etc., in which case the attraction and repulsion conditions cancel out. This is the neutral condition. The third phase where like poles are lined up, produces the condition of repulsion.

Now the important thing about the analogy is, if the rig is analogous to the atom and the rotation of the magnets so high as to render measurement impossible, we should have no idea of the slight changes in relative speed and the corresponding phases which were taking place. Therefore we might just as easily attribute several different causes to explain the various phenomena, which in this case is change in relative motion only. Of course the same experiment can be performed by two simple electromagnets, but this would hardly serve to illustrate the point in mind. Past experience warns me that some flying saucer enthusiast may be tempted to interpret the above example in terms of UFO behaviour, but obviously I do not intend this.

There is still quite a lot of land to survey and we are not quite finished with magnetism, which I like to think of as a kind of selective gravitation.

In modern relativistic thinking, there is no room for the outdated conception of an all pervading ether and of course physics is carrying on quite nicely without it by the substitution of algebraic abstractions. But if our philosophical rule requires simplicity, then for a while we shall keep our ether, for let us remember we started from square one deliberately to see if we missed any ground.

The following account will not be found in any elementary text book on magnetics, this may be a pity, for quite a number of students might have redirected their thinking had it been so.

At the 1881 Electrical Exhibition held in Paris, Professor C. A. Bjerkness, of Christiana, demonstrated some most interesting experiments, showing that it is possible to imitate most of the well known actions between currents and magnets by means of bodies pulsating and vibrating in water and other fluids. At that time it was felt that the experiments offered a possible clue to the nature of the mechanism which transmits electric and magnetic forces through space.

The apparatus used consisted of a glass trough, filled with water, in which the vibrating and pulsating bodies were placed. These consisted of little drums, with elastic diaphragms at the ends, which were made to pulsate by drawing air in and out by means of pumps. The pumps had no valves, and were constructed like a child's ordinary squirt, so that the air was drawn in and out at each motion of the piston. There were two pumps which actuated the pulsating bodies whose mutual actions it was desired to study. These were driven rapidly by hand wheel, and by altering the position of one of the crank pins, they could be made to work either in the same or opposite phases. In the experiments, one pulsating body was held by hand and the other pivoted on a stand, so that it was free to move like a compass needle under the attraction or repulsion of the first one.

PHASES. Two pulsating bodies are said to be in the same phase when they both expand or both contract at the same instant, and in opposite phase when one expands while the other contracts. It was found that there is a close analogy between the mutual actions of pulsating and vibrating bodies, and of magnetic poles and electrified bodies, but that in all cases the analogy is inverse. The force in all four cases varies inversely as the square of the distance between the attracting and repelling bodies.

Attraction of Light Bodies and Soft Iron

If the suspended body is disconnected from the pump, then a pulsating body will repel it in the same way as an electrified body attracts light objects, or a magnet attracts soft iron.

Attraction or Repulsion of Compass Needle

A body oscillating along a horizontal axis, which is free to turn, will follow or fly from a pulsating or vibrating body, just as a compass needle will follow or fly from the pole of a magnet. The direction of the force depends on the phase.

Two Magnets of Unequal Strength

If two magnetic poles of the same polarity, but of which one is much stronger than the other, are placed a little distance apart, they will repel, but if brought near together, they attract, as the large one induces in the small one a polarity opposite to its own and stronger than its natural polarity. Similarly, two pulsating bodies moving in the phase which produces attraction, and one which is much larger than the other, will attract when they are a little distance apart, but repel when they are near.

Diamagnetism

Faraday suggested that many of the phenomena of diamagnetism may be accounted for by supposing all bodies to be paramagnetic, but of different strengths, and that the apparent repulsion observed with bismuth and other bodies is only due to the stronger attraction exercised on the medium in which they are immersed. It is probable, however, that this explanation is not sufficient to account for all the phenomena observed.

The analogous case in the Bjerkness experiments is that the actions on a body are opposite, according to whether it is lighter or heavier than the medium in which it is immersed. Bodies heavier than water are attracted by a pulsating body, bodies lighter than water are repelled. In each case we consider the body heavier than water as a type of diamagnetic body, remembering that all the phenomena are inverse; that the one lighter than water to be similarly the (inverse) type of a paramagnetic body. Thus the body heavier than water is acted on like soft iron, the one lighter like bismuth.

If two magnetic poles of the same name be placed a little way apart, a piece of iron will be repelled from between them; if they are of opposite names, it will be drawn in. Similarly, if two drums are placed a short distance apart, a body lighter than water will be attracted to the centre if the drums vibrate in the same way phases, and will be repelled if they vibrate in opposite phases.

Lines of Force

Professor Bjerkness succeeded in tracing out the lines of force in the water due to the various pulsating and vibrating bodies experimented on, in a form which enabled him to compare them with the corresponding lines of force displayed by magnets and currents as traced by iron filings. The apparatus consisted of a heavy metal ball, supported on a stand by means of a light steel spring. When this was placed at various points, it oscillated along the direction of the line of force at that point, *i.e.* along the direction of the wave in the water. A fine rod attached to the top of the device projected from the water, carrying a camel hair brush, which recorded the direction of vibration on the under side of a piece of smoked glass. A series of curves were obtained, which show that the lines of force between pulsating and vibrating bodies in water are exactly similar in form to those between converging electric, magnetic and electro-magnetic forces.

A Mr Strob since repeated Professor Bjerkness' experiments and reproduced nearly all the phenomena by means of sound waves in air. He caused air, vibrating by sound, to transmit the forces in the same way as Professor Bjerkness transmitted them through water.

The summary went on to state, 'In these researches we see opened a possibility of explaining some of the mysterious mechanism of electric and magnetic attractions, without the necessity of supposing any force to be at work other than those with which common experience makes us familiar, for we see them all reproduced by vibrations of material fluids, which differ from our supposed ether only in the superior quality of elasticity and smaller density of the latter, that is, they differ from it only in degree and not in kind.'

5

Gravitation and the Ether

THAT gravitation is caused by an all pervading ether was long contended by earlier thinkers; we shall now examine a theory which helps to partly reconcile this old idea with the modern concept of wave mechanics. Further, it offers an explanation for the belief that gravity is a 'push' from without rather than a 'pull' from within. It suggests that both concepts are incomplete, that it is just as unsatisfactory to consider gravity as an inward attraction force of unknown origin (comparable to the tension of elastic bands 'pulling' several bodies together) as it is to imagine an unknown source of 'pressure' acting from without, Fig 13.

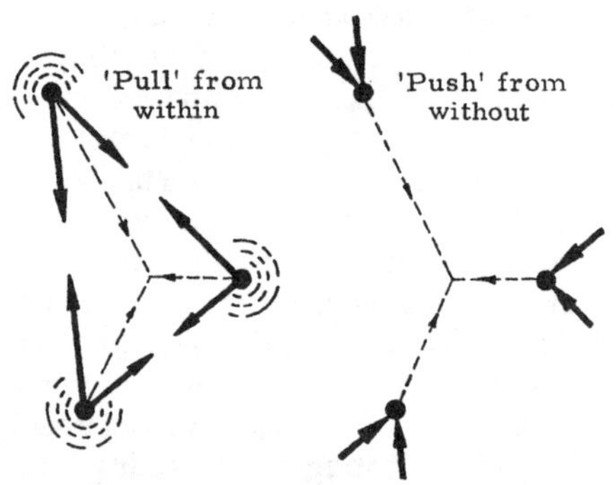

Fig 13. *Is Gravity a 'push' from without or a 'pull' from within*?

It is argued, that if we could reconcile the apparent 'pull' of unknown origin and the equally unknown 'push', the problem has been clarified to some extent, but if in addition a possible working hypothesis can be offered which is borne out by known facts, then we have done more.

To consider a boundless space bespangled with matter in varying degrees of formation upon which is exerted a 'pressure' from without, leads to the obvious question: what is the origin of this pressure and how is the layman to visualise it, free from mathematical entanglement?

The theory offers a possible explanation, the nature of which conforms to a general pattern displayed by nature. It is circular motion.

We have seen how some of the greatest teachers have postulated the unity of the universe. Faraday again and again expressed his conviction of it. Dr Albert Einstein tried to establish it in his unified field theory. Then in one of the most fascinating and complete works beyond the scope of this book, an Englishman, Antony Avenel of Yorkshire, developed a theory called *The Unity of Creation Theory*. The present author enjoyed the privilege of using a condensed version of the theory in *Space, Gravity and the Flying Saucer* and, as this is now out of print, I feel it is important to re-quote this remarkable theory for the benefit of those readers who may be unacquainted with it.

Unity of Creation Theory

The recent correspondence in technical magazines seems to show that many readers feel the need for something less coldly mathematical than Einstein's 'Theory of Relativity' and his subsequent theories. Few suggest that Einstein's brilliant calculations and theories are faulty, yet by themselves those essays in pure logic are not comprehensible to the average person.

One cannot gain the 'mental picture' of Einstein's theory because the theory is not in a form which leads to a mental picture. If you read the test performance figures for a new aeroplane, you will know a lot about what the plane will do, but you will be unable to visualise whether its lines are beautiful or ungainly, or anything of its appearance.

Einstein's theory was before its time. The calculations of, for instance, the precise amount that a material contracts along the direction of its movement seems to be out of place before it has been explained why the contraction occurs.

I suggest that there has been too much mathematical jig-saw puzzle making and solving, and that formulae have been put forward which, though probably correct, are by themselves without very much meaning to the intellect.

The Michelson-Morley Experiment

This experiment aimed at finding the speed of the earth through the ether. Scientists had assigned to the ether descriptions ranging from an elastic solid to a rarefied gas. If our speed through the ether could have been determined, it would help us to understand (among other things) whether we are labouring through a mass like black treacle, or wafting our way through a substance as thin and delicate as perfume.

Those who rely on this experiment, or on similar experiments make an assumption which I believe to be false, that is, that the ether is a three dimensional substance—such as gas. Only if the ether was a material substance would the passage of the earth through it cause an ether drag, or an ether wind, which could be measured.

The result of the Michelson-Morley experiment showed (apparently) that either there was no ether, or if there was an ether the earth was not moving through it. Neither of these conclusions seemed to be probable; it would be unlikely that the earth should remain stationary in space when all other observed heavenly bodies were moving. Nor was it likely that there was no ether, for how else could the passage of rays through space be explained?

As neither of these conclusions could be welcomed, it was later suggested that the Michelson-Morley experiment really did show a positive result, but that the measuring rod in the direction of the earth's movement through space contracted by an amount exactly sufficient to remove the positive result from being apparent (the Lorentz Contraction). The proposition was of course that all materials contracted in the direction of travel; the supposed contraction was not confined to the measuring rod in the Michelson-Morley experiment.

The Lorentz Contraction at first sight seems to be an artificial and far-fetched theory, yet I think that those who have studied the calculations, and those who care to do so, will agree that the contraction must be accepted as something which actually does take place.

The Theory of Unity suggests reasons why the Lorentz Contraction takes place.

Theory of Unity of Creation

The interest shown lately in the physical world prompts me to offer an outline of that part of the theory which affects this subject. The theory suggests, among other things, why the phenomena forecast by Einstein's theories take place. It is unsatisfying to be told that time slows when you travel through space, and even to be informed of the precise amount by which it slows compared with your velocity, before any attempt is made to explain what time is and why it is capable of slowing.

The following statements and arguments are set out in rather a dogmatic and over-simplified form, which I hope will be excused, in order to try to offer an outline of the theory which can be followed without undue effort.

The theory anticipates the ultimate result of the fact that research discovers one unity after another in physical phenomena. One is led to expect that before long it will be proved that there is one basic building material for the whole universe. I do not pretend that there is sufficient data available at present to prove the theory fully, but there are many indications that it is an anticipation of what will be proved, by, let us say, the year A.D. 2000.

The theory that I put forward is that the ether and space are the same, and that space is formed out of nothing by a grid of extremely high frequency rays (probably having a wavelength of less than 10^{-13}cm). Space must be distinguished from 'nothing'. Space—even if it is empty—possesses the qualities of length, breadth, thickness and time. 'Nothing' has no qualities whatsoever, and cannot support any material or ray. In other words, creation of the universe takes the form of making space out of 'nothing', and the method adopted for making space is a network or grid or rays, which I call 'creative' rays.

Outside the Universe

Taking 'the universe' to mean all created space, there is 'nothing' outside the boundaries of the universe. The old problem of imagining the boundaries of the universe, outside which stretched empty space—which space must have boundaries, and what was outside that?—should not arise. 'Endless space' is a contradiction in terms. Space has dimensions and boundaries and cannot be endless. The hand of creation has not touched the 'nothing' outside the boundaries of the universe, and that 'nothing' has no dimensions and therefore no boundaries.

To put it in another way, space is positive creation, while 'nothing' is the absence of space, and thus purely negative.

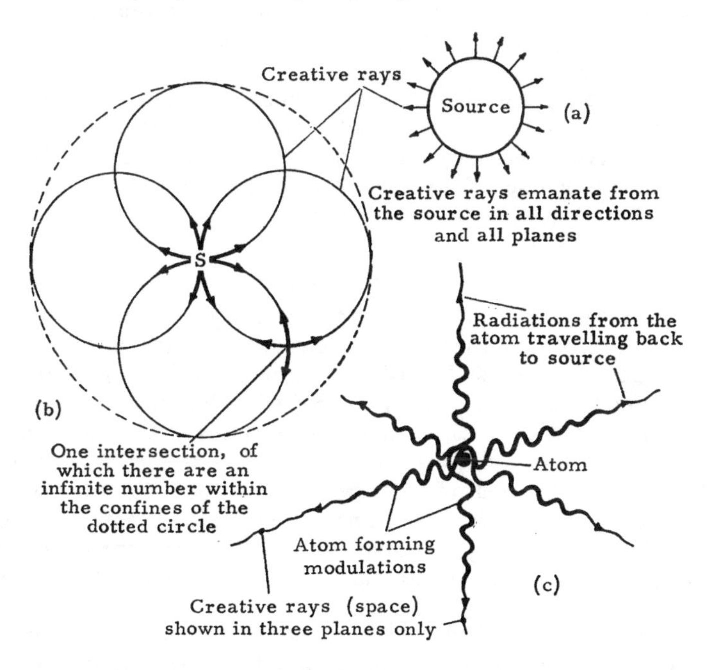

Fig 14. Formation of space and matter according to the Unity of Creation Theory.

You cannot visualise 'nothing' for obvious reasons; it has to be accepted. If anyone particularly wishes to try to relate it to human experience, it could be said that he has had more of it than he has had of space and time. It is what he experienced, or did not experience, before he was born.

Space or ether is formed by the creative rays which emanate from one source in all directions and in all planes, Fig 14(a). Each creative ray covers a circuit from source back to source, and each circuit is probably the same size. In this way space with boundaries of globular shape is built, and whatever point is taken in space, creative rays travel in all directions towards the source, Fig 14(b) and Fig 15(a).

By the word 'source' I do not imply that the formation of the creative rays operates in only one direction in each circuit; the action may be alternating.

Light is a Modulation of the Creative Rays

All rays of whatever frequency, visible or invisible, detectable or undetectable, are modulations of the creative rays, in the same way as a high-frequency radio wave is modulated by a musical note. As a radio carrier wave can be modulated by a number of separate notes, so can the ether carry between the same two points any number of waves of differing frequencies.

It would appear that rays or modulations are always caused by a disturbance in three-dimensional material, and that they are only of consequence when they encounter other such material. When a ray travels through space it is merely a slight modulation or disturbance of the creative rays and of no importance.

Material Objects

The atom is the building material for all solids, liquids, and gases, and each atom is composed of a nucleus round which revolve electrons at distances from the nucleus which vary with the type of atom. I submit that the atom is not solid fundamentally, but that it is composed of modulations of the creative rays in three planes. Although a modulation is normally a ray which travels in all directions from *its* source towards the source of the creative rays, the chord of modulations forming an atom are locked together in three planes. This lock prevents the modulations travelling in opposing directions as rays. Does not the release of atomic energy show the very close relationship between atoms and rays? Fig 14(c).

The main point which I want to make is that rays and atoms are both modulations of the creative rays, the former being simple modulations, the latter being complex and static ones.

An atom could in some ways be compared with a ripple caused by a stick in a smoothly flowing stream of water. It remains the same in appearance yet it is formed from a constantly changing medium. If this is correct, the universe is made from the same medium throughout, and what appears to be empty space between the Earth and Mars is in reality a connecting medium.

Time

I suggest that time is the effect on our minds of the frequency of the creative rays. If the atoms out of which our brains and bodies are made are formed out of the creative rays, we cannot but be aware of the alternation of the creative rays. We cannot escape from time unless we also escape from space, or, in other words, cease to exist.

It is impossible to look either backward or forward in time from a fixed position in space. If we could travel at the speed of light and thus 'keep up with time' we should probably cease to be three dimensional which would not assist our observations! In any case, we should find ourselves in a different position in space, so we cannot by any means foresee what is going to happen, or look back on what has happened, on earth.

Alteration of Time

If we were to travel at a very high speed—a substantial proportion of the speed of light—the frequency of the creative rays in the direction of our travel would be increased, because we would be travelling relatively to the pulses of the creative rays. It can be envisaged that something akin to the Doppler effect would take place, with the result that our basic time would be increased in frequency. We should not be aware of this, because the frequency of the creative rays is our only standard of time, and there is nothing *nearby* against which we can test this standard. But a stationary observer could, by rays of light, calculate the difference between our time and his time; he would say that our clock was going slow compared with his clock, or that our basic time frequency was quicker than his.

Clock time is our way of counting the number of pulses of basic time. If basic time frequency increases, clock time still counts as one million pulses what are now, say, two million, and the clock time appears to be going half speed.

Some space travel enthusiasts consider that if you could travel fast enough in a space ship, you could spend twenty earth years away from our planet and come back only a year or two older than when you left. If this is calculated using basic time-space, it is found that the effect on your body, and the impression on your mind, is exactly twenty years' worth of earth time, and that you could not therefore enjoy almost perpetual youth by this very inconvenient method.

Contraction of Length

If an atom moves along the creative rays, the increased frequency referred to before results in a shorter effective wavelength of the creative rays, which decreases the measurement of the atom in the direction of its travel.

For the purpose of simplicity, take it that the material length of an object is formed by the wavelength of the creative rays, while basic time is the frequency of the creative rays, then wavelength $\times$ the frequency of the creative rays will remain constant at whatever speed the object travels, because as the frequency increases the wavelength decreases. The product of the length of the object and the basic time is unaffected by the velocity of the object, and it is this product which gives to our minds the impression of time and of the proportions of the object.

The creative rays' present existence or the possibility of existence, and time and space are a division of that presentation. In whatever proportions the division is made, the whole remains unchanged.

Rays and Materials are Temporary

I would now like to meet the objection of those who say that it is just as difficult to believe that the creative rays travel through 'nothing' as it is to accept that light travels through space without an ether to carry it. My reply is that the theory holds that the creative rays create space not casually, but permanently: their cause is not casual, like the cause of a ray. The theory proposes that rays and materials are casual and temporary modulations or disturbances of the creative rays. It would seem unreasonable to believe that a special act of creation is necessary every time you choose to switch on an electric torch. The theory of unity holds that you, by switching on the torch, are able slightly to modulate the creative rays, which are permanently present, and that the casual phenomenon of visible light is the result.

Gravity

It is usually accepted (to put it basically) that if in space two masses exist, they *attract* one another. I suggest that this idea is wrong, and that it is impossible for a material object to emit rays which *pull* another object. Nor is there anything other than a ray which could exert the supposed pull. Rays can exert a small amount of pressure on an object in the direction of the ray's travel, but they cannot pull.

An alternative theory is that gravity is due to an increasing velocity, and the analogy of a lift rising at constantly increasing speed is often used. If a person in this lift released a pencil, it would appear to that person to fall to the floor of the lift and he might well consider that the pencil was attracted by the floor. If this is the explanation, why does gravity act in more than one direction? It requires adjustments which seem to me to be very artificial to answer this.

The theory of unity explains gravity as the material version of the natural travel of a ray towards the source of the creative rays. The modulations forming an atom tend strongly to split up, to break their three dimensional bond, and to travel in all directions, like ordinary rays towards S.

Referring to Fig 15(b), S represents the source of the creative rays, and the circle represents the circuit of one creative ray. While X represents the place at which an atom is formed by the intermodulation of the creative ray shown on the diagram with the creative rays in other planes; the latter cannot clearly be represented on paper, nor of course, can an attempt be made to draw to scale.

The tendency of X to act as a ray and to travel to S via Y and Z, and via other planes is nullified by the three-dimensional strength of X. The modulations of the creative ray start for practical purposes at Y and Z, but they do not interlock with modulations in other planes until X is reached. These preliminary modulations in ray form I will call extension modulations; some of them are of measurable frequency, others are of a frequency too high to be measured by a material device.

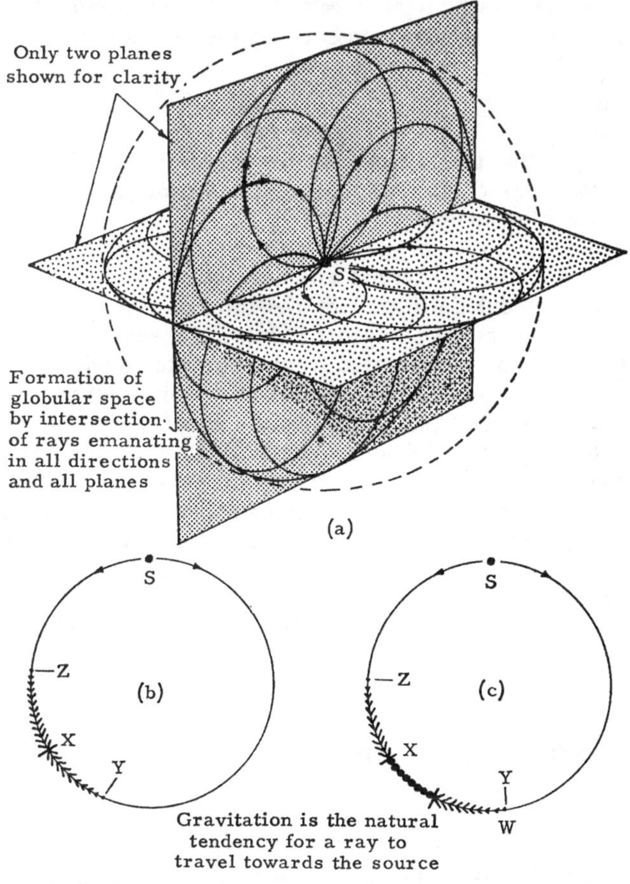

Fig 15. The formation of globular space and the function of gravitation according to the Unity of Creation Theory.

So long as X is undisturbed, it remains still in space, the tendency to travel to S via Y being balanced exactly by its tendency to travel to S via Z. Its tendency to travel to S in other planes is also balanced. But, referring to Fig 15(c), if in the position W (before the extension modulations of X have for practical purposes faded) another atom is formed, the extension modulation of X is interfered with and unbalanced between X and W. The extension modulation of X in the direction of Z is unchanged; XZ and XY are now no longer balanced, and as a result X moves towards W; W also moves towards X, according to the laws formulated by Newton, or approximately so.

Although X moves towards W, it is not *attracted* by W, any more than light from the sun is attracted by the earth. (Here I am ignoring the almost negligible element of gravitation between a material object and a ray: the reason why light travels from the sun towards the earth is not because of mutual attraction between the light and the earth) X moves towards S via W.

Magnetism

I suggest that it is not possible for the north pole of a magnet to emit rays which attract the south pole of another magnet, and repel the north pole of another magnet.

The travel of one magnet is not towards another magnet but towards S. Some atoms of iron are arranged, or can be arranged, so that the extension modulations are not the same in all planes. This lack of symmetry can be encouraged by electrical means. It is quite possible that a single magnet removed from a powerful gravitational field would move through space of its own accord. A single magnet on or near the earth is prevented from moving by the gravitational field of the earth—that is, it is prevented from moving through space of its own accord. If another magnet of opposite polarity or a piece of iron is placed near the first magnet there is apparent attraction, but what actually happens is akin to gravitation. The first magnet moves towards S until it reaches the second magnet or piece of iron. The strength of the magnet probably depends on the number of atoms in the magnet which have unbalanced extension modulations, the degree of lack of symmetry in each atom remaining constant.

Electricity

I suggest that this is a general disturbance of the extension modulations.

Flying Saucers

True 'flying saucers'—that is those which are not the results of the imagination of the observer—are vehicles which are based on the principle of unbalancing the extension modulations of material carried in the vehicles.

Reality

The question arises: 'Are these changes in time and space real, or are they only *deemed* to happen'?

The answer to this is, I think, what you and I and everyone else are concerned with is basic time × length representing the whole effect of both the frequency and the wavelength of the creative rays. In judging reality before our eyes, we are not concerned with the division of time × space into time and space.

If you want to listen to a concert on the radio it makes no difference to you whether the programme is carried to you by a 500 metre carrier wave or a 1,000 metre carrier wave, and you could detect no difference in the reality of the reception. You might then say that there was no real difference; but an engineer who is more interested in the method of your hearing the programme than in the programme itself would say that one programme was the result of modulating a carrier wave of 500 metres wavelength and frequency of 600kc/s, while the other programme was brought on a carrier wave of twice the wavelength and half the frequency. To the listener who was unable to go further into the problem than to hear what came out of his loudspeaker, the programme would be the same.

The answer then is, shortly, that although the change does actually take place in time and space, it is not real in the sense that it could be observed by a human being living within the sphere of the change, for such a person has not the means to measure basic time or basic length as an engineer can measure the wavelength and the frequency of a radio carrier wave!

The present author would once again like to thank Mr Avenel for the re-use of this remarkable thesis, without which I feel this present work would be incomplete. As many readers may automatically associate the pseudonym *Antony Avenel* with the *Unity of Creation Theory* I feel it will cause less confusion by referring to him thus, rather than Antony Parker.

Inertia

As in my earlier work, I would now like to take up the theory with the following: If gravitation is the most enigmatic force in the universe, surely the property of inertia runs a close second, in fact Einstein has already indicated the two are inseparable. If we can question the existence of the property of inertia (that is in the form accepted by the world of science), and thereby indicate that there is something akin to a higher 'octave' of matter in Space (or as Mr Avenel says 'creative rays') we will be in a more receptive frame of mind for the descriptions which are to follow.

To say that there is no such thing as the property of inertia, seems to be a denial of the long accepted and proven laws: to make such a statement may amount to something like scientific blasphemy, but we must not blind ourselves to the fact that the term 'inertia' was simply invented for a condition we just do not understand. What is inertia? Can we handle it or buy it by the pound? Has anyone seen it? The text book tells us that it is the property which a body 'possesses' by which it shows a reluctance to be moved from a condition of rest, or the reluctance to be brought to rest when 'possessing' momentum.

Tell some physicists that there is such a thing as a higher 'octave' of matter and we will be received with raised eyebrows, yet no one shows any alarm whatsoever when we say a body 'possesses' such an intangible thing as inertia. In actual fact the one is just as intangible as the other. Let us reflect; in the first place a body is said to be at rest when it has no relative motion to another body, and so *ad infinitum*, but is there such a condition? Not that we know of, because we cannot point to a single body in the whole of the known universe that has no relative motion of some kind.

Consider a body in space, there being no planets, suns, nothing save the one body. It is obvious such a body can be moving only in respect to something else, but there is nothing else, so the body is not moving. Still the physicist tells us the body 'possesses' inertia. It will stay where it is unless a 'force' is brought to bear on it.

Now the body is surrounded by nothing; we will accept for the time being the modern trend of thought that there is no such thing as an ether; how then can we accept the hypothesis of a body possessing of itself a reluctance to move unless it is fixed, or part of something else? Surely we must come to realise that the surrounding space *is* that something else? Then it follows that if the body of matter (the term given by us to a collection of non-material 'forces' called atoms) is affected by space, then in some way that space must be composed of a higher 'octave' of matter.

Let us examine another simple analogy, for in it we may find several interesting possibilities as far as inertia is concerned.

Fig 16(a) shows a plate upon which impinges two opposite and equal flowing jets of water, thereby holding the plate in a state of equilibrium. It is quite clear that the force acting on the plate is only proportional to the mass flow of water issuing from the jets and it follows that any increase or decrease in the jet velocity will result in a greater or lesser force.

Should we try to move the plate in the direction of one of the jets as in Fig 16(b) it follows that there will be a relative increase in velocity on that side and a corresponding relative decrease in velocity on the other side. Therefore the plate will experience an unbalanced pressure and will show a 'reluctance' to move (inertia).

But a body in space will only show a reluctance to move initially; thenceforth, it will continue to move. Therefore we must modify our analogy to accommodate this condition. We can do this by making jet A in Fig 16(c) controllable by means of a solenoid operated cock, and the solenoid in turn controlled by an electrical contact formed between the wheels of the trolley and the rails.

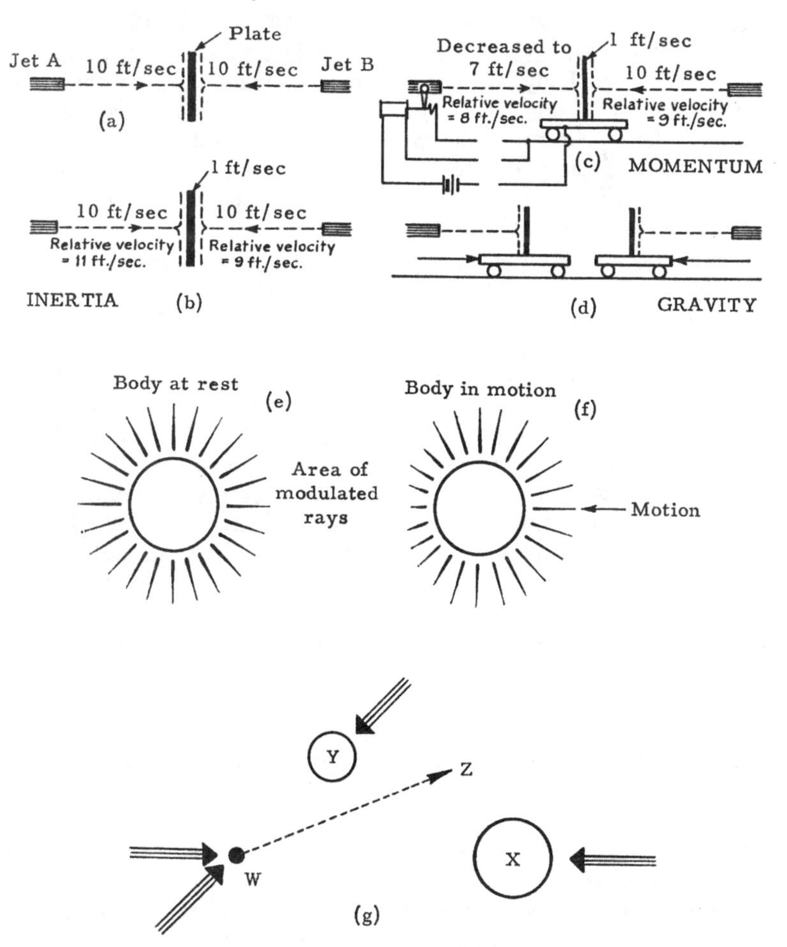

Fig 16. Motion, inertia, and gravitation are manifestations of modulated space.

Now, when we try to initiate a movement, there will still be the reluctance to move as in Fig 16(b), but immediately after this stage has been reached, the wheels complete the electrical circuit and the velocity of jet A is decreased and, as will be seen, the relative velocity of jet A will still be less than the relative velocity of jet B. Therefore the plate after suffering its initial 'inertia' will continue to move of its own accord. The relationship of inertia to gravitation is shown in Fig 16(d).

The analogy is limited and would be improved only at the risk of further complications; which in any case is unnecessary, for we are now in a position to analyse the condition of inertia in what may well be its true state. Fig 16(e) shows a body in free space, the radial lines indicating the etheric, or as Mr Avenel calls them, creative rays in one plane.

It follows that any reluctance the body exhibits to movement is not caused by the body, nor by the rays in themselves; rather that the phenomenon is common to both, comparable with the water jet and the plate. As we controlled the source of the jet in order to obtain continual movement of the plate, so by modulation of the creative rays—by an applied force—we bring about an unbalanced condition resulting in continual movement of the body, Fig 16(f).

We know that for a 'given mass' and a given applied force over a 'given time', a body will acquire a certain velocity. The theory suggests that this velocity is simply the result of the degree of the unbalanced condition, and that inertia is simply the resistance set up by the rays to being modulated. The greater the applied force, the stronger the modulation, and therefore the greater the resistance set up. Furthermore, it will be apparent that only those rays opposite to the applied force are modulated, therefore the body takes on a movement in the direction of that applied force, and what is equally important is the fact that the body will continue to move along the modulated rays (a straight line to our reckoning) according to Newton's First Law. The same reasoning can be applied to all aspects of our known mechanics. Even the phenomenon of so-called centrifugal force is now made clear.

Referring to Fig 16(g), W is a body which has previously received motion by the 'gravitational' or unbalancing effect of X. But it will be observed that there will be disturbance and therefore a corresponding unbalancing with respect to Y. The result of an intermodulation of this kind has the effect identical with the unbalancing effect brought about by X and Y, a *new* 'set' of rays is modulated in the direction indicated by Z.

The body has a new 'set' of modulated rays to move along and it will try to do this, but due to increased unbalancing effect of Y, again there is intermodulation.

The process is now continued until either W spirals into Y or, as in the case of a satellite, where the forces or modulations are balanced, it will continue to orbit indefinitely.

The whole point is, that because the body is forced continually to modulate a new set of rays every degree of arc, there is shown a reluctance which we call centrifugal force, or rather the reaction to centripetal force. Much the same sort of condition may exist in the atom, the proton 'catching' the electron by an unbalanced condition. The repulsion of electron by an electron is possibly due to an unbalanced condition in the opposite sense.

The great importance of the theory lies in the fact that it is not difficult to modulate the etheric rays. We do so every day of our lives, by moving our bodies or any inanimate object. Mr Avenel has told us that we probably do the same thing by pressing the button of an electric torch. It is very simple, but strangely enough we are not aware of it. It may be just as simple to move a vehicle from our planet's surface other than by the blasting method of the rocket. At present in order to 'lift' the vehicle we exert a force in the direction we desire it to move, that is, upwards, and in doing so we modulate the very rays which were keeping it in its original state (gravity). Had we an understanding of the etheric rays, we might be able to modulate them by other means and achieve the same result with far less expenditure of energy and discomfort.

We are now in a position to elaborate a little on Mr Avenel's basic theory of gravitation. It is possible of course that in some respects my own theory may differ slightly from Mr Avenel's, but the difference (if any) will be of detail, and of no fundamental importance to the theory.

Although the subject of gravity is of prime importance, the developed theory is beyond the scope of this book. However we can briefly correlate the known aspects of gravitation to the general theory, in the summary which follows:

Inertia. The resistance offered by the etheric rays to being modulated. The greater the mass the greater the number of rays to be modulated, therefore the greater the resistance set up.

Velocity. Rate at which modulations are transferred by 'resonance' to intersecting rays.

Momentum. Number of rays modulated in direction of motion times the rate of transference of modulation in intersecting rays.

Acceleration. Unbalanced modulations by a continued interference or applied force.

Gravity. In the case of two equal masses V and Y, Fig 17(a). Mutual unbalance by interference resulting in continual increase in velocity (acceleration) which is evenly shared.

In the case of two unequal masses W and Y, Fig 17(b). Greater number of rays being modulated and greater mutual unbalance by interference, therefore continual increase in velocity which is not evenly shared, due to 'inertia' resistance set up in W. If the mass of W is now

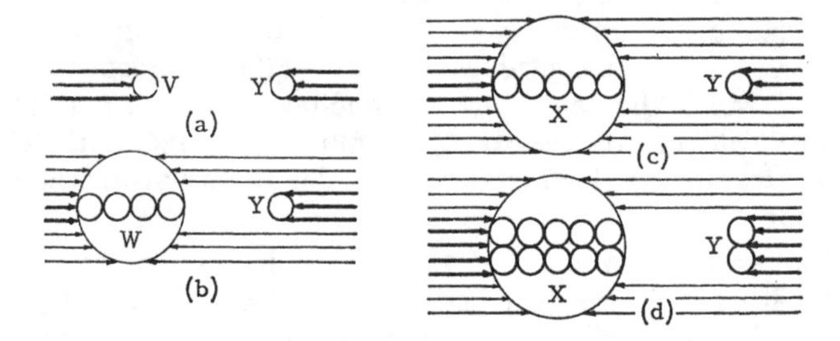

Fig 17. *Gravitational phenomenon expressed in terms of Unity of Creation Theory.*

increased to X, an even greater number of rays are modulated, therefore the higher will be the degree of unbalance and the greater the acceleration, Fig 17(c). If now the mass of Y is doubled as in Fig 17(d) it will still move towards X with the same acceleration as it did when it was only half the mass, in accordance with Newton's laws.

All these facts then can be observed in everyday mechanics and although the comparisons just made are deliberately simplified and can only be improved by complication, the basic argument for further speculation is there. If we can show that there is some truth in the theory, then we may be in a better position to judge the possibility of employing other means of space ship propulsion.

Here and there throughout this work I would like to emphasize the uncanny, though perhaps significant similarity of conclusions of independent researchers, working from completely different standpoints, often with completely different objectives in mind. Few, for instance, would care to attempt to postulate a simple, all embracing underlying pattern of the cosmos, such as Mr Avenel has done in the *Unity of Creation Theory*. Dr A. Einstein tried to establish something like it quantitively in his unified field theory and we shall see more overlapping of ideas as this story unfolds. But now in order to examine the state of the art as it stands today, we must venture a little here and there into some fundamental facts concerning nuclear physics. So from the mechanistic ranges just contemplated, we must cross the bridge where the terrain is somewhat more tedious. To those of my readers who are unacquainted with some fundamental physics, I would hasten to reassure them by saying, it is not always necessary to be acquainted with the local scenery in order to follow a country lane.

Now before we consider some of these facts, one other relevant thought on which to ponder: according to relativistic thinking, if two observers separated by a great distance in a remote space and moving on a collision course, were to send reflected pulses of light at set intervals towards each other, they would be able to measure the diminishing time lag and thereby determine the relative velocity, Fig 18(a). Observer A would say that observer B was rushing towards him at velocity X and because of the constancy of the velocity of light, similarly observer B would calculate that A was rushing towards *him* at precisely that speed. Now according to Einstein and Mr Avenel, a beam of light would theoretically travel a vast circle in a cosmological sense, finally bending back on itself. Of course this isn't demonstrably practical, but it serves to illustrate the issue.

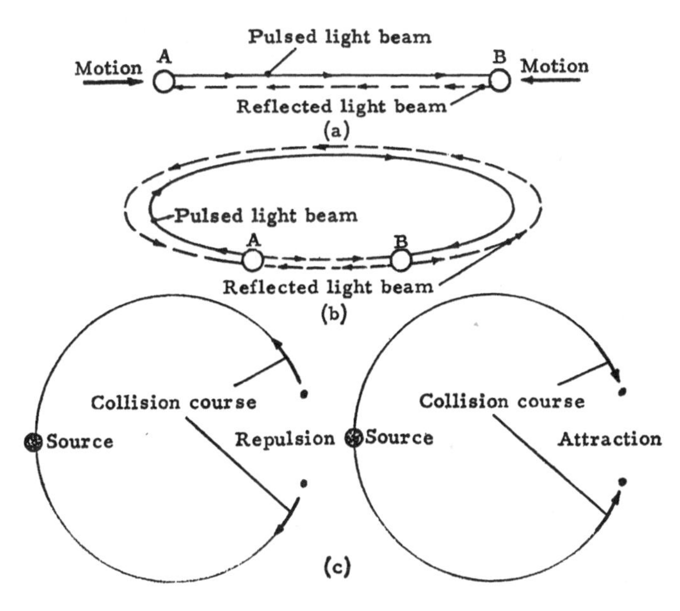

Fig 18. Whatever their relative motions may be, two moving bodies are on a collision course when considered from the source.

Therefore let us imagine that one of the observers, say A, turns through 180 degrees with his back towards the approaching observer B, and while in that position sends a stream of timed light pulses directly ahead as before. This time the light will travel a much greater distance in a circle before reflecting back again, but this time, by calculating the *increasing* time lag, observer A will postulate that he is now receding from observer B by precisely the same amount as the previously measured impact speed! in a word, according to the measuring technique alone, both bodies are simultaneously approaching and receding from one another; which is absurd, Fig 18(b).

Similarly then with attraction and repulsion of two masses, in both these conditions it could be said that all we observe is motion back to the source, and that in either instance, two such repelling or attracting masses are still on collision course with each other, when viewed from the source, Fig 18(c).

6

The present state of the Art

NOW let us cross the bridge to examine the problem of gravity in terms of present day acceptance, and see if there are any signposts which confirm the route Mr Avenel has taken.

Well at the present time, the aim of most technologists is the generation of energy from thermo-nuclear reaction, such as the transmutation of deuterium from 'heavy water' into helium. For example it can be shown that the fusion of one gramme of hydrogen into helium will yield some 200,000 Kilowatt-hours in the form of heat and radiation. This is the result of what is known as the 'mass defect', the minute difference in mass when the helium nucleus is formed. It is this minute difference which is liberated as energy.

Although it may be romantic to do otherwise, we shall be more realistic if we anticipate that although progress in scientific research will widen our knowledge of the laws of nature, no discoveries or changes can be expected which will cancel any of our present fundamental laws. We may see some aspects of nature in another light, but they will function and be predictable as before.

Thus it is logical that any exploration of other avenues in physics should not involve any conflicts with the present laws. Even phenomena which sometimes appear to conflict with these laws, on closer appreciation, will be found to obey them. Therefore we shall be wise to adopt this reasonable limitation in making a sober survey of future possibilities in astronautics. But first let us take a quick glance at the present scenery.

Cosmologically, universal gravitation devoid of polarity, is related to the expansion of the universe (as established by the red-shift in the light from distant stars), that is to the continuous creation of space.

According to F. Hoyle, this expansion causes a corresponding creation of mass, as required by the constancy of the universal gravitational constant.

P. Jordan offered practical evidence which supported Dirac's hypothesis, that the gravitational constant is in fact very slightly time-variable. Both these hypotheses can be correlated by the following microcosmic conceptions, but first the reader should remember that technicologically, in recent years, much hope has been placed on the possible anti-gravitational properties of antimatter and other strange particles such as the neutrino. It was expected that antimatter would be gravitationally repulsive towards ordinary matter, but this was not established experimentally.

P. J. Wyatt claimed that antimatter might exist in bulk in the universe on the grounds of certain mysterious traceless meteoric impacts, with the possible accompaniment of annihilation of antimatter. It was pointed out that if this be so, such antimatter meteors would have been subject to gravitational attraction. Indeed this appeared more logical, as positrons, antiprotons, antineutrons etc., exhibit the property of inertia.

In 1948, Gamow suspected the elusive neutrino to be the 'graviton', that is the quantum particle of gravitational radiation. The idea was favoured by the neutrino's exceptional ability to pass through the most dense matter. It could for example penetrate a layer of liquefied hydrogen a hundred million light years thick, with a very low capture risk. Perhaps this is more readily appreciated when we remember that the neutrino has a rest mass of less than 1/2,000 of that of the electron, in other words, zero rest mass for all practical purposes. Gamow's hypothesis was discarded due to the lack of experimental proof, in fact it is now accepted that the elusive particle is subject to gravitation as are all the other subatomic particles. Even particles of absolute zero rest mass which possess merely velocity-acquired, or inertial mass (such as the photon, which is the quantum particle of light) are subject to the force of gravity, so there can be no anti-gravitational solution in the subatomic domain. And present knowledge indicates that it would be futile to hope for the discovery of any particle devoid of inertia and/or not subject to gravitational attraction.

Some scientists are of the opinion that any conversion of mass into energy, is bound to be accompanied by the release of gravitational energy, in some form. They believe that the lack of evidence on such energy release might be due to the exceedingly weak nature of gravitational energy (as we saw earlier) which renders detection impossible by present day methods, compared with the relatively easy task of measuring other forms of energy released during the conversion of mass.

It might be said in fact, that experimentally the position towards such effects is analogous to the inability of researchers to observe undeniable losses in mass, for example those which occur during exothermal chemical reactions, or those during which heat is being released. During such reactions, a minute part of the reacting mass is converted into heat. Even so, the mass lost is by far too minute to be detected by even the finest present day instruments. The late A. R. Weyl writing on this very subject said: 'The great inferiority in the field strength of gravitation suggests two important conclusions:—

(a) It largely explains our ignorance of the nature of gravitation. Conversions of gravitational energy into other forms of energy (and *vice-versa*) are obviously beyond our powers of observation and detection. We may be confronted with them in everyday life, but we could not notice them! The position is the same as for the terrestrial proof of the consequences from general relativity.

'This inability leads one to suspect that information on the nature of gravitation might be at hand, but will escape notice until our instrumentation is immensely refined.

(b) 'A neutralisation or reversal of gravitational effects should need merely a tiny expense of the elementary energy at our disposal [this is borne out by the undeniable ability of some mystics to levitate, author]. This demonstrates how grossly inefficient our present methods of energy conversion for air and space flight are!'

Some scientists are inclined to the belief that the existence of gravitons —by analogy with photons of light—must be presumed if indeed gravitational fields are subject to the quantum laws, although it is true to say that this has never been theoretically or experimentally established. At the time of writing, the latest advances involving the theory of gravitational fields, has not revealed if in fact they are subject to quantum mechanics or not. Therefore theorists are inclined to view the existence of gravitons with doubt.

So also is evidence lacking as to whether the velocity of light is variable in the presence of gravitational fields and that such fields are also subject to quantum laws, as postulated by J. Mandelker.

Max Born also suggested that the origin of gravity might be hidden somewhere in the microcosmos, as a kind of 'incomplete compensation in otherwise polar fields'. The same hypothesis was put forward by F. Ba Hli and M. A. Gerardine when they independently suggested that gravitation might have its root in electric fields. That the secret to the enigmatic force was in fact the prevalence of electric attraction over electric repulsion. And the minute difference detected was what we in fact call gravity.

From explorations in Einstein's Unified-Field theory, H. Hlavaty derived the hypothesis that rest mass density and gravitational fields have electro-magnetic character and that the 'primary and elementary field is the electro-magnetic one'. Without it, Hlavaty postulated, gravitation does not exist. But both these fields are interdependent and each can be expressed in terms of the other. However, one kind of electro-magnetic field is incapable of creating gravitation; and that is plane waves of light.

Weyl goes on to say: 'Dr Einstein always held that mass, gravitational and electro-magnetic fields, were closely inter-related; interchanges and combinations between the three on the basis of relativity were much in his mind.

'Accepting Hlavaty's important deductions from the Unified-Field theory of Einstein's, it seems obvious that during every conversion of mass into energy, all or some of the gravitational energy inherent in mass will manifest itself as radiation. [The reader will note how closely this follows UFO phenomenon, author]. In what form this energy release will take place is unknown. It may be as gravitons, as gravitational waves without quantisation, or as pure electro-magnetic radiation.

'Einstein's theory applies macrocosmically. It neglects the quantum theory of fields and existence of non-electric nuclear forces. For closer information on the nature of gravitation, an extension of the theory is needed.'

In recent years work has been done by mathematicians in Britain and practically every country in the world on theoretical research into gravitation from relativistic considerations. Largely the work concerns the properties of gravitational waves; that is the result of the acceleration or deceleration of masses in gravitational fields.

Both Einstein and N. Rosen concluded from their investigations into gravitational field equations of the relativity theory, that cylindrical waves of gravitation might exist. Then later Rosen was able to prove that the theory excluded the possibility of plane polarised gravitational waves, but his conclusions were challenged by mathematicians in Britain.

F. A. E. Pirani of King's College, London, attacked the problem with a new theoretical approach in which he abandoned Einstein's analogy with the theory of electro-magnetic fields, since in physical effect these types of field differ, as for example, the principle of equivalence between heavy and inertial mass. Pirani's work did much to enhance the view that gravitational fields do exist, 'and not merely spurious phenomena in the supposition of abstract mathematical theory'. Indeed Einstein and his collaborators during their later years of research accepted this hypothesis.

One interpretation of Pirani's theory predicts how gravitational radiation affects the motion of tiny particles and investigation of the relative acceleration of two such particles in the presence of gravitational radiation, would establish the laws governing the transfer of such gravitational wave energy.

Pirani proposed an experiment which consisted of two masses connected by a spring or piston-cylinder device. He reasoned that on the passage of gravitational radiation the masses would receive different accelerations which would convey to the spring a net force which could be measured in magnitude and direction, in this case, by the heat developed during the deformation of the spring. In a similar way the energy of gravitational waves might be investigated.

Then H. Bondi, also of King's College, London, was able to calculate the possible energy transfer from plane polarised gravitational waves, confirming Pirani's conclusions while contradicting the views of Rosen.

In 1958, L. Marder produced a theory involving the generation of gravitational waves, which theoretically permits a controlled transfer of gravitational energy through space. Therefore some have claimed, if such waves were physically possible, they would carry energy and this energy would be physically transferable. Other researchers believe this would open new possibilities for space travel and signal transmission through space and dense matter.

Yet another researcher, K. Stanyukovich, related gravitation to low temperature physics, in which he developed mathematical expressions of the laws of gravitation. He suggested that the force was dependent on temperature and predicted a lessening of gravity near the absolute zero. There does not appear to be any evidence that this was established experimentally, despite many experiments which approached the absolute zero very closely. There are of course many interesting physical phenomena exhibited at such temperatures, such as superconductivity and the superfluidity of helium He 11, below 2.19 degrees K, when it loses its viscosity and tends to creep over the rims of containers. But in no case was there a hint that gravity was the least bit affected.

Other researchers have held out hope in electro-magnetic fields, but here again there have been negative results. Strong magnetic fields as high as several millions of gauss can be produced, but this in no way affects the force of gravity to any degree measurable with modern equipment. Similarly attempts at shielding gravity have also produced negative results.

Professor J. Allais claimed to have detected small fluctuations in the oscillation of his Foucault pendulum during the total eclipse of the sun on 30 June, 1954; he claimed the fluctuations were as much as 13 degrees, but refused to conclude that the effect had anything to do with the shielding of the sun's gravitational field by the moon.

B. S. DeWitt, H. E. Salzer, B. Heim, H. J. Kaeppeler and other theoretical physicists of academic and high professional standing have developed extensions of the four dimensional relativity theory, to even broader field theories comprising six dimensions. Needless to say such exceedingly mathematical treatment renders the theories even more complicated than Einstein's field theory. As mathematicians of the highest degree find this difficult to interpret physically, it is obvious that such complex theories have little meaning for minds of lesser calibre.

Even so it is said that in spite of these obvious limitations, conclusions gained from such 'elaborate orgies in tensor analysis' indicate the possible existence of an *intermediary* field, distinct from the electro-magnetic and the gravitational fields. (Note author.) According to these theories, this third field is interchangeably connected with electro-magnetism and gravitation; that is, it acts as mediatory between both. Nevertheless, it must be recorded as a *field* in itself. The reader will notice the close parallel with which this follows the 'Unity of creation' wave theory.

We have seen how all elementary particles known to science are subject to the force of gravity. Therefore it is logical to conclude that such an intermediary field must be found in the sub-nuclear domain.

Other thinkers suggest that this is the reason why nuclear researchers have failed to detect intermediary field conversions predicted by the above theories.

It is interesting to note that H. E. Salzer ascribed relativity to 'uniformly accelerated accelerations' and to 'changing gravitational fields'. Accordingly he deduced a purely mechanical concept involving interrelations in a 'dynamic intermediary field'. In this respect we notice the similarity of John W. Keely's 'inter-atomic ether'.

Almost lost among the galaxy of theories involving gravitation, space and matter, the faint whisper of the psychic researcher now and again can be heard. He and his fellows know only too well the reality of the phenomenon they investigate, yet for some strange reason scientists generally do not wish to know. Yet it is absolutely ludicrous for them to deny that psychic phenomena does occur, as even the barest examination of poltergeist activity, for instance, will illustrate. And although such research may not reveal the nature of gravitation, it does establish all too vividly the possibility of an anti-gravity state.

The long list continues. We can mention only a few more. H. J. Kaeppeler is said to base his purely micro-physical theory upon nuclear phenomena. While St. Deser and R. Arnowitt strive to integrate the quantum theory of fields into the theory of relativity.

H. Hlavaty has conducted research into a multi-dimensional general field theory which seems to agree very closely with the formalism of Heim's intermediary field concept. Burkhard Heim, extremely handicapped by war wounds (he is blind, armless and nearly deaf), has evolved a six-dimensional theory containing the general relativity theory, and the quantum theory of fields as special cases. Which means, that the formalism of these two theories is derived by neglecting certain members in Heim's field equations. Therefore Heim's theoretical approach has a more universal character than any other presently treated theoretical approach. In addition it bridges some fundamental contradictions between general relativity and the quantum theory, which tends to restrict them to the macro and the microcosmic domain. The reader will not be surprised therefore to learn that B. Heim's six-dimensional theory yields results which predict physical phenomena mentioned elsewhere in this book.

Other great thinkers are known to have expressed some doubt concerning established micro-physical conceptions. L. de Broglie, the originator of wave mechanics on the basis of Planck's quantum theory, appealed for 'a broader view' in this domain, and proposed a critical re-examination. He was particularly concerned with the theoretical interpretations of the quantum theory. In this respect it is interesting to note that neither Planck, nor Einstein, nor Schroedinger considered that this theory furnished a complete description of physical reality. 'The introduction of a 'sub-quantum' milieu, deeper than the level of current micro-physics . . . might lead us farther than the current theory will go'. Concerning this, A. R. Weyl said, 'Quantum electro-dynamics, that is, field theory on the basis of quantum mechanics is quite able to describe the interaction of electrons with electro-magnetic radiation, but fails to apply to the sub-nuclear domain (to meson fields). This has been demonstrated from present theory and confirmed empirically: the experimental data do not agree with those established from the perturbation theory.'

H. Yukawa confirmed that the atomic nucleus is bound by strong forces which couple neutrons and protons together, yet which are non-electrical in nature and so strong that they quite easily overcome the powerful electric repulsion between the two protons. But unlike electric forces they act only over the very short sub-nuclear distances. It can be shown that when momentum is exchanged from one particle to another during the process of binding, it sometimes involves an exchange of electric charges between the particles, from which phenomena Yukawa suggested the existence of a special category of particles, the mesons. No less than seven different varieties of these have since been discovered experimentally. Physicists have shown for instance, that the pi-meson of nuclear binding while having a free life of the order of a few 10^{-8} seconds, has 276 times the rest mass of an electron. Although experimentation on this is still difficult, nevertheless it is the foundation of the 'meson theory' of nuclear binding forces.

Fundamentally the theory holds, that every neutron or particle of an atomic nucleus, is surrounded by a meson field: this causes the particles non-electrical interaction with other nucleons, whether they carry electric charges or not. Therefore mesons are the carriers of energy quanta in this particular type of field, and can be regarded as equivalent to the protons in an electro-magnetic field. Even so meson fields still possess the permeating characteristic of gravitation.

Eddington, Schroedinger and Born worked on hypotheses involving a relation of Yukawa's meson fields with gravitation. While Pauli and Blackett suggested that the life of mesons varied inversely with the root of the gravitational constant.

Now we have seen that the universal intermediary field predicted by the multi-dimensional field theories, differs from the electro-magnetic type field, and apart from gravitational fields the only other elementary category of field of which there is experimental evidence is that of the nuclear binding non-electric, short range forces.

Weyl suggested that this justifies an attempt to identify Heim's intermediary field with the meson field of the nuclear binding forces. He therefore conceived that three types of elementary fields govern all natural phenomena 'between mass and energy' and also constitute them:

(a) the electro-magnetic field;
(b) the meson field of nuclear binding as intermediary field; and
(c) the gravitational field

where the electro-magnetic field is dominantly quantised and micro-macrocosmically universal in range, whilst the most powerful meson field is entirely microphysical and restricted to the ranges of the order of a few Fermis. (1 Fermi= 1×10^{-13} centimetres.)

I would remind the reader that among his notes John W. Keely often refers to three such forces. In assessing the following typical example, it would be well for us to remember that Keely demonstrated anti-gravity experiments to eminent scientists of that time, many of whom acclaimed him a sadly maligned genius. Among other things, he is known to have said: 'The action of Nature's sympathetic flows regulates the differential oscillary range of motion of the planetary masses as regards their approach toward and recession from each other. These flows may also be compared to the flow of the magnet which permeates the field, existing between the molecules themselves, sensitising the combined neutral centres of the molecules without disturbing, in the least, the visible molecular mass itself. In the planetary masses—balanced as it were in the scales of universal space, floating like soap-bubbles in a field of atmospheric air, the concentration of these sympathetic streams evolves the universal power which moves them in their oscillating range of motion to and from each other. This sympathetic triple stream focalises and defocalises on the neutrals of all such masses; polarising and depolarising, positive and negative action, planetary rotation etc., etc. It is thus that all the conditions governing light, heat, life, vegetation, motion, are all derived from the velocity of the positive and negative interchange of celestial sympathy with the terrestrial.'

Burkhard Heim has not only developed a mathematical theory of gravitation which sounds remarkably like A. Avenel's *Unity of Creation Theory*, but he has also predicted a space ship which is suggested by that theory, and which sounds remarkably like our own.

In order to illustrate more emphatically this similarity of independent conclusions, I shall now let A. R. Weyl take up more of the story for a while. In the following discussion he interprets Heim's intermediary conception of gravitation.

'Let us now see how such a daring trinity conception could help in understanding the qualitative relations between mass and energy.

Einstein suggested that mass might be an inter-relation between electro-magnetic and gravitational fields. No evidence exists to contradict this conception.

Modern developments in nuclear physics have led to the idea of the existence of a common basic 'original' substance ('Urmaterie'). From

Plate 9. *Lee-Richards annular wing monoplane.*

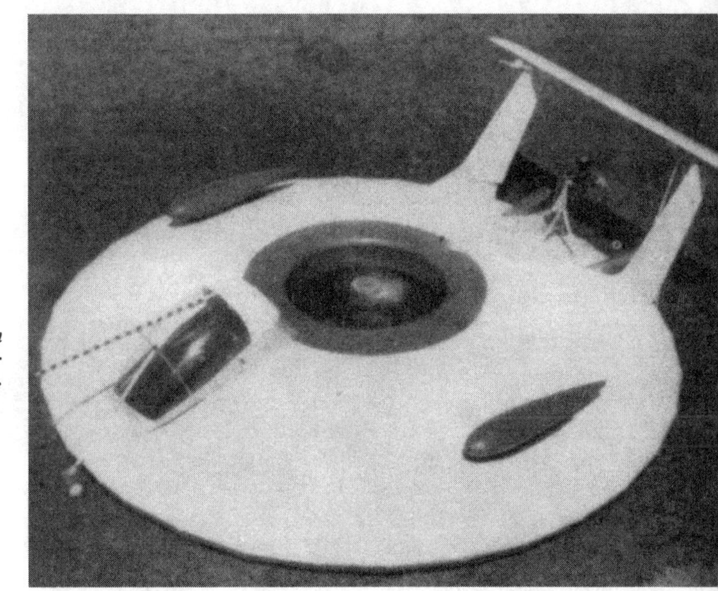

Plate 10. *Princeton University X-3C circular wing hovercraft.*

Plate 11. *French aerodyne designed by René Couzenet. Had a diameter of 27ft powered by 4 engines.*

VTOL model in flight.

6ft dia. part structure illustrates simple construction of the circular wing.

Sectioned model showing th main inherent aerofoils.

Plate 12. General view of the author's models. Above is a view of the variable cascade and test wing on rig.

this, all mass particles in the universe are thought to be constituted, matter and antimatter alike. This substance cannot, of course, be anything like mass; it would be latent energy, that is, some form of dormant energy occurring throughout the universe wherever mass is formed.

Such a conception would seem to be in conflict with the relativistic idea of velocity-acquired (inertial) mass. This argument disappears, however, when the 'original' substance is conceived as an all-pervading all-present ether. Speculating upon this idea, we may ascribe some other properties to such an ether-like substance of latent energy.

By definition, it can be neither mass nor dynamic energy. It can have no size, nor dimension, nor shape (W. Heisenberg's uncertainty principle). It will be all-permeating through the entire universe. *Indeed, we might identify it with space.*

As it will exhibit no radiation nor drag it will remain undetectable, and it will also not be subject to quantum laws. It has static-field character.

As a tentative model conception, 'original' substance might also be described as an *ether* of latent 'meso-static' energy consisting of an infinitely large number of infinitesimally small meso-static fieldlets each possessing an infinitely small individual range of influence.

Against such a definition the objection could be raised that the non-existence of a world ether has been proved experimentally.

However, the Michelson-Morley experiment merely relates to the propagation of light waves. Light waves have lengths much longer than the ranges of the stipulated fieldlets of the 'original' substance. An ether of the kind under consideration is incapable of forming a carrying medium for light waves and could, hence, not effect the propagation of light or, for that matter, of any conventional electro-magnetic radiation. Thus, for the Michelson-Morley experiment, our ether-like 'original' substance would not exist. P. A. M. Dirac has, from electron theory, also postulated the existence of a '*world* ether', although this might invalidate certain suppositions of relativity theory.

One step further in our hypothesis leads us to ascribe to this undetectable 'Urmaterie' ether another important property. It is that of *gravitational repulsion* between the fieldlets.

The absence of any polarity in gravitation has always been puzzling. Magnetic, electric, and nuclear-binding forces (mesons) have a polar character, with attraction *and* repulsion. The existence of exclusively attractive forces in gravitational fields remained a mystery for rational physics.

If we conceive our 'original' substance ether as filling the universe or, more explicitly stated, as 'space', or 'vacuum state of negative energy' (Dirac), and bear in mind the evidence for an expanding universe, one must logically ascribe to such ether the tendency to 'thin out', that is, to create more space. Such gravitational repulsion between the fieldlets of the ether would thus naturally compensate for the universal gravitational attraction between masses. This property of repulsion would also be responsible for the creation of new mass during the expansion of space.

When P. A. M. Dirac interpreted his relativistic wave equation for the electron, enlisting W. Pauli's 'exclusion' principle, he argued that the absolute vacuum of space was represented by quantum states of electrons having negative kinetic energy, or negative inertial mass, respectively, *observable* effects would occur when an electron of the said vacuum absorbed sufficient energy for transition into a state of positive kinetic energy. The unoccupied level in the vacuum (= negative mass or energy) thus left, should lead to the simultaneous appearance of a positron. This particle was indeed discovered three years after Dirac's postulate. Electron-positron pair production is today an accepted fact of nuclear physics.

Recently, Dirac confirmed his conviction that, that which we are wont to call the 'absolute vacuum of empty space' should be considered as a store of negative energy. This conception covers precisely our ether hypothesis if the negative kinetic energy postulated by Dirac is interpreted as latent energy incapable of observation.

The fact of pair production, incidentally, is by no means restricted to electrons; it applies to all other nuclear particles, for example, to protons (antiprotons); neutrons (antineutrons), to mesons, and so on, all of which have physical existence.

Mass can be conceived as the result from an interaction of appropriate electro-magnetic fields with the 'original' substance ether defined before. This complies with Einstein's conception of mass [and reminds UFOlogists of the formation of the mysterious substance called 'Angel Hair', author].

During such interaction the fieldlets of the said ether would absorb energy—as postulated by Dirac—and assume the dynamic state. This state would become manifest as meso-dynamic (= nuclear binding) energy and also invert the gravitational ether property into gravitational attraction: both constitute mass. Electro-static charges are imparted, with or without spin; magnetic properties would arise. From inner resonance of the newly created particle, quantum characteristics would result, and with them stability requirements, that is, the need to form discrete mass particles. Within these, dynamic phenomena prevail; they cause fluctuations in the exchange of electric charges and, hence, meson exchanges. Other manifestations of this activity of quantised elementary mass is the emission and absorption of orbiting energy quanta (meson particles).

From the engineering point of view it is important to speculate on what kinds of electro-magnetic fields might be able to produce mass.

Fields of such high energy transfer must necessarily be waves which are sufficiently short and powerful to be capable of interaction with the infinitely small fieldlets of the ether. They too, cannot be plane waves of light. This leads one to stipulate the necessity for 'ultra-gamma' radiation of an energy which has hitherto neither been observable nor experimentally achievable, with wave lengths of less than say, 10^{-2} Fermi. Rays down to 1 Fermi ($= 1 \times 10^{-13}$ centimetres) are presumed to occur in cosmic radiation (observations by Kohlhoerster, Haas and Millican, respectively). Could they be responsible for the creation of mass particles in this radiation? Thus, technologically, one of the principal problems of the near future in this domain would be to achieve and to control such powerful ultra-gamma radiation.

When this problem is studied, possibilities will be discovered how to transform meso-dynamic energy directly into electrodynamic manifestations (by, for example, the production of electron streams), and finally, how to oppose gravitational attraction 'at its root' at a minute expense of mass or electro-magnetic radiation.

Gravitational fields also link the 'original substance' ether to mass. This inter-relation would also lead to a new conception of inertia. One might visualise this as an increase of intensity in the field of gravitational attraction during acceleration of a mass [or resistance of the 'C' rays to being modulated, author] on account of interference by the fieldlets with the powerful dynamic fields which are inherent in mass. The inter-relation would, by inversion, produce an increase of the moving mass.

This conception not only allows one to understand the relativistic mass increase with velocity; it also provides an explanation of the apparent contradiction to which G. Burniston Brown drew attention in his criticism of the mathematical approach in natural philosophy; the inertial mass of a body is about 4,000 times greater than its attractive mass; yet according to the theory of relativity, both masses are identical.

During the experimental approach, a formidable problem might be the penetrating character of ultra-gamma radiation. It is foreseen that this could unlock much elementary energy, *with the formation of heat and other biological side effects.* It might, however, be possible to bunch and to guide such radiation so as to constitute focused beams for experimentation, without affecting the surroundings. In this manner too, protective weapon rays can be conceived, of indefinitely higher effectiveness than photon beams. Also, the conversion of mass *directly* into usable electro-magnetic or mechanical energy should be within the reach of practical possibility'. [Italics are mine, author]

As I sum up this chapter on gravitation, I feel sure that my thanks (co-mingled with a sense of friendly criticism) will be kindly received by the shades of Mr A. Weyl, Dr A. Einstein, Sir Isaac Newton, and the many others, when I say to the lofty scientific sceptic, none of these great souls have yet explained to humanity, exactly what the phenomenon of micro-macrocosmic *attraction* and *repulsion* really is. John W. Keely may have come very close to it, I think Mr Avenel may be even nearer, while Mr G. De la Warr demonstrates the cause; the much disputed ether. Ether is space, space is ether.

To the many people from various corners of the world who wrote to me after the publication of *Space, Gravity and the Flying Saucer*, concerning their own particular views on the nature of gravitation, I would take this opportunity to thank them all additionally, for their willingness to share ideas, and although an explanation is not really necessary, I feel I should add that the above latter views on gravitation expressed by men of science, carrying more weight as they do, have been quoted here for the subject's sake to the exclusion of all other material available. Many of you good people, I feel, are very near to the truth, each may be thinking his particular theory quite original, and in the sense that he conceived it, it is. I am sadly aware of the fact that many will never be publicly recognised. Having experience of these things I would say, take cheer, you are in good company. I would ask further indulgence to remind you that two thousand years ago, a great soul speaking to a mortal few said: 'Unto you it is given to know the mysteries of the kingdom of God, but to others in parables, that seeing they shall not see, and hearing they shall not understand.'

The sun is rising, we have crossed the bridge, explored the foothills and talked through the night. But we are rested and it is time to move on our way. Just ahead, among the receding shadows we can see the first faint outlines of the next signpost, we have covered a lot of territory, but in another sense we are but just beginning. . . .

7

Gravitational Space Ship

WHEN *Space, Gravity and the Flying Saucer* was first published in 1954, the gravitational field propulsion theory, which was supported by the *Unity of Creation Theory*, was as far as we could venture, for to a large extent we were on our own in the no-man's land of borderline exploration. Perhaps just a little too far out for those unacquainted with technical matters, but who nevertheless want to know the truth about flying saucers, while some of the ideas expressed were probably not taken seriously by a number of professional people. I expect, to some of those, many of the conclusions on natural phenomena were perhaps tolerably amusing, that is, if they bothered to read them at all, while the correlation of flying saucers probably caused many a raised eyebrow. This I accept in all good will. Even so, it is nevertheless encouraging to know that an ever increasing number of scientists the world over are beginning to look more closely to reports of UFO sightings. For, as most students of the subject are aware, many scientists have now recognised the technically corroborative pattern in the ever growing list.

In *Space, Gravity and the Flying Saucer*, as indeed elsewhere throughout this subsequent publication, I have tried to strengthen the case for the flying saucer by stressing the almost uncannily similar conclusions which completely independent researchers have arrived at, frequently stemming from totally different points of view or interests. For instance, my original interest in gravitation was no doubt greatly influenced by witnessing a genuine levitation phenomenon. It made me think, as such phenomena have made thousands, who will at least condescend to investigate, think also. No doubt some sceptics may see fit to use such information as additional fuel, and this is regrettable, for with all humility, I know it to be their loss.

Mr Avenel conceived the *Unity of Creation Theory* because he felt the need for such a unifying completion, which basically did not clash with relativistic thinking. Lieutenant Plantier's conception of the flying saucer was originated from the point of view of an airman, and I must confess, his ideas which were brought to my attention after the publication of my book, caused me a mingled reaction of disturbing surprise on the one hand, then more realistically, pleasantly confirmative admiration on the other. Such repetition or overlapping of ideas prompts the argument in favour of pooled research, but although there is much to be said for this (albeit practical difficulties could be overcome) the private enterprise of the individual will always go on, as indeed artists will always paint. Such duplication is still occurring and I pen these words fully aware that at this precise moment some unknown inventor somewhere, may be on the very threshold of a discovery which will help throw light on the mystery of the UFO. Someone else, who has never heard of A. R. Weyl, A. Avenel, Lieut. Plantier, B. Heim, the present author, and many others.

In the last chapter we took merely a glance at some of the theoretical work being done on gravitation and I should now like to correlate this phenomenon with space ship propulsion.

Perhaps we can compare our present relationship to gravity with a little boy racing up a descending escalator; he has only to cease his efforts and he is promptly brought down again. Gravity, like the downward moving escalator, is being continuously 'expended'.

Our future relationship to gravity may be more attractive and perhaps comparable to the same little boy suddenly discovering yet another escalator, which will give him a free ride to the top, whereupon he changes back to the downward-moving escalator to have an equally free ride down again. The whole point being that there is no comparison between his first frantically exhausting efforts to climb up and his leisurely stroll from one escalator to the other to achieve the same journey.

There is reason to consider that we are in the position of the little boy who has begun to believe in the existence of another escalator and that if we pursue this belief we, too, may find, not one, but an infinite number of such 'gravitic' escalators which we may harness at the expenditure of a comparatively small amount of energy.

Modern aeronautics has reached the stage when serious consideration should be given to the fact that gravity is but another natural phenomenon waiting to be explored, understood and put to useful purpose as have light, sound and electricity before it. From the aeronautical point of view, one of the first observations we might make, is the fact that gravity displays a unique and most enviable quality in its make-up, that of uniform acceleration.

It is known that a body experiencing acceleration due to a gravitational field is under no strain whatsoever, for it is moving in a uniform field where each particle or atom forming its structure experiences a force equal to its neighbour. Therefore there is an absence of progressive acceleration experienced by the particles. This condition is readily observed in the case of a falling egg, Fig 19(a). During the free fall every particle of the egg experiences an equal force to its neighbour, therefore there is no relative movement between them. On coming in contact with the ground however, the egg shell is *decelerated before* the yolk, which still possesses momentum. Therefore, there exists a relative movement between them and, of course, collision with resulting structural failure, as in Fig 19(b), deceleration in this case being identical to acceleration, that is, change of velocity.

Perhaps the point is more effectively illustrated by imagining that if we were 'falling' towards the earth, that is, moving in the earth's gravitational field, we would be accelerating at 32.2ft per second per second. Should the earth's mass then be suddenly increased to that of the dwarf star Sirius B, we should experience a vastly increased acceleration, in fact something to the order of 20,000g, but we should suffer only a condition of weightlessness during this drastic change.

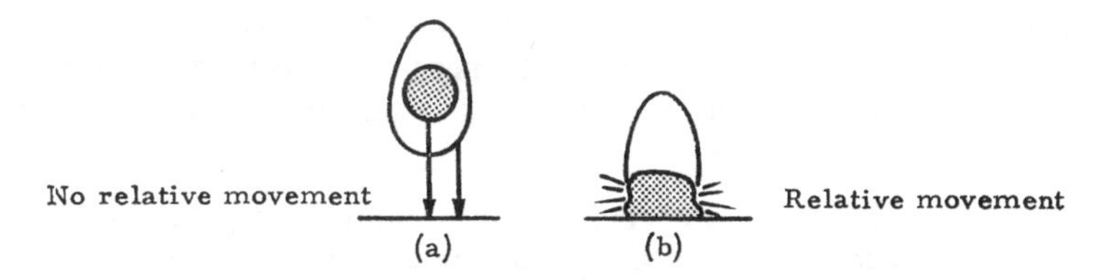

A falling egg offers a simple example of acceleration and deceleration, when all the molecules in the former are accelerated uniformly and decelerated progressively on impact.

A progressively conveyed force producing structural failure.

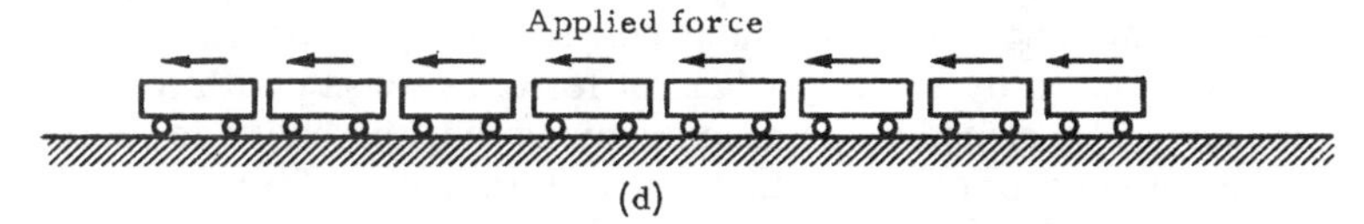

A uniformly applied force shared by the structure. By this simple principle there is no limit to which the structure may be accelerated or decelerated.
Fig 19.

Now if matter *is* a three-dimensional bond of carrier wave intermodulation and gravity is only a mutual unbalance of that bond, or perhaps more aptly, a condition of plane polarised space, then it should be possible to encourage this condition by other means. Once this was achieved, a space ship could move through space, perpetually 'falling' into its own part polarised gravitational field with a minimum of energy expenditure, analogous to the little boy crossing from one escalator to another.

Moreover, it must be borne in mind that this condition could be brought about not only in outer space, but near the Earth's surface, it being possible to move a material object by this means in any chosen direction. Should the unbalance be caused in the opposite sense to the Earth's gravitational field, then a condition of 'negative weight' would be obtained, the strength of which would depend only on the degree of unbalance.

Due to the nature of the field produced, a gravitically-propelled vehicle and its occupants could be moved at fantastic orders of acceleration and deceleration, neither would its function be restricted to the Earth's magnetic field as some suppose. It would be capable of extreme man-oeuvres and even being violently stopped; the effect would be exactly as 'falling' in any gravitational field, the occupants would be completely unaware of any change.

On the other hand, the rocket, like an automobile shooting suddenly forward, receives all its thrust via the motor and the astronaut is pushed rudely along by the seat of his space suit. In the case of the automobile, first the wheels turn, ejecting as it were a 'jet' of roadway. This thrust is conveyed to the chassis millimicroseconds later, and millimicroseconds later still, the seat of the car is pushed also. Next it is the driver's turn and his body receives its dose of acceleration, but even so, the blood-stream, muscles, and internals want to stay where they were before the rude interruption (inertia). The driver's nerve endings record this as an uncomfortable jerk and he is on his way. Usually this jerk is a fraction of a g, this may serve to illustrate what, say 10g over some minutes would be like!

All this step by step build up of velocity (or loss of it) with its resulting discomfort and sometimes fatal consequences, would be absent with a vehicle moved by a uniform field of force, as we saw with the falling egg analogy. A glance at the accompanying diagram serves to illustrate the difference of principle further. The little railway trucks in Fig 19(c) resemble molecules in a structure. In this case a force has been applied on the first truck giving it motion, but the other trucks stay where they were until the motion is conveyed by impact, truck by truck, or as the case may be, molecule by molecule. In other words, there exists relative motion between them. But as in Fig (19(d), should the original force be applied separately to each individual truck (or molecule) at the same instant, then there will be a *uniform* motion shared by the trucks, and no *relative* motion. Such a condition might be imagined if all the trucks were made of iron being attracted by a powerful electro-magnet.

The velocities theoretically attainable are colossal compared with any vehicle we have today, for the equivalent of the 'C ray pressure' acting on matter probably goes into millions of tons per square inch, yet the energy input required to unbalance an extremely useful portion of this 'pressure' may be comparatively small.

Some astronautical scientists are apprehensive of meteoric bombardment in outer space, but here again, as a natural consequence of gravitic propulsion, the chances of small meteorites puncturing the vehicle's skin must be considerably reduced, for on entering the surrounding field they would be violently decelerated, ricocheting off back into space, as in Fig 20.

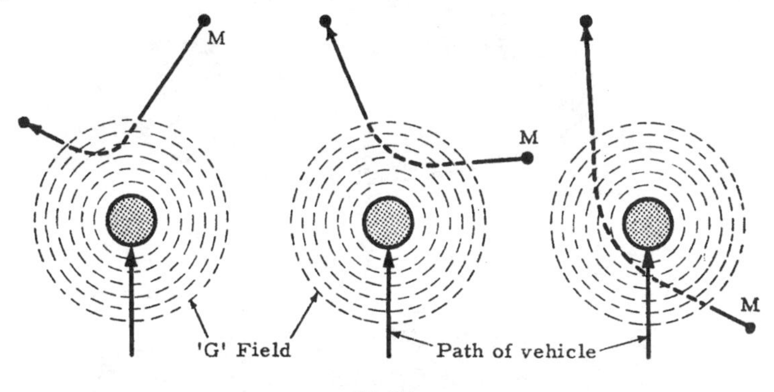

Fig 20.

It is paradoxical that although the harnessing of this motive source may produce velocities undreamed of by us at the present and yield powers in industry beyond our imagination, our lives may become less prone to mishap than they are today. Surface vehicles propelled by these means could approach head on, and under the influence of a reverse field brake to a stop without the occupants experiencing the least shock or discomfort.

One of the most interesting projects involving flight devoid of stress concerns the work of an American, Townsend T. Brown, who has been engaged on research into electro-gravitic propulsion of aircraft for more than thirty years. Involving as it does a little understood phenomenon, Townsend Brown has conducted pioneer work comparable to some of the most notable researchers.

It is claimed that the work is largely based on the principle that a highly electro-statically charged condenser tends to move in the direction of the positive electrode. In this, there appears to be evidence of an interaction between electrical and the terrestrial gravitational fields, in a similar manner to Fleming's left-hand rule applied to electro-dynamics. This is also in sympathy with Michael Faraday's work, and like him, Brown suggests there may be a similar relationship between the condenser and the gravitational field, as there is between the coil and the magnetic field. We have already seen that Einstein was trying to establish a similar relationship in his Unified Field Theory. Fig 21 shows Townsend

Brown's free flying condenser, shaped like a disc. If the two arch shaped electrodes are highly charged, negative and positive, the disc will move in the direction of the positively charged rim. The higher the charge, the greater the interaction between gravity. It is claimed that with a potential of several hundred kilovolts, the disc would reach several hundred miles

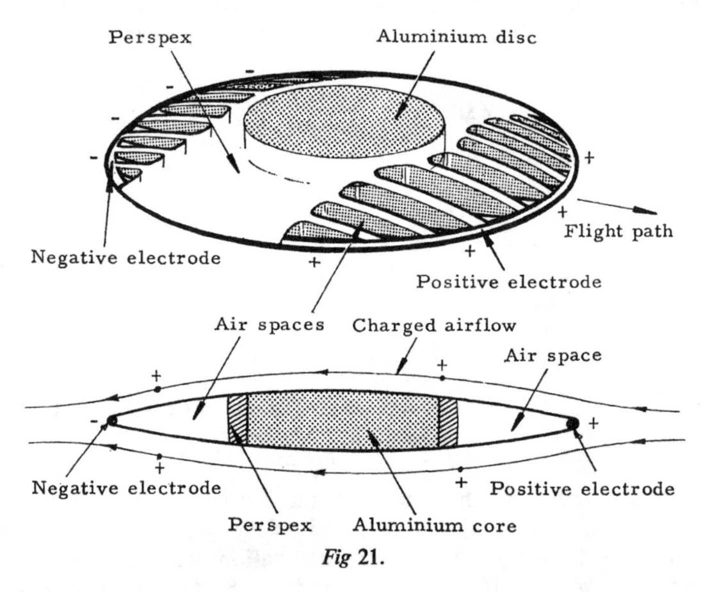

Fig 21.

per hour. As it is, model disc aerofoils, two feet in diameter, and charged at fifty kilovolts with a total continuous energy input of some fifty watts, achieved speeds of seventeen feet per second, moving in a twenty foot diameter circle. Later these were stepped up to three feet in diameter and run in a fifty foot diameter aircourse under a charge of one hundred and fifty kilovolts, and quotes one report, 'with results so impressive as to be highly classified. Variations of this work done under a vacuum have produced much greater efficiencies that can only be described as startling'. One can only wonder what results were achieved with the fifteen million volt flame-jet generator, developed to supply even more power.

In the electro-static condenser shown in Fig 21 the centre of the disc is formed in aluminium, while the solid rimming at the sides is in perspex. In the leading and trailing edges are the electrodes separated from the aluminium core chiefly by air pockets. The electrodes take the place of the plates in a simple two plate condenser, which on reaching full charge, normally loses its propulsive force. So in order to continue moving, the device has to give off packets of energy. In the configuration shown, the air between the supporting wires is also charged, so that as the disc moves forward from minus to plus, the charged air is left behind, therefore the vehicle moves into new uncharged air. Thus both the charging process and propulsive force are continuous.

From this it will be seen there is an additional propulsive force due to the rearward ejection of the charged airflow, similar in fact to the plasma rocket.

Although some UFO researchers will be quick to associate this fascinating device with flying saucers, I should point out that the gravitic interaction does not provide a lifting force, the resulting force being at right angles to gravity, that is, horizontally. Even so it could be argued that the high potential saturating the disc would propel it, and more important, everything in it free from relative accelerating forces, in exactly the same way in fact, as the gravitationally propelled vehicle, and for the same reason, the propulsive effect of the field acts on the individual particles in the structure.

Now at this stage, it might be a good idea to stop for a while and take stock of the present position as far as the signposts have thus far guided us. We have foreseen the limitations of the aircraft as a purely aerodynamic vehicle and we have soberly assessed the rocket as being the only means at our disposal for space ship propulsion . . . for the present. In the last two chapters we took a look at the nature of gravitation and came to the conclusion that such a principle would offer much as an alternative means of space ship propulsion. With this in mind then, let us take another look back and evaluate the position further in terms of modern technology.

To begin with, the law of the conservation of mass-energy requires that it is pointless to pursue levitation or contra-gravitic states without energy expenditure of some kind. This is fair enough, but it must be stressed that the idea of anti-gravity is by no means absurd. I shall illustrate this with another analogy.

Imagine the permanent magnet in Plate 14(a) to represent the earth, to which is 'attracted' the small disc representing a space ship. The fact that this *is* attracted to the magnet by a circumferential iron coil need not lessen the validity of the analogy one bit, though naturally I do not intend the experiment to be interpreted literally.

Now as every schoolboy will have learnt, soft iron, normally paramagnetic, will not be attracted to a magnet when heated bright red, and for that matter a magnet will lose its magnetism, or become diamagnetic if heated. So now let us imagine the magnet to be the earth on to which was held the little space ship. If we now heat the circular iron coil by passing an electric current through it, it will become diamagnetic, and to illustrate this, the small disc has been suspended on a counter-balanced arm, so that it is free to move upwards when the diamagnetic state has been reached, Plate 14(b).

If in our imagination we care to endow the disc with a soft iron crew which would be heated up by the coil, then they too would become diamagnetic, which in the analogy represents weightlessness. In the experiment shown, an external power source was used to heat the coil, but the little ship could have carried its own independent supply and the analogy would be complete, *i.e.* a space ship neutralising the pull of a parent body by means within its own confines. At this stage it may be of academic interest only to note that heat is the visible component of molecular vibration, indeed a magnet will become diamagnetic when given a series of hard blows; note also that vibration of an extremely high order was the basis of John W. Keely's strange levitating force.

Now by this I am not suggesting we build a space ship and violently agitate it and its occupants to pieces to obtain anti-gravity, but, if magnetism is as I suspect, a kind of selective gravitation, then it is certainly reasonable to use the above simple but applicable analogy. If there is a means of producing a diamagnetic state, then there must also be a means of producing the gravitational counterpart.

It will be apparent that by substituting a small electric solenoid for the disc and suitably adjusting the direction of current flow, we could obtain not merely a neutral condition, but a repulsive one also. We shall see later that in a sense this might be equally true of gravitation. Again it is also true that if a small magnet is placed near a very large and powerful magnet, then no matter what the polar relationship may be, the powerful magnet's pull will predominate and the small magnet will be subservient to its greater strength. But if the small magnet is removed some distance from the larger one, with like poles facing, then repulsion will take place between them, the smaller magnet taking most of the movement due to its smaller mass. The same can be said of two electrostatically charged bodies, in fact some have suggested that this is gravitation, and when a space ship was removed far enough away from the earth, it would reach the repulsion state. But again I must remind you that none of these techniques would provide a horizontal thrust component for travel over the earth, but as analogies they serve to establish the point.

At this stage I would with all due regard, point out to the lay reader and many UFOlogists at large, that there simply isn't any point in trying to interpret UFOs as being electro-magnetically levitated or propelled. In fact it is a pity that some otherwise well meaning contributors to the subject still insist on offering pet theories to the public which are often not only fundamentally inaccurate, but immeasurably harms the subject as far as otherwise interested scientists are concerned. One is presented with vague theories on 'crossing the terrestrial lines of magnetic force', 'electro-magnetic drive', 'electro-static propulsion', 'ionic and plasma propulsion' etc., etc., and as I have tried to show, none of these have the faintest hope of doing what the perpetrators would like to claim.

That there are electro-static and electro-magnetic fields directly associated with UFOs is quite obvious from the very nature of sighting reports, but it is foolhardy to make lay interpretations and I would in all humility ask that we keep to the little facts and correlate as many of them as possible to tell even bigger facts, but in the light of reason let us leave technical evaluations to the expert, for to do otherwise is to trespass irresponsibly on his domain.

In this respect it is equally true that I also am offering a pet theory, but I would emphasise that it is theory built on *facts.* A man sifts through many parts of a jig-saw puzzle and finds many pieces which fit. In all truth then we might say the man is in more of a position to say 'the part I have found looks like the sky in the picture' rather than the man who has picked up a few random pieces.

It is with neither sense of reproach or self esteem that I say, to some degree I feel rather like the man who has the good fortune to find many pieces which fit. And like that man I have to tell you now, I *do not* really know if in fact I have identified my completed part correctly, I *think* I recognise the part of the jig-saw, but I cannot say for certain this is true. So by all means let us theorise to our heart's content, it is jolly good mental exercise. But we would be wise not to interpret UFOs as being electro-magnetically powered simply because there are electro-magnetic effects, for we would then be comparable to the man who has picked up just one piece of the puzzle. On the other hand we have several pieces to examine. First the largest piece which represents *gravity.* Secondly, strange craft which are visiting our skies seemingly disobeying the law of *gravity.* Thirdly we know of no field of force which will propel *machine and crew* at *uniform* accelerations as would the *gravitational field.* And many pieces more. Why start with the electro-magnetic jig-saw piece when we know it will *not* fit the rest? For while a magnetic field might accelerate a metallic device, it will not accelerate the crew. Indeed they would have to suffer crushing accelerations just as do the crew of the modern rocket! The energy required to produce such a magnetic or electro-static field would be enormous, on the other hand gravity is extremely weak.

Not only may it be possible to produce a contra-gravitic state, but as we shall see later on, it would require such a minute fraction of the available elementary energies, given a really efficient liberation technique, as to be of little consequence. That this should be so is borne out by the history of technical development, but in this respect we still have a very long way to go.

We have seen how our present heat engines, including the modern rocket are afflicted by low thermal efficiencies or heat losses, which is further increased by poor mechanical conversion into work. A piston driven airscrew for instance, is considered pretty good if 25% of the available heat energy is converted into propulsive force, while the remaining 75% of the heat energy is wasted. True, we have progressed from the days of the steam locomotive, which usefully converts no less than 4% of the available heat energy, the remaining 96% being dissipated into smoke, exhaust steam, conducted and radiated heat, incompletely burned fuel, frictional losses etc., etc. While on the other hand the direct conversion of mass into energy required to neutralise or reverse gravitational effects, would represent merely a tiny expenditure of the elementary energies available.

Bearing this in mind it becomes obvious that we must look to the very core of gravitation if we are ever to accomplish real space flight. No doubt it is such considerations as these which led Burkhard Heim to develop his 6 dimensional field theory. His results concern in particular 'mutual relations between the gravitational force and the matter which generates it'. Heim postulates that since electro-magnetic waves are special cases of material fields, then electro-magnetic fields must be accompanied by gravitational ones. Heim's theoretical investigations have shown that the 'meso-field' may exist in two states, 'contrabaric' and 'dynabaric'.

When 'contrabaric' it is able to transform electro-magnetic waves *directly* into gravitational ones, it could induce the *acceleration of mass* from a direct conversion of *electro-magnetic waves*. Heim believes the energy required for this phenomenon could be derived directly from nuclear processes.

Although hitherto Heim focused attention chiefly on the consequences for space flight, it becomes obvious that such an identical conversion process could be employed to generate kinetic energy from electricity without any wasteful intermediate thermodynamic process, that is, practically free from loss.

In the 'dynabaric' state, the 'intermediary field' is inverted so that electro-magnetic energy is directly liberated from matter, without accompanying heat or other waste. Fantastic prospects would result if engineering techniques were available to harness the dynabaric state.

In addition, it must be stressed, Heim's approach does not conflict with known laws of nature, in fact it strictly agrees with the quantum theory. UFO researchers will note the parallel with electro-magnetic saucer effects here.

Heim's theory goes on to predict mutual interactions between inertia forces and electro-magnetic radiations. Einstein said they were equivalent to gravitational forces, Avenel in effect says the same thing.

As an open minded scientist, A. R. Weyl's following comments will be encouraging to UFO researchers. Writing on Heim's work he said:—

'In Heim's theory the member which represents electro-magnetic radiation is related, by way of an 'operator' (that is, an instruction for carrying out definite manipulations of computations in an ordered manner), to a 'space density variable in time'. It supposes an inertia force of such a kind that the operator's treatment of the electro-magnetic radiation produces effects upon the inertia, from the radiation.

'If it were possible to realise this 'operator' physically, that is to devise means which actually carry out the formalistic manipulation of the theory, the direct transfer of electro-magnetic waves (for example, light) into mechanical force (gravitational waves) would become reality, subject to some efficiency factor, of course. Also, mass could be directly converted into radiation without the production of heat.

'B. Heim intends to render proof for the validity of his theory, by way of conclusive experimentation.

'Despite his financial handicap, he has made some tentative experiments with a primitive preliminary assembly. This experimentation had to be discontinued for the want of proper instrumentation. Besides, the arrangement which had been designed for 3.2cm waves was run with a coarsely adapted generator. The strict proof of the theory would require two new instrumentations using shorter waves and having higher efficiency. Heim is convinced that he can prove, for example, that nuclear radiation energy is directly transformed into mechanical energy. Till now, alas, Heim has not been able to scrounge together more than about £170 for all his experimentation—not much evidence for official scientific eagerness in Federal Germany!

'Heim's intermediary field (represented by the spirit which motivates his 'operator') would also show the possibility to neutralise or to reverse gravitational acceleration (=gravitational waves) at *direct* expense of mass-energy, by way of an electro-magnetic/gravitational conversion which would impose practically no loss. A dynabaric state of the intermediary field should also be capable of producing the direct conversion of mass into electro-magnetic energy, without heat or the formation of waste products.

'If Heim's conclusions from his comprehensive theory should prove realisable, certain fantastic consequences somewhat of the kind usually ascribed to 'flying saucers' (including their alleged immunity from the effects of rapid accelerations) would become attainable.

'It would also appear possible to propel space vehicles from external sources of natural energy, that is, *from the conversion of light or other electro-magnetic radiation.*

'It must be said again, that, unlike many other physicists dealing with comprehensive field theories, B. Heim is willing to prove the validity of his conclusions from his theory'.

Earlier Weyl said 'Leaning upon his present deductions from his theory, Heim has claimed possible air and space travel by exploitation of the contrabaric state of his 'meso-field'. Beyond that, he states to have discovered the fantastically sounding possibility of immunising the occupants and the structure of such vehicles against any effects from accelerations of the vehicle, however great and violent. He claims to be able to achieve this unique property by an appropriate arrangement of his 'field inducers' which provide the acceleration of the mass of the complete vehicles.

'If Heim were right, the amazing properties commonly ascribed to the mysterious 'flying saucers' would be in fact, *sound physics and proper engineering*! Hitherto, no respectable engineer could have imagined as possible the idea of vehicles, dashing about at fantastic supersonic speeds and in all directions without 'visible means of sustentation', with their crews not being subject to the effects of acceleration.

'Heim is definitely not a flying saucer fancier, being as serious and sober a scientist of academic qualifications as anyone would imagine a German physicist to be. It is rather doubtful if he knows much about the flying saucer stories. As a blind scientist, his time is devoted to concentrating his attention upon his mathematical theories, the consequences from it, and upon planning his experimentation'.

Not until *Space, Gravity and the Flying Saucer* had gone to press, did the present author hear of Heim's work. In fact some readers of that book may remember a stop press quote was made concerning him. In summing up this chapter, I could wish for no finer nor worthier piece of evidence than from such a courageous independent scientist. As to whether Heim's conclusions are coincidental to UFO behaviour, I shall let the reader be the judge.

We have stood looking thoughtfully at this particular signpost for a little while and now we must move on. Over the horizon, rays from the rising sun crack incandescently between the distant hills. There, caught as it were in our imagination we see the thing, indistinct at first, as it glides soundlessly, swiftly towards us. Then for a brief moment we see it clear, a round thing, a shimmering beautiful thing. In a split second it is gone.

We hesitate for a moment breathless, staring back unbelievingly, then our pace quickens in the direction from which it came.

PART TWO

Analysis of the Technically Corroborative Evidence

8

The G. Field Theory

AS some readers may remember, the above title appeared as Chapter 9 in *Space, Gravity and the Flying Saucer*, and the repetition of it here is by way of an inevitable continuation of the material in that earlier publication. In the twelve years which have elapsed since then, the numbers of UFO sightings has increased so much that we are now able to make a more concise observation on the behaviour of flying saucers.

For the moment however let us interpret the basic conclusions arrived at in the last chapter in broader terms. We might profitably do this in the form of a study into the kind of space ship we could expect to evolve from such gravitational manipulation, and what kind of phenomena we might expect to accompany it. Appropriate technically-helpful UFO sightings are included to lend some interesting support to the theory, later we shall analyse more fully some of these sightings from the purely technically corroborative point of view. First it must be made quite clear, our space ship is *not* a so-called anti-gravity machine, in the sense that it is a device which cancels out or *shields* off the earth's field. For even if this were possible, the machine might at best only sail aloft due to displacement to the limits of the atmosphere. Without additional lift and directional propulsion of some kind such as the Townsend Brown effect we have already examined, certainly it could hardly be classified as a space ship, Fig 22.

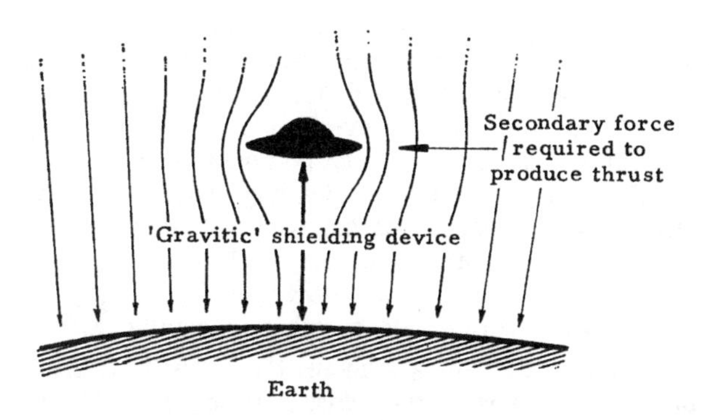

Fig 22. A purely anti-gravitational device could never produce a directional thrust, other than an aerodynamic one.

So we assume our space vehicle can generate a field which does not cancel out the earth's field, but rather *opposes* it. In Chapter 5 we saw how gravity might be a modulation of the rays that created space, or 'unbalanced inertia', and later it was suggested that the condition might be encouraged electrically or by other means. Now as we are primarily concerned here with the pure mechanics of using this technique to operate a space ship, we will start from the beginning with the premise that our space ship creates such a condition in the space around it. So in order to become 'weightless' on lift-off, then we assume that this G. field is created above the ship. Now it is obviously difficult to imagine a massless point source into which a material device is 'attracted'. But the difficulty can be removed once the general conception of what we know as space is modified. I have sometimes found it helpful when talking at lectures to describe the condition in the following broad terms. This at least enables the student to follow the general idea. Essentially it is what we have said earlier, but put in a more analogous form.

If modern physics can talk of matter as being 'something' and space as being 'nothing' (which in itself is a contradiction in terms) and then go on to tell the student that space is 'curved', then with all good grace we can equally ask him to accept that space *is* something and what is more, space and matter are one and the same thing, their relationship to any examination by man is different, that is all. Now being synonymous they must interact with one another, we have considered this interaction in the *Unity of Creation Theory* as modulations of creative rays. A simpler way for the student to visualise gravitation may be to consider that particles of matter do not 'attract' each other, but in a way we can regard it as a case of matter 'attracting' space and *vice versa*. A ray which formed space a moment ago is changed into a different frequency to be borne into our detectable electro-magnetic spectrum an instant later. The two cannot help but be 'attractive' to one another. Fig 23(a) and (b).

Now if matter 'attracts' space, and space 'attracts' matter, we could regard the condition of gravitation between two masses as one in which the masses are attracted to the denser space in the region of the masses. Therefore the conception of gravitational space takes on a new meaning as in Fig 23(c). Of course it must be remembered when considering this state of affairs that although matter is gravitative towards space, it will not disintegrate, because matter itself is nothing but 99.9% space anyway. Therefore there is just as much space within the particles of matter and the interstices between them as there is anywhere else, matter in its normal state cannot help but aggregate and coalesce.

If we can consider this argument, then it will be easier to accept the idea of locally increasing the density of space (creating a gravitational field) without the formation of an equivalent mass, be this enormous and tenuous, or small and very dense. Bearing this underlying factor in

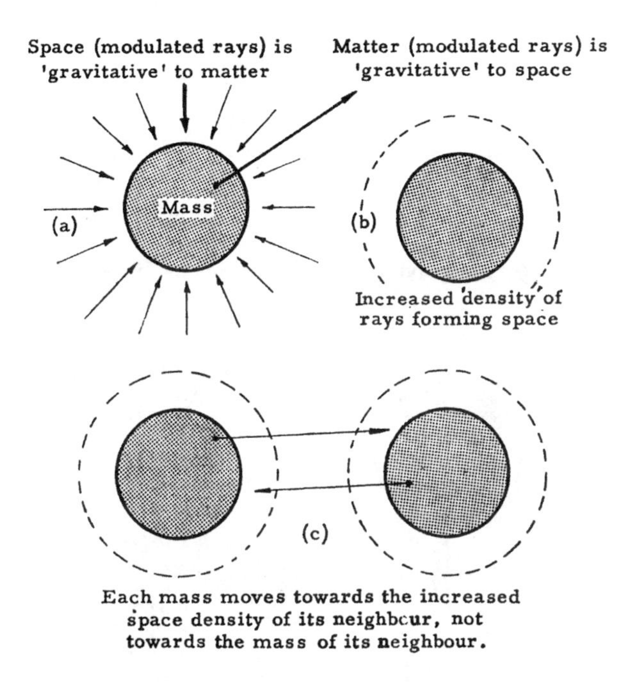

Fig 23. Analogous representation of the mutual interaction between space and matter.

mind then, we can proceed by first considering the idea in terms of actual masses.

Fig 24(a) depicts a machine placed above the earth where it is subjected to the 'downward' acceleration of 1g. if we could now imagine a duplicate of the earth situated above the machine and at the same distance as the earth from it, then the craft will experience an 'upward' acceleration of 1g. also, which of course would cancel out, and the craft would be immobilised in space. Of course such a condition could hardly exist, because the two larger masses would rush towards one another with equal acceleration and not much would be left between! So for the moment we shall imagine the two masses to be separated by some physical means, for instance, they could be rotating about each other.

Now in order to take the next step, we must bear in mind some of the governing factors about gravitation. For instance we know that the gravitational field strength between two bodies obeys the Newtonian formula;

$$F = K\frac{M_1 M_2}{R^2}$$

Where M_1 & M_2=the weights of the two bodies in lb respectively.
R=the distance separating the bodies in feet.
K=the Newtonian gravitational constant 1.09×10^{-9} $ft^3/lb.sec^2$.
F=gravitational force exerted between the bodies in lb.

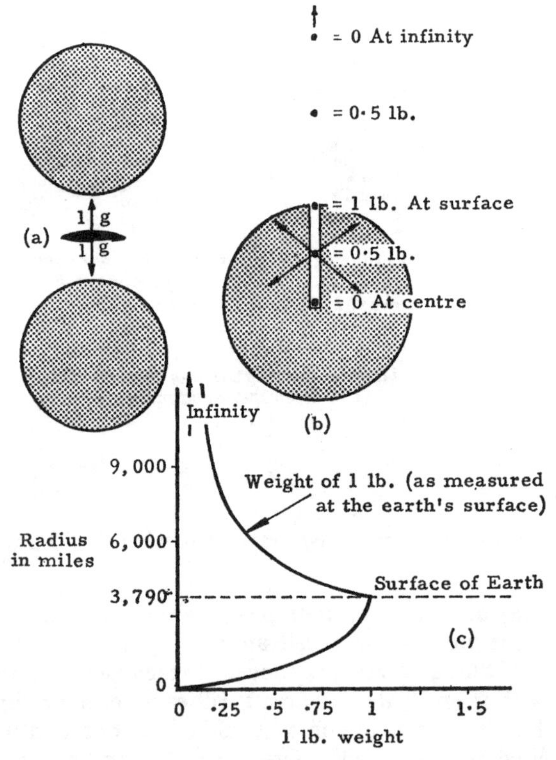

Fig 24. Gravitational attraction is a function of masses and the radius between the mass centres.

Also, while it is true to say that the gravitational pull exerted by a planet on a smaller body, *increases* the nearer the body gets to the planet, it is *not* true to say that the weight of the body will continue to increase if it could descend a great pit towards the centre of the earth, Fig 24(b). For it will be apparent that the moment it goes below the surface, then the surrounding mass is exerting a force from all directions on it, until we can visualise a condition when on reaching the centre of the earth, the weight of the body would be zero, according to the graph Fig 24(c) or nearly so.

In order to find the exact gravitational differential over a measured height, a group of scientists from the Henry Krumb School of Mines studies at Columbia University, led by Dr J. T. F. Kuo, recently took readings at the top and bottom of the Empire State Building. After correcting their sensitive gravimeter for the building structure, geographical features within a 10 mile radius and lunar and solar tidal effects, the team established the gravitational differential over the 1,250ft building as being 0.01%.

Now it is equally true, that according to the formula $F = K \frac{M_1 M_2}{R^2}$ if the earth could shrink while retaining its same mass, *i.e.* become more dense, then a 1lb weight situated on the earth's surface would steadily get heavier as the radius diminished, Fig 25(a). It follows therefore, in

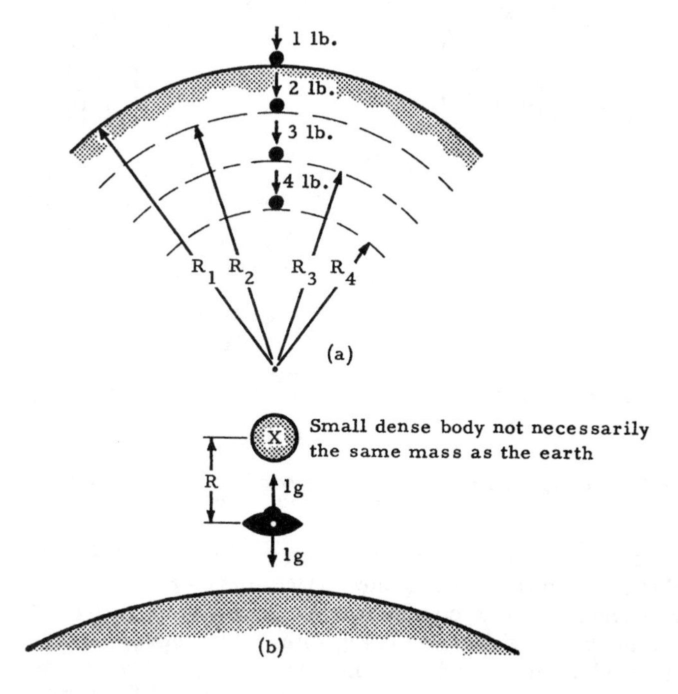

Fig 25. Development of the massless point source conception.

order for the 1lb weight to remain constant as the radius diminished, the earth would have to lose a proportionate amount of its mass, so we should arrive at a condition in which a small, very dense earth would still produce 1g. on a body placed near it. So we can now modify our analogy Fig 24(a) to look like Fig 25(b), *i.e.* a small dense earth X situated a comparatively short distance above the machine to produce an upwards acceleration of lg. on it.

But the previous objection raised in the case of Fig 24(a) still applies, *i.e.* the two earths would move towards each other, unless otherwise sustained, but in this case, the velocity would *not* be evenly shared, due to the fact that the small one has lost so much of its mass. Now it is a

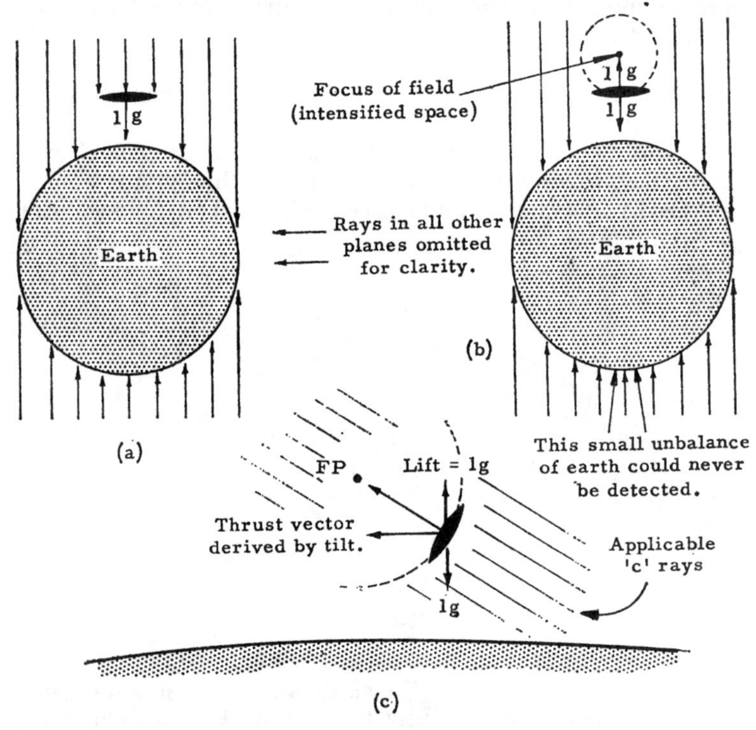

Fig 26. The logical development of a controllable 'G. Field' idea in terms of the Unity of Creation Theory.

comparatively short though by no means simple step to visualise that if matter is a form of modulated space, then we can shrink our already dense body X to a condition where it becomes virtually non-existent, together with the distance R, Fig 25(b). Our analogy is almost complete and can be summarised by saying: in so far as X has now become a pure field, it will not be accelerated towards the mass of the normal sized earth.

The situation might be comparable to a plate raised above the floor, and by some means the plate be able to cause a vacuum or implosive effect in the air above it into which it promptly pops. If split seconds later the procedure was repeated, we can imagine the plate lifting or hovering according to the frequency of cycle employed. Fig 26 (a, b & c) illustrates what may well be a more correct diagrammatic conception of the true state of affairs with a controllable G. field.

From this it will be seen that should the craft be tilted, thereby directing the focal point to one side, a thrust component will move the machine forward in any chosen direction, while a change in tilt angle, and/or power, offers helicopter type manoeuvrability in all planes. In forward tilt thrust attitude the occupants of such a device will of course experience no acceleration, though 'down' to them would not necessarily be vertically towards the earth, instead it could be at right angles to the plane of the ship. In other words, the floor of the craft would be the true horizontal, and looking out of their porthole windows, the panorama of the earth below might well appear to be tilted up towards *them*. The situation can be more readily grasped by considering diagram Fig 27, which shows the forces acting on the nearest counterpart, the human centrifuge; with one exception, in the case of the centrifuge, the rider would experience *increased* weight, while the pilot of the saucer on the other hand might well experience a loss of it, unless, as we shall see later on, other factors are brought to bear to remedy this.

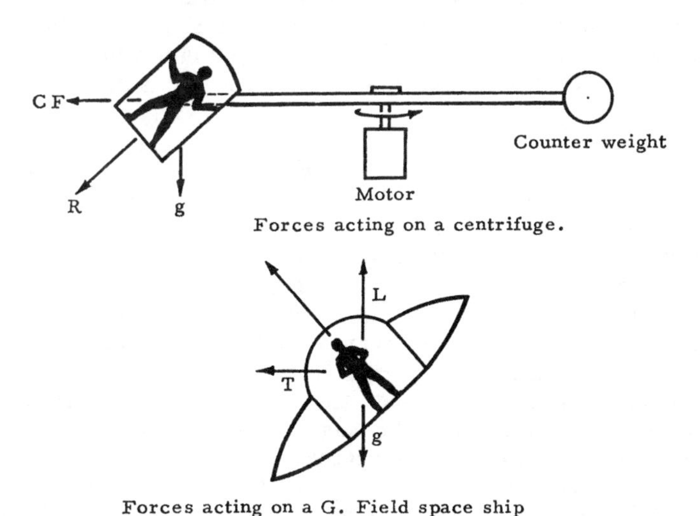

Fig 27. Conditions in the modern centrifuge produce the nearest counterpart for crew environment in a 'G. Field' propelled space ship.

Now, so far we must restrict ourselves to this somewhat limited and perhaps not strictly accurate conception of G. field propulsion but it will at least enable us to proceed with our enquiry into some associated phenomena we might expect to accompany such a technique.

However, if at this point the reader should doubt the feasibility of the generation of such a powerful field purely on grounds of energy expenditure, it may be as well to remember that two protons of 1.66×10^{-24}gm weight can at 1cm distance, by virtue of their masses, attract each other with a force of $(1.66 \times 10^{-24})^2 \times 66.8 \times 10^{-9}$ dynes $= 18.10^{-56}$ dynes, according to Newton's law of gravitation. At the same time they exert an electrical repulsion of $(4.774 \times 10^{-10})^2$ dynes $= 22 \times 10^{-20}$ dynes. The electrical force is therefore roughly a *sextillion* (10^{36}) times as large and of opposite sign to the gravitational pull. For the layman's sake 10^{36} looks like this: 10,000,000,000,000,000,000,000,000,000,000,000,000. While as regards the feasibility of a massless point source, I would equally recommend some doubting scientists to look more seriously at some genuine cases in metaphysics. They may perhaps learn something about matter that at the moment they are at a loss to understand.

Stability and Control

Of the possible various techniques which might be employed to stabilise a gravitational device, probably the most obvious which comes to the mind of many a UFOlogist is that of revolving masses. In fact, because of its more obvious use, I propose to give it only a mention in this chapter, sufficient to say here that we would in all probability develop such a technique for stabilising our space ship, as in fact are some modern satellites. Control and orientation brings with it other problems which due to its technically corroborative nature is best dealt with later on.

Air Displacement

If the G. field was low in intensity, as might be at the beginning of a slow take off, then similarly the extension of the field might be restricted within the confines of the craft's structure, *i.e.* low intensity field, low extension of field as with a bar magnet. Therefore when the machine took off, surrounding air would be displaced as it is with a conventional aircraft, and if an observer were to be close, he would detect a swishing noise. This has frequently been the case.

Many readers will remember that particular February morning in 1954 when Stephen Darbishire, he was 13 years old then, climbed the hill near Lake Coniston with his eight year old cousin, Adrian Myer, to take one of the most controversial photographs of a flying saucer.

At 11 a.m. Adrian was looking towards the mountain; Stephen was looking away in the opposite direction, when suddenly Adrian clasped his arm and gasped, ' Look at that thing'!

Down from the direction of the sun a strange, silvery, round object was descending. It came to earth about 100 yards away and disappeared behind a bit of rising ground. A few seconds later it came into view again. Suddenly it tipped up on its side and shot up into the sky with a deep swishing sound, but until then it had been completely silent. In a few seconds it had disappeared into the clouds.

Just before it went down behind the rise, Stephen had succeeded in obtaining a photo. As the object came into view again he took another. Here we have the swishing noise and notice the tilting motion on take off. We shall hear more about this later on.

As we are here trying to establish our case by dealing with the fundamental behaviour of a hypothetical space craft, it is necessary to be more generous. Here then is a parallel case to that of Stephen Darbishire, and although it occurred much later, in the year 1962, the similarity is most striking. One report of the event contained the heading ' Riddle of The Thing in the Midnight Sky '.

A wife spoke last night of a strange object which, she said, hovered over her husband's car during a midnight drive.

The woman, Mrs Myra Jones, of Norrishill, Moira, Leicestershire, said the object appeared as they drove along country lanes on the Leicestershire-Derbyshire border on Thursday, 13 September.

Mrs Jones went on, 'I saw a light over the car, leaned forward and looked up through the windscreen. Right above the car at about the height of the telephone poles was a grey luminous object bigger than the car. It was curved underneath and the top was domed like a child's humming top. There were dark spots around the rim of the base and the whole thing seemed to be *tilted* slightly and to be revolving. I was absolutely terrified. I thought it was going to land on the bonnet. Then it made a swishing noise like a rocket and disappeared'.

Note again, the sighting was accompanied by a swishing noise suggesting evidence of aerodynamic displacement, which due to the comparatively low speed of the vehicles and the resulting low intensity of the G. field, is required by the theory.

Here also we have the tilt of the device, and as the thousands upon thousands of UFOlogists throughout the world will know, there are many, many similar sightings, hundreds of them, 'on take off, a *tilt* and a *swishing noise*'.

The following two cases are somewhat different in so far as the witnesses claim to have seen beings with the landed saucers, and the inclusion of these typical incidents here is required by the dual purpose we have in mind. In this instance my intention is to point out the technical similarity between widely diverse sightings which sceptics and some UFOlogists are all too ready to otherwise doubt.

The following amazing incident happened in November 1958, when the Territorial Army were on a weekend excercise in Scotland. Such exercises are quite frequent and this particular one took place a few miles outside the village of Tarland which is about 60 miles from Aberdeen. It also lies between Braemar and Ballater.

When the exercise got under way, two young lads were left to guard a small hilltop for a few hours. They had full equipment with them and they had also dug a trench to lie in.

It was some time in the very early hours of the morning, when the first rays of the sun appeared over the eastern horizon that they heard a strange 'gurgling' sound coming from behind some trees only a few hundred yards from their position.

As the noise persisted, they decided to investigate. They started to walk towards the trees. Then suddenly two large figures came stumbling towards them.

As the two boys stood, rooted to the spot with fear, they noticed that the beings were at least seven to eight feet tall, and that the 'gurgling' sound was being caused by the two strangers talking to each other in a guttural tongue.

The beings were dressed rather peculiarly and found difficulty in walking across the rough hilltop.

The two young soldiers turned and fled down the hillside to the main Tarland road to seek civilisation. On reaching the road they headed for a small hut which was temporarily housing a few Post Office engineers. As they ran along the road, one of the boys turned to see what was causing a *swishing* noise behind them. They saw a huge brilliant object, disc-shaped, coming up behind them down the road and only a few feet from the ground. The lads started to run as fast as they could and then the large craft swooped up over their heads, pulsating and leaving a *shower of sparks* in its wake. It soon disappeared from their sight.

The boys arrived at the hut and thumped on the door for someone to let them in. The engineers took the two boys in and made them rest for a while. Naturally they were considerably shaken by their experience.

Now *is* there a world wide conspiracy? Do people suffer from shared delusions? Or did these events really happen?

Here is a translation of an article from the *Corriere Milanese* for 19 December, 1962. The date of the occurrence was 17 December. 'Martians near the Porta Magneta at Milan'?

A 37 year old night patrolman of the Milan police was confronted with this problem. His name is Francesco Rizzi and he lives at 1, Viale Berengario at Milan. His story goes:

It was exactly 2.20 o'clock at night when it happened. The night patrolman was about to fulfil his nightly round and at this precise moment he entered the premises of the mill 'Filatura cascami di seta' in

Santa Valeria Street, a few steps away from the Santa Ambrosia Square. It was his job to clock in and in order to do this he had to walk across the vast courtyard of the mill. 'Exactly in the middle of the courtyard I became aware of a *swishing noise* to the right of my neck', said Rizzi later. 'At first I thought it was inside my ear, but then I noticed that the swish grew stronger and stronger. I therefore turned around and beheld an incredible thing! Of course, I had read about flying saucers and Martians coming to visit us and spy on us, and also about messages coming from other worlds, but I would never have imagined that it would fall to my lot to see such an object under my very nose. Yet there it was—just in front of me, hanging about three feet above the ground.

'It was of a clear metal, perhaps of aluminium with silvery reflections on it; its diameter may have been 12 to 15 feet. On the top there was a turret around which were a number of dormer windows, lighted. I was paralysed and tried hard to believe my own eyes, when suddenly the noise stopped. At the bottom of the disc a door was opened through which a little man came out, a little man of about three feet and a few inches. I could not see him very well because it was very dark in the courtyard and the man's head seemed to be absolutely black. To make matters worse he wore a kind of luminous overall which made it hard to see him clearly.

'The man had perhaps no hostile intentions. He pointed one finger towards my hand and with his other hand he gave me a sign to come nearer and not to be afraid. Yet I was completely unable to move. Soon after, another man jumped out of the disc, submerged in a blue haze. With a commanding gesture he made a sign to the other to re-enter the disc. Suddenly the door closed behind them both, the *swishing noise* started again and the disc disappeared in a cloud of *white smoke*'.

Only after the disappearance, the policeman regained command of his legs and he made good use of them. He rushed out of the courtyard to tell everything to another night patrolman. This man gave him the advice to report at once to the commander, and Rizzi did so. In order to be co-operative, a search patrol was sent at once to the spot, but the 'Martians' had been careful not to leave a trace. Note in some cases the noise may well be caused mechanically from the machine itself.

But if there are many of these cases where the witness claimed to have heard a swishing noise, then the numbers of sightings quoting saucers in a high speed hop making no noise whatsoever, are prodigious and there is no point in spending more time by quoting them. Instead let us go on a stage further with our hypothetical craft.

Aerodynamic Effects

When the propulsion field was high in intensity as when the vehicle was making a high speed run, then we would expect the same rule to apply, *i.e.* high intensity field, high extension of field and therefore we could expect it to noticeably extend beyond the confines of the craft's structure. And therefore we would also expect the surrounding air in contact with the surface of the disc to receive the same 'thrust' as the vehicle itself and move along with it. As the field strength fell off at greater distances from the originating source, we would expect the local air in these regions to receive proportionally less 'thrust'. Therefore there would be formed a molecular shearing effect, or velocity gradient. We would also expect that the vehicle would suffer no aero-dynamic heating through friction, although undoubtedly there would be a temperature rise due to the shearing effect, which in turn might produce an effect similar to sonoluminescence, the well known laboratory phenomenon sometimes produced by ultrasonic experiment. Even so one would expect this large volume of continually shearing and changing air to act as a heat sink. In general terms then we could say, high speed flight, no noise. But, and this is important, should there be a power failure at high speed, then the tilted machine would virtually collide with an unyielding wall of air, which would almost certainly cause it to break up, or even volatise should it be travelling at hypersonic velocity. In which case there would be generated an enormous shock wave, heard over several miles of countryside below. I do not have to remind my well informed readers that there are on record sightings of sky phenomena which may well explain such occurrences.

18 February, 1948. Stockton, Kansas, U.S.A. A terrific explosion in Northern Kansas rocked buildings, broke windows and terrified local people. Origin unknown. A farmer near Stockton said he saw a flying saucer before the explosion.

7 January, 1954. An unknown object like a disc with a wake of crimson light streaked across the North French sky, exploded, and fell with a violent noise into the Channel off Dieppe. French military authorities say that it was neither a 'plane nor secret missile'. Houses were shaken over a wide area. The blinding flash, white then orange, was seen 80 miles away, and for three seconds was like the noon-day sun at night. A French trawlerman, about 30 miles off the coast, said he saw 'a tremendous bowl of fire flash over the sky in the direction of Dieppe. It left a wake of sparks'.

8 November, 1961. The *London Evening Standard* reported that Hertfordshire police were probing the mystery of an explosion in the sky. At about 9 p.m. on the evening of 7 November, police stations in Hertford and Hatfield received calls that a burning object was passing overhead. A second report said that it had exploded near Hertford. Police patrols searched the area without success and the mystery remains unsolved.

There are of course many more and it may be wiser not to explain them all away as natural phenomena, meteors, etc.

We have seen that another aerodynamic behaviour we can predict is the pendulum motion, for should the craft be hovering in a weightless condition and the pilot to close down the lift factor to something less than 1g. then the craft will gently fall, the rate of fall depending on the value of the diminished lift. As it falls the machine will be subjected to aerodynamic forces, which unless compensated will cause it to descend like a 'falling leaf' as in the Topcliffe incident. This we shall examine more closely in the next chapter. Sufficient to say here we would expect it to happen.

Now if the air does move with the craft, then there should be a local drop in atmospheric pressure in the near vicinity of the device, accompanied by a drop in temperature, and again we do not have to search far for verification. Here is a typical case. Remember these are witnesses' own unprompted testimony taken from the UFO files in my possession. In this particular instance we have to allow for the witnesses' excited state, but in one or two paragraphs the facts we require are there.

The *Dublin Evening Express* in its issue of 6 August, 1963 carried the following account of a youth's extraordinary experience: 'A teenager said a strange white light chased his car at speeds up to 120 miles an hour early yesterday morning. Ronnie Austin (18) of Wayne City, Illinois, told authorities the light followed him and Phyllis Bruce (18), 10 miles as they drove home from a drive-in theatre at Mount Vernon, Illinois.

'Austin claimed the light stalled the car's engine as it passed over and caused the radio 'to go crazy'. At one time it approached as close as 100 feet he estimated. He said it made a humming sound and had a *cooling effect* as it passed overhead'.

Wayne County Deputy Sheriff Harry Lee, one of the officers called to the case, said he saw the light at a distance. He said it was 'three or four times bigger than a star and was moving, but not twinkling'.

Here is another. 18 September, 1954. M Güitta, of Casablanca, was driving along the coastal road, when suddenly he saw in his rear mirror a grey thing diving down at him. Gripping the steering wheel tightly, he ducked instinctively, as a few seconds later the grey object passed on the left almost at ground level and 'at terrific speed'. A violent gust of *cold* air followed the passage of the object, which in spite of M Güitta's effort to hold the wheel steady, created a strong suction which carried the car to the left. There was no noise. M Güitta then caught a glimpse of the object disappearing on the horizon in front of him. He said it looked like a small grey disc.

And again, beyond Montlevic, at Jettingen, near Mulhouse (Haut-

Rhin), a landing case was reported during the night of 7-8 October. Here is the story as told by the witness, M René Ott, a railway official.

'I was riding a motor scooter along Route D-16, between Beeantzwiller and Altkirch, when just this side of the village of Jettingen I saw in my headlight, very clearly, an object in the meadow to the left of the road, only nine or ten feet away. It was shaped like the top of a mushroom, or a low hut, and it might have been ten feet from tip to tip. In the cupola shaped silhouette I could see a lighted rectangle, like a door, about four and half feet high by two feet wide.

'The thing frightened me and I speeded up. But maybe 60 feet farther along, I was caught in a fierce white light from behind me, that shone at least 200 yards ahead and that seemed to be coming closer at terrific speed. In fact, the machine or whatever it was passed over my head no more than 15 or 20 feet above the road, and I distinctly felt a very strong current of air'.

M Jean-Pierre Mitto of Toulouse, technical representative for an industrial firm, was driving at considerable speed along Route N-631 near Briatexte, when 'I suddenly caught sight—as did my two cousins who were with me—of two small creatures, the size of children of eleven or twelve, who were crossing the road in front of my car. I stopped instantly. But before we had time to get out we saw a red glowing disc taking off straight from the meadow next to us, and saw it disappear in the sky a few seconds later'.

In the course of the police enquiry conducted on this case, M Mitto emphasised that the machine had left the ground at a tremendous rate of ascent, sucking the air up from beneath it in the process. At the place in the meadow where he said the machine had rested, the police found brownish spots of some unfamiliar viscous matter.

Submarine Capabilities

Continuing speculation on some of the side effects caused by the field of our space ship, we might ask the question, 'If the surrounding air would move along with the craft, can the same be said of water, if so might this enable a space ship to operate as a submarine craft, and is there any evidence of the latter to substantiate the flying saucer hypothesis'? Yes indeed there is, the following are typical.

March 1945. Fourteen men on the U.S.A.T. Delarof, an attack transport, operating near the Aleutian Islands, saw a dark spherical object rise out of the water, circle the ship and fly off at speed. So says an official report sent to Washington.

The sceptic may read nothing significant into this one, but UFO students will recognise the pattern. From Thor Heyerdahl's *The Kon-Tiki Expedition*, he tells us, 'and on one single occasion we saw the sea boil and bubble while something like a big wheel came up and rotated in the air'.

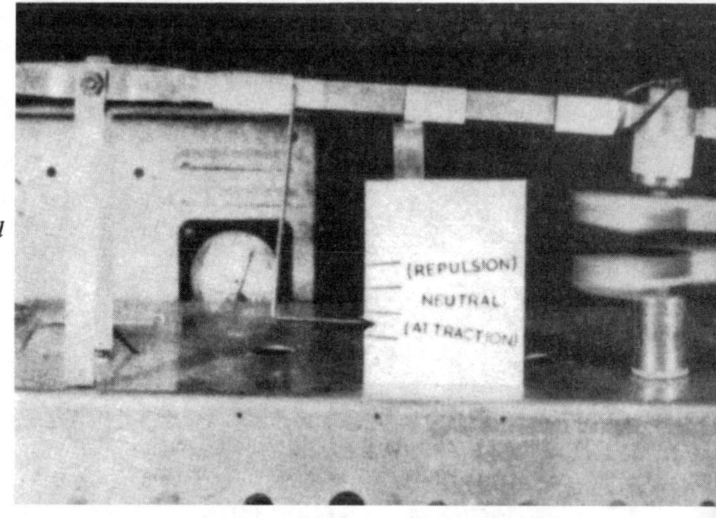

Plate 13. *Magnet rotary phasing analogy of electrical and magnetic phenomena.*

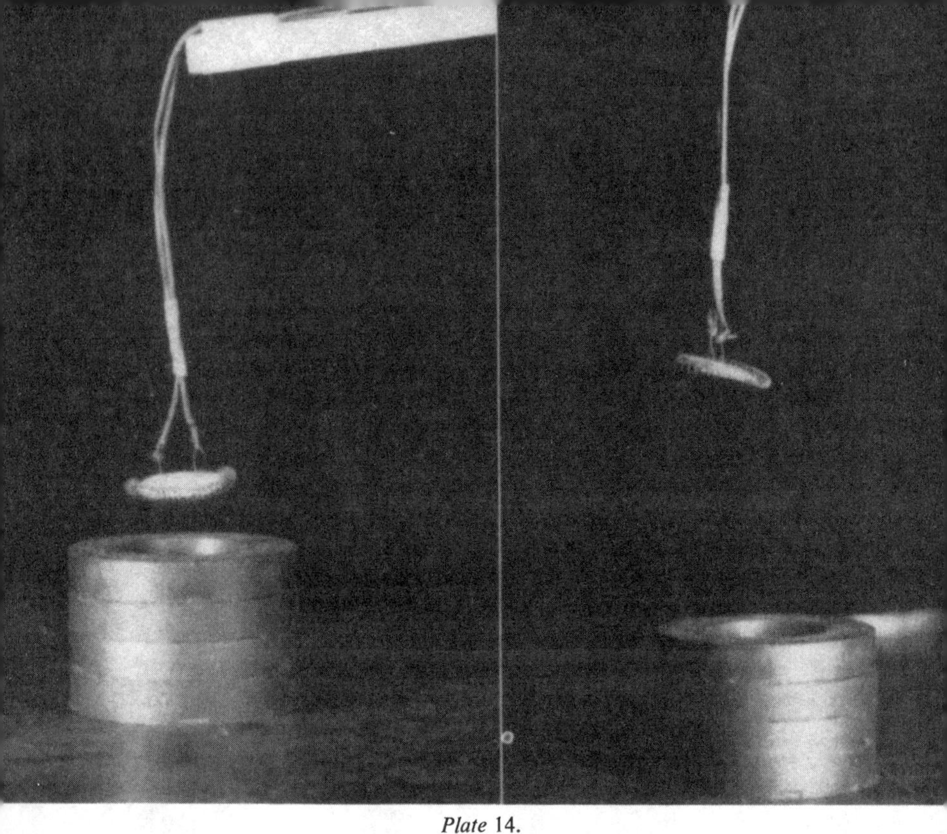

Plate 14.

Magnetic analogy of 'anti-gravity'. Iron coil attracted to magnet. (*a*)

When the coil is glowing hot the magnet has no further hold over it. (*b*)

Plate 15. *Convair Sub-plane.*

From the salty waves of the Baltic come reports of these strange events. The first took place in 1957 at a place called Koszalin when two mysterious objects were seen to emerge from the depths. Then in April 1957 at Kolobrzeg, coastguards and civilians saw the sea suddenly become agitated. Then the waves in a particular stretch of water swelled, and out shot a triangular object of 4-5 metres in size. It rose swiftly, encircling the locality, then finally rushed upwards disappearing into space. Army units were rushed to the spot and made an exhaustive search of the brushwood on the seashore. The next day a team of divers did the same in the water.

But the strangest case of all occurred in September 1961, 125 kilometres eastwards—towards a place called Leba. This is a pleasant fishing harbour and seaside resort situated on the open sea. There are also coastal lakes in the vicinity. Mr Czeslaw K. Kawecki, a twenty eight year old textile technician, had just spent his holiday there and this was his last day before boarding the train back to his hometown, Lodz. In order to have a last good look at the sea, he had decided to walk back to the hotel. He took his shoes and socks off and strolled along the sandy dunes that separated Lake Lebsko from the sea, and stopped for a while on the seashore to gaze at the waters lit splendidly by the full moon. Reluctantly noticing the time by his wristwatch (11.35 p.m.) he had turned to move on when 'a sudden noise of rushing waters made me turn towards the sea again and right in front of me, about 300 metres from the shore, the surface was rising in one spot. It looked like a round hill—pushed up from beneath. Then splashes of water gushed from the top, and like fountain-jets, fell around the 'hole' in the waves. From this opening in the water emerged an object which at first I thought to be an elongated triangle . . .' With a mixture of uneasiness and excitement, Mr Kawecki said he watched for further developments. Then, 'the object rose a few metres and hovered above the same spot and there was now a *whirlpool* of water rushing inwards with a loud sucking and gurgling noise. The object itself was black and silent.

Suddenly there appeared a belt of steady white light segmented by a number of convex dark streaks. This light made glowing reflections on the lower rim of the object. It also lighted considerably the upper rim and all the rest. Now it became apparent that 'the thing' had the shape of a huge funnel with two rims, separated by a belt of segmented light. About half way up the upper part was a thin strip of something whiter than the rest, of a rather dark body. The slim end of the 'funnel' had a rounded top, from which protruded a stump, thinning upwards, and bent in the middle on one side.

The stillness of this object lasted about a minute, then there appeared the glow of a second light under the object. Also a white one, but much stronger and sharper than that emitted by the segmented belt and almost immediately the 'funnel' tilted slowly northwards revealing the bottom. After remaining in this position for about half a minute without changing, it glided about 50 metres eastwards, stopped but soon glided back and stopped again. All the time the bottom of the object was visible and consisted of a dark circular perimeter corresponding to the lower (and wider) rim of the 'funnel'. Towards the centre was a wide ring of strong white light, with a number of dark, hook-shaped streaks upon it. Next was a dark ring with three evenly spaced triangular spikes, which protruded over half the width of the lit, streaky ring. Finally, there was a central disc which looked as if it was made of highly polished silver or crystal. It reflected the light with great brilliance.

Mr Kawecki said he was 'certain that there was some rotating movement involved. I could not make out whether the spikes were moving or the dark streaks gyrated under them. But I had no doubt that one or the other rotated. The light now became bluish and more intense. Then the object moved towards the north and upwards at an angle of about 45°, with a speed not exceeding that of a jet. It became just a diminishing spot of light until it finally disappeared. There was no sound. The entire observation lasted not more than four to five minutes'. Mr Kawecki estimated the width of the object at about 5 metres, and its height (without antenna) at about 6 metres.

Recovering from his surprise, he then realised that standing only a few paces from him was a young couple—gasping with amazement. They were Mr and Mrs J. H. and A. Poniewicz—who were also returning to their hotel. Mr Kawecki started to talk with them, when they were joined by two men who arrived hurriedly from the opposite direction. Both were local and rather scared. They did not want to give their names, and went to inform the coastguards about the happening.

While we are on the subject of UFOs and water, it is perhaps appropriate to quote the following case here.

31 October, 1963. 2.30 p.m. on the banks of the Peropava River, in Sao Paulo Province, Brazil. Several people suddenly heard a loud roar in the sky and looking up saw a 'shiny disc shaped object'. It was flying at low altitude, so low in fact that it struck a glancing blow on a palm tree which stands by the side of Mrs Elidia Alves de Souza's house. The blow was only sufficient to gouge the tree near the top, but the effect on the disc was fantastic, for it immediately began a series of thrashing, erratic manoeuvres, which obviously spelt trouble of some kind. Seemingly unable to regain its flight attitude, the object plunged into the river near the shore opposite the de Souza house.

Witnesses said later, that when the UFO came into contact with the surface, it was as if a hot iron had fallen into cold water. For the water at that spot began to bubble and surged up, then 'became exceedingly muddy and continued to boil'. At that point the Peropava river is about 12 feet deep, with about 15 feet of clay and mud at the bottom.

Nine year old Ruth de Souza, daughter of Elidia, was the closest witness. The object had flown directly over her head. She had looked up on hearing the loud roar and had seen the thing in collision with the palm tree. On the other bank, local fishermen, including Japanese Tetsuo Ioshigawa witnessed the whole episode. The police questioned the witnesses and established that the disc, which resembled an 'aluminium basin' was about 5 metres in diameter and 1 metre thick. It was extremely bright, 'almost luminous', and it travelled at a fairly slow speed.

Since then, numerous attempts to recover the object have failed. There is talk that the disc may have been 'retrieved' during the night before the search started.

If as a natural consequence to its mode of operation, our dream ship could function as a submarine craft, then it would be fulfilling the dream of inventors the world over, and only recently information has been released concerning a design study for such an idea called a 'Sub-Plane'.

One writer pointed out that the surface of the sea is an interface between air and water which remains one of the only physical boundaries not thoroughly mastered by man. The submarine can cruise on the surface and move freely beneath the sea; while flying boats use it as a platform from which to launch themselves into the air. But we still have no craft which can fly and then penetrate the air-water interface and move equally at ease in the ocean depths.

The public announcement said that the idea of a craft which could approach its target through the air as a seaplane, then dive underwater to pursue it as a submarine, is now feasible, practical, and well within the state of the art. The work is a result of a study made by the Convair division of General Dynamics, leaders in nuclear submarine and advanced military aircraft development. Plate 15

It is said the craft would have a seaplane hull with conventional wings and tail for airborne flight. Three engines would be installed, one on top of the hull to provide thrust for flight while the other two would be used for take off. The engines would be sealed off for submerging with alternative means for underwater propulsion. The flight radius would be between 300 and 500 nautical miles at speeds between 150 and 225 knots and it is claimed the vehicle would be able to operate submerged at depths up to 75ft and speeds of 5 knots. It is a pity that once again the emphasis has to be a military one. (*Author.*)

9

G. Field lift Effects

Craters

NOW let us see what might happen if the pilot of our space ship decided on a rapid take off. Well, naturally this will call for more lift which can only be attained by a greater degree of unbalance and we have already seen that the greater the intensity of the field the farther it will extend. Air in its immediate neighbourhood will tend to move along with the craft, while at take off it might be necessary to keep the machine a short distance above ground level, for should the vertical thrust vector be generated high for a quick take off, then the negative 'g' effect would lift any movable object beneath the vehicle. There are cases where saucers have taken off violently, appearing to suck up the ground beneath them as they did so. Many eye witnesses have claimed to have actually seen such ground effect taking place, the following is a typical case.

It was about 8 p.m. on 4 October, 1954. The place Poncey-sur-L'Ignon, France.

'About twenty yards from the house, in M Cazet's meadow, a luminous body was balancing itself lightly in the air, to the right of the plum tree, as if preparing to land. As well as I was able to judge, this object was about three yards in diameter, seemed elongated, horizontal, and orange coloured. Its luminosity threw a pale light on the branches and leaves of the tree'. The witness had made a noise which seemed to disturb the strange vehicle, for it had immediately taken off '*forward and upward* at *prodigious speed*, leaving a gaping hole in the ground over which it had hovered'. Over an area, a yard and a half long, 27 inches wide at one end, 20 inches at the other, the ground appeared 'to have been sucked up'.

'On the fresh soil of this hole, white worms wriggled. The earth that had been torn out was scattered all round the hole, in clods ten or twelve inches across, over a radius of about four yards. On the inner edge of the hole, similar clods hung down; the earth had been pulled out in such a way, that about half way down, the hole was wider than at ground level. In short, it looked just as if the mass of earth spread over the surrounding grass had been sucked out by a gigantic vacuum'.

Some readers will remember this description is almost word for word identical with that used by Mr Blanchard, when interviewed about the crater which appeared in his barley field at Charlton, England in July 1963. We shall hear more startling facts about such craters later on.

Yes, we would expect the vertical thrust vector of our hypothetical vehicle to 'suck up' any nearby objects if the field intensity was high. Furthermore, if the machine was given a simultaneous forward thrust vector, we would expect the hole it might leave in the ground beneath, to be elongated, narrowing down as the vehicle rose higher. Plate 19 shows an exact simulation by a magnet on iron filings. I feel it is vitally important, even at the risk of inciting some impatience among my readers, and I feel I might have to do this again and again, to remind you once more that the above, though perhaps a small technically corroborative point was the claim made by an untrained person who had nothing to gain from this story, other than derision and ridicule for her pains.

Here is yet another case in which the field of a flying saucer was seen to have an effect on the surface beneath it, in this case, water. It occurred at 6 a.m. one day in April 1958 on the Atlantic sea-board of north-eastern Brazil. Senhor Wilson Lustosa, a Brazilian jeweller was travelling along the beach from the port of Maceió to a place called Parapueira. At a spot called Saude, the witness stopped to ask some fishermen what they were looking at. They said that it was a flying saucer. He could see nothing at first, but soon he heard a humming noise, which grew rapidly louder, and perceived something which seemed to be falling out of the sky towards him, from the direction of the sea.

When the machine was at a distance of some 40 metres from him and the group of fishermen, and about 15 metres above the water, it began to rock sideways, and then stopped and hung there. It was from 15 to 20 metres in height, and 'its width was approximately that of a travelling circus'. It seemed to have three distinguishable parts: the upper half was the colour of aluminium, like an inverted bowl, and on top of it was a small protuberance or dome, with a light as bright as that of an electric welding arc. The lower part was also a bowl, of the same size as the top, but dark in colour, and around the widest part, where the two bowls met, was a band with a number of square portholes, from which came a reddish light. The portholes nearest to the onlookers were darkened, 'as though there were people looking out through them'.

Beneath the machine, the water seemed to be boiling, or being *sucked up*, but without actually touching the under part of it, note this might well have been in part an aerodynamic effect. A faint humming could be heard at brief intervals and from the under part of the machine a number of things like leather thongs were *hanging* motionless. No, this latter observation does not constitute an inconsistency, in fact, as we shall see later, it is entirely corroborative.

Undercarriage

We have seen how the force field of our hypothetical space craft would extend beyond the confines of the structure, both in the horizontal plane as well as the vertical, and therefore we would expect the lifting component of the latter would also tend to make weightless, or 'suck up' anything lying beneath the machine should it come to rest, and earlier I said it might therefore be desirable to have some marginal clearance above the ground while hovering. Indeed this would be necessary even on take off, and we might design an undercarriage to obviate this problem. Our designs might call for a multi-legged landing gear, or for simplicity and for other reasons, even one central pad of sufficient supporting area. It is a practical solution to the problem, but once again it looks as though we have been beaten to the post.

Dealing as they do with claims of contact and near contact with space people similar to ourselves, it is of the utmost importance to give fair consideration to the technically corroborative value of the following cases.

The first is the sensational story that in the Spring of 1952 caused a great deal of controversy and no doubt a great deal of scepticism; such stories are usually difficult to accept, but we must be patient; more shocks than this may be awaiting us before the flying saucer mystery is solved.

The man who first brought to light this story is grey-haired 48 year old ex-mayor Oskar Linke of Gleimershausen, near Meiningen. He had escaped from the Russian Zone with his wife and six children. In the company of West Berlin officials, Herr Linke with his eleven year old step daughter Gabriele, swore this solemn affidavit before a judge.

'I was riding home on my motor cycle, with Gabriele on the pillion, when a tyre burst near the village of Hasselbach.

'As we were pushing the machine towards Hasselbach, Gabriele pointed to something about one hundred and fifty yards away. At first sight, in the half light, I took it for a young deer.

'I left my motor cycle by a tree while I approached the 'deer' cautiously. I was now about sixty yards from it.

'I then realised that my first impression had been incorrect. The thing I had noticed was really two apparently human figures now about fifty yards from me.

'They appeared to be clothed in a kind of shimmering metallic substance, and were bending down and studying something on the ground.

'I wormed my way to within about thirty feet of them. Peering over a small ridge, I noticed a large object, which I judged to be about forty to fifty feet across, though it was hard to say exactly. It looked like a huge warming pan.

'There were two rows of holes along the sides, about a foot in diameter. Each row was roughly about a foot and a half from the rest.

'Out of the metallic object rose a black cylindrical 'conning tower' about ten feet high', Fig 28(a).

Linke went on: 'I was now alarmed by a call from my daughter, who had remained some distance back. The sound must have reached the two figures, for they rushed back to the object, clambered rapidly up the side of the conning tower and disappeared inside.

'Previously I had noticed that one appeared to be carrying a lamp on his chest. The lamp flashed on and off at regular intervals.

'The outer edge of the 'warming pan' in which the holes were sunk now started to glow.

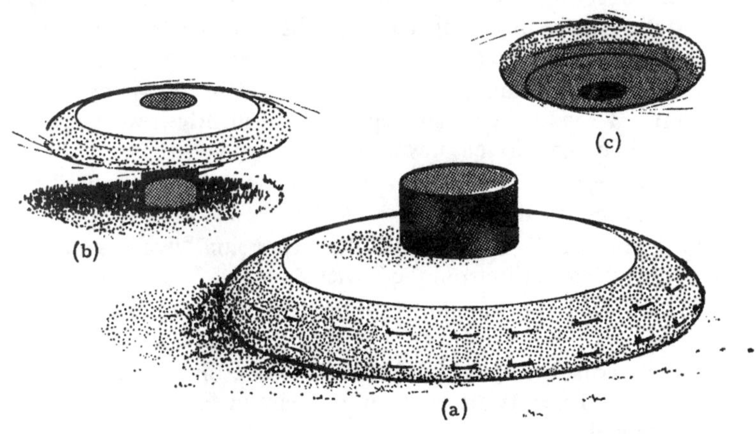

Fig 28. Artist's impression of the three stages in the saucer take off as seen by Herr Linke.

'The colour at first seemed green, then changed to red. At the same time I heard a slight hum. As the glow and the sound increased, the 'conning tower' was retracting into the centre of the 'warming pan' and the whole object rose slowly from the ground.

'From the swirling effect of the glowing 'exhaust', I got the impression that the whole object was spinning like a top.

'It seemed to be resting on the cylindrical piece which had sunk through the centre of the object and was now protruding from the bottom and standing on the ground, Fig 28(b).

'The 'warming pan' with its glowing outside ring of flame, was now some feet off the earth.

'Then I noticed that the whole object was rising slowly from the earth. The cylinder on which it had rested had now disappeared inside the centre and reappeared again through the top, Fig 28(c).

'The rate of ascent now became much greater, and at the same time my daughter and I heard a whistling sound, rather like the noise made by a falling bomb, but not nearly so loud.

'The object rose in horizontal position, swerved away towards a nearby village and disappeared, still gaining height over the hills and forest towards Stockheim'.

Several other people in the area later told Herr Linke they had seen something which they took for a comet. One, a shepherd, Georg Derbst, who was about a mile and a half away, said he thought a comet had bounced off the earth.

A sawmill watchman told Herr Linke he had seen what he thought was a 'low flying comet' flash away from the hill where Herr Linke saw the object.

After appearing before the judge, Herr Linke said: 'I would almost have believed that my daughter and I had dreamed the whole episode, were it not for one thing.

'When the thing had gone, I went to the place where it had been standing. I found a circular depression, evidently freshly made, where the earth had been driven down.

'This was exactly the shape of the 'conning tower'. I realised then that I had not been dreaming'.

He continued, 'I never heard the expression 'flying saucer' until I escaped to West Berlin from the Soviet Zone.

'When I saw the thing first, I thought it was a new Russian war machine.

'I was terrified, for the Soviets do not like one to know about their goings-on, and people are shut up for years in East Germany for knowing too much'.

Now this is not a unique case in which the central pedestal is mentioned, far from it, but we must restrict ourselves to one more.

It was 10 April, 1962, and Signor Zuccalà was returning home to San Casciano, Val di Pesa, from nearby Florence, where he goes to work every morning: it had been a day just like any other for Signor Zuccalà.

He arrived at San Casciano by the coach service SITA about 9.15 p.m. After he left the coach, he walked on three or four minutes and met a friend of his who went with him on a motor cycle as far as the street in the district of Cidinella which, passing through the wood of Cidinella, leads him home. He started walking by himself about 9.25 p.m. along the street and before 9.30 p.m. he arrived at the 'carbonaia' (coal cellar)—an open ground from which two streets lead, one which goes to the house of Signor Zuccalà, the other which goes on beyond. The sky was covered with stars, with a small moon, and the air was hot and still.

While he was busy going across a small canal which flows across the street, he felt himself struck and 'lifted up' slightly by a sharp gust of wind.

Turning round to the left, he stood stock still and terror stricken; 6 or 7 metres above the earth an object was hovering. The object resembled two bowls put one on top of the other, ashen in colour and of a diameter thought to be about 8.50 metres. The object passed over Signor Zuccalà and settled 6 or 7 metres away from him at a height of 2.50 metres from the earth. A *cylinder* of diameter about 1.50 metres was let down from the lower side of the machine until it touched the ground.

Signor Zuccalà had the impression that the cylinder, once it had touched the ground, re-entered the machine again, leaving exposed one side of the cylinder in which a door opened slowly, while two small doors were gliding towards the outside. There may have been two cylinders gliding within the other, Fig 29.

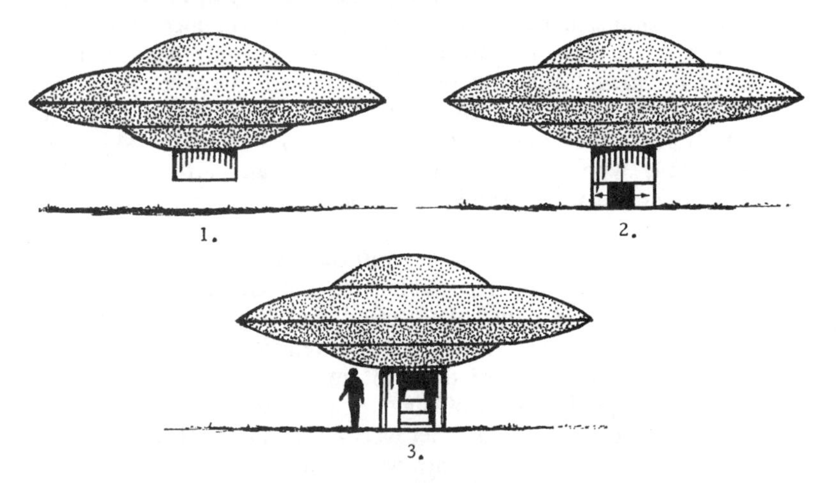

Fig 29. Artist's impression of the landing sequence of the UFO seen by Signor Zuccalà.

From the door thus opened there appeared an empty space lit up by a diffused light of a brilliant white colour. Inside three steps about 40 centimetres high could be seen.

Two beings came out of the opening and he described them as being in height about 1.50 metres (Signor Zuccalà is 1.63 metres high). Their bodies resembled ours in so far as they could be seen, *i.e.* as to exterior form, because as for the rest they were completely covered by an 'armour' of shining metal. Two antennae came out of their helmets. With the help of Signor Zuccalà a drawing of these beings has been reconstructed.

These two little men took hold of him gently under his armpits and took him inside the object. Signor Zuccalà went up three steps and went inside. The interior was empty and shining all over with the same light which he had seen from outside. Signor Zuccalà did not notice any detail in the interior of the object.

Did these witnesses really have these experiences, are they lying or did they have identical hallucinations? Some of the information in this book will help the reader to decide, but as yet there is a lot more ground to cover.

I have tried to show that our design for a space ship has inborn disadvantages, one of which we have to overcome by employing an undercarriage. But even then we have seen how the ground beneath the saucer will be lifted by an excessively strong field on take off, *i.e.* powerful field, greater extension of field. In which case, it follows on occasions our craft of the future would be taking aloft all sorts of materials, viz. soil, grass, stones, potato crops, yes, even some poor unfortunate domestic cat, in fact anything which happened to be beneath it on a rapid take off. Once aloft we would only have to reduce the power for a moment, and our unintentional cargo could be jettisoned and somewhere below, much mystified earth peoples would be scratching their heads about this unaccountable debris from the sky, on which incidentally whole volumes have been written in the past. We shall satisfy ourselves with just a little of it later on.

Other Lift Effects

Remember the basic requirement of the G. field theory allows for something like a highly concentrated and localised gravitational field which can be developed from within the vehicle itself, and this field could be unidirectional to move the craft in all possible planes. We have considered several aspects of flight phases, such as when the ship is in motion and/or hovering, now let us consider some other local conditions which might accompany a hovering saucer, some side effects of the levitating force or vertical thrust vector.

As in the forward thrust condition, every molecule of the craft's structure would experience an upward acceleration simultaneously, and we know that in the hovering condition above the atmosphere of a planet, there would be an upward acceleration of 1g. to balance out the earth's downward 1g. acceleration or 'pull'. But hovering a few hundred feet above ground level in atmosphere would require a small adjustment, *i.e.* the vertical lift component must be slightly less than 1g., otherwise the craft might behave similarly to a conventional lighter-than-air craft or balloon due to aerodynamic effects. As this lift complement might only be in the neighbourhood of several hundred pounds, depending on the volume of the machine, it will help to simplify the issue if from now on we consider the upward vector to equal 1g.

As in forward flight, the supporting field of the hovering machine would extend radially outwards, producing a lift gradient all round the vessel. Therefore, the air molecules immediately in contact with the periphery would also be subject to an upward acceleration of 1g, this becoming less the further out from the craft they were situated.

This is important. For it implies that there would exist round the machine a belt in which a mass would be rendered weightless. If the occupants, for some reason or another, jettisoned, say, a gaseous substance—or a liquid for that matter—then on leaving the craft if would immediately coalesce into a spherical shape, exactly as it would do in fact if it were in outer space, or as in an astronaut's capsule when in orbit.

Moving slowly outwards away from the ship, it would continually be subjected to ever-decreasing regions of the G. field and of course would begin to disintegrate, falling as it did so. Readers may remember it was a glowing ball of light which rendered scoutmaster Desverges unconscious in Florida in 1952. Going to investigate, the scoutmaster saw 'an object large enough for six or eight men to stand in. It was about ten feet high in the centre and about thirty feet in diameter and shaped like a half rugby ball, tapering down to three feet thickness on the sides.' He believed he got near it for about three minutes. It was only ten feet from the ground and it made a hissing sound like a tyre going down. Desverges said that from the object a *ball* shaped flare was shot at him which seemed to float slowly towards his face. 'When I awoke,' he added, 'I had no sense of feeling'. A deputy sheriff reported that the hairs on Desverges' arms had been singed and tiny holes burned in his cap.

While reports of such phenomena as this continue to fit the theory, they will be accepted by the author until, under the scrutiny of newer evidence, they break down.

Accounts of the Sheffield sighting have been published elsewhere, but in view of its importance here, I have repeated it so as not to miss the salient points.

In February 1962, Alex Birch, a fourteen-year-old schoolboy, was taking snapshots with his box camera of his dog, in the back garden of his home at Moor Crescent, Mosborough, near Sheffield, when he spotted the objects shown in Plate 21. Later Alex said: 'I suddenly noticed five objects in the sky—about 500 feet up. They were not moving and they made no sound. The possibility that they might have been flying saucers did not cross my mind at the time'. With him were his friends, David Brownlow aged twelve, also of Moor Crescent, and sixteen-year-old Stuart Dixon of the British Oak Inn, Mosborough. 'I think it was for about four seconds, when suddenly there appeared dazzling balls or blobs of light from the region of the objects, which appeared to move outwards, then began to change shape and fall. Then other blobs appeared and the same thing repeated itself. I thought of my camera. It seemed as if seeing the lights urged me on to try and snap them, which I promptly did and as I did so the lights seemed to dim and die away.

Suddenly the objects seemed to move, as if gathering speed, then shot off at a terrific rate in a north easterly direction over Sheffield. There was no sound at any time'.

As we correlate the above sighting to the G. field theory, I would ask the reader seriously to consider the witness's age, which in this case was a mere fourteen years. On the grounds that Alex Birch and his friends were not suffering hallucinations and had not misinterpreted a natural phenomenon, as was proved by the photographic evidence, are we to consider that these boys have taken part in a childish hoax?

These, of course, are the arguments of the scientific sceptics in that order. Clearly there can be no others, and the first two are invalidated. Therefore I propose to answer the only remaining argument to the point and with the utmost conviction at my command.

Can any serious and responsible person really consider that these boys, having perpetrated such a hoax and produced a fake—which the Air Ministry is satisfied the photograph is not—then go on to dream up effects which exactly fit predictable behaviour in the G. field theory? We can just imagine them inventing 'glowing spheres', but for the youngsters to also invent the break up of such formations, the break up born of those weakening fields, is this not asking a bit too much? Or if they care to listen at all, will the scientific diehards once more prefer to take their stand on a hoax favoured by chance? The same chance in a million, in fact, which appeared to have favoured Stephen Darbishire over a decade ago? I ask the reader to soberly bear these facts in mind as we continue.

10

Accompanying Phenomena

The Private Cloud of a G. Field Craft

IT is now generally known that high flying conventional aircraft passing through upper strata of the atmosphere, frequently leave a 'vapour trail'. What is not generally known is the common cause of the phenomena, which briefly is this. At a certain moisture content in the atmosphere, just before the formation of clouds, droplets of water vapour can be encouraged to form prematurely by pressure disturbances caused by the passage of a fast moving aircraft. The droplets so formed coalesce into larger droplets, or, they form a cloud. Thus the formation will frequently be seen generating behind the wingtips and behind the engines of high flying aircraft where eddy vortices provide more turbulence for the phenomena to develop. Note, the aircraft is continuously leaving this artificial 'cloud' behind.

In developing its lift field, our aircraft of the future will set up a similar disturbance in its immediate vicinity which will cause similar effects in the atmosphere. This will be further augmented by the uprising air currents caused by the field, and as with the aeroplane, so we would expect the generation of a local cloud-like formation in super saturated regions of the atmosphere. But with this difference. Due to the extension of the disturbance radiating around the craft in *all* directions, we might expect the cloud to completely envelop the machine, which if it were hovering, would indeed look no more than that from the distant earth, a cloud! A strong gravitational field supporting a vehicle would also locally condense the atmosphere, therefore as the density increased, we could expect it to drop below the dew point, producing fog.

Should the vehicle be moving, we would expect it to retain its convenient camouflage-like covering whilst trailing behind the more conventional vapour trail with which we are already familiar. So much for theory, what about fact? Well, as any UFO researcher knows, cloudlike saucers are common, the records are loaded with them. Here are a few descriptive quotes: 'A strange circular cloud crossed the skies over the town moving at high speed. It was seen by a number of observers. They described it as having a kind of luminous focus at the centre. At that same moment, in Iguatu, a strange luminous smear was seen in the sky, moving from west to east at very low speed. As it approached the town, it became evident that an object was inside the smear, glowing with a very intense blue light. Dozens of persons witnessed this sighting'. And again, 'The light was orange and was encircled by a kind of halo—a luminous and misty halo which looked like a transparent, luminous cloud, circular in shape and several times the size of the full moon'. Yet again, 'almost disappearing, reduced to a great cloudy, luminous ball'. Compare these descriptions of comparatively recent times with the following extract from the log of Captain F. W. Banner, skipper of the British ship 'Lady of the Lake', while in the North Atlantic.

'22 March, 1870, in Lat. 5° 47′N., Long. 27° 52′W.' (the position would be west of the coast of Rio de Oro, N.W. Africa). 'My crew reported a strange object in the sky . . . I saw it. . . . It was a 'cloud' of circular form, with an included semi-circle divided into four parts, and a central shaft running from the centre of the circle and extending far outward and curving back. The thing was travelling against the wind. It came from the south and settled right in the wind's eye. It was visible for half an hour, much lower than the clouds'.

I can hear many a sceptic murmuring impatiently, 'ball lightning—natural phenomena', but let us remember we are not here evaluating one aspect of the G. field propelled machine individually, rather all the aspects collectively. Sceptical readers will be wiser to reserve judgement until the end of this hearing.

Radar and Optical Effects of the G. Field

In the course of writing and lecturing on this subject, the author has been constantly aware of, and to some extent in sympathy with the many natural reactions and objections to the interplanetary or interstellar flying saucer hypothesis, particularly among scientists, as well as the general public at large. But if I have a sympathy for them, then I have the greatest respect and compassion for those who in one sense, have the misfortune to witness a saucer at close range. Such a witness might describe an incident to his or her friends which sounds ludicrous to their lay minds and perhaps even outrageous to many a scientist. That this should be so is entirely consistent of course, because it is quite obvious even the modern scientist is a kind of specialist in his own right, in other fields, completely unexplored by him, he is also reduced to the ranks of the layman. I say this with all goodwill and feel sure that any intelligent person worth his salt will be the first to admit this fact. The whole point being that the general public look so completely to the world of science and modern technology for guidance over cases like the UFO about which some of these bodies are completely unqualified to venture an opinion.

In a way perhaps it is a good thing that many scientists choose to adopt the ostrich technique over some anomalies in science—perhaps it wouldn't do to have all our mysteries solved too quickly, for it is a well known fact that man creates and changes his environment and in turn is changed by it, so in a way we are largely dependent on change. Change of thought, change of living, ultimately we may depend on change of dimension. So by presenting this case for the UFO, who knows I may succeed in arousing the attention of a few who might otherwise have remained unaware of certain facts. Then perhaps the environment will be changed a little more, *ad infinitum.*

In the course of my investigations, many people have asked me if saucers are ever detected by radar. I have always been painfully aware of the true, though unsatisfactory answer. Yes and no. Sometimes they are detected and sometimes they are not. But even this apparent anomaly will be reconciled before we reach the end of these pages. Here is a case in which radar did not respond, involving 35 passengers and the crew of an airliner. It is quoted here in full.

Thirty five passengers watched in wonder as three illuminated flying saucers escorted their American Airlines four-engine DC-6 airliner for 45 minutes! Captain Peter Killian, a pilot of twenty years' experience with a total of more than four million miles to his credit, was at the controls. The co-pilot and First Officer was John Dee, also an experienced airline pilot.

The airliner had left Newark Airport at 7.10 p.m. on 24 February, 1959, on a non-stop flight to Detroit, and the DC-6 was over Pennsylvania when Captain Killian spotted the UFOs.

'We were flying at about 8,500 feet, between Phillipsburg and Bradford (Pa). It was 8.45 p.m. when I looked off to the south and saw three yellowish lights in a single line formation.

'At first glance I thought they were stars in the Belt of Orion. Then I took a second look and saw both Orion and the objects. Orion was considerably higher; also the objects were brighter and of a different colour'.

Killian pointed out the lights to his co-pilot, John Dee, the other crew members and the passengers. The crew assured him he was not 'seeing things'—that the objects were there right enough. Killian even radioed two other American Airlines' planes flying in the vicinity to make sure 'I wasn't seeing lightning bugs in the cockpit'. Both other captains called Killian back to assure him he wasn't—they saw the things too.

'I have never seen anything like it', Killian said.

Among the passengers was Mr N. D. Puscas, general manufacturing manager of Curtis-Wright Division in Utica. He said the strange objects were in 'precise formation' and seemed to dance in the sky. 'They were roundlike, and every now and then one would glow brighter than the others as if it had moved closer to the plane'.

Both Captain Killian and Mr Puscas agreed that the sky was cloudless above the airliner. 'There was no chance of mistake', said Captain Killian. 'Though there were broken clouds below us, at 5,000 feet, all the sky above that layer was absolutely clear. We had a visibility of about 100 miles.

'At first I estimated that the objects were not over a mile from us. Since I didn't know their size this was just an impression. I believe now that they were not that close. However, at intervals one would move in closer, then fall back into formation'.

To enable the passengers to have a better view of the objects, the stewardesses, Edna Lagate and Beverly Pingree, turned out the cabin lights and everyone watched the fantastic 'out of this world' spectacle for the next forty minutes.

Only one passenger was somewhat scared. 'I told him', said Killian, 'if there was any danger of attack I'm sure they would have done it long ago. *While the objects were in sight I kept watch on the radar screen but saw nothing on it*', *he added.* [During the London Airport sighting of 26 February, 1959, the yellow disc which hovered there for 20 minutes was not tracked on radar either.—author.]

Captain Killian radioed a report of the incident to American Airlines' communications at Detroit Airport before coming in to land. The three objects became lost in a low-altitude haze.

In addition to the crew and 35 passengers of his own plane and the two other American Airlines' planes he had contacted by radio while in flight, the UFOs had also been sighted by the crews of three United Airlines' planes! All the pilots and flight engineers agreed that the lights were on separate vehicles which were in formation.

Further confirmation now came from ground sources. The Akron UFO Research Group reported several local sighting reports between 9.15 and 9.20 p.m. describing three glowing objects.

According to a report published by NICAP, Lt.-Colonel Lee B. James, an Army missile expert associated with Wernher von Braun, in the Army Ordnance Missile Command at Huntsville suggested that the objects came from outer space! Speaking before the Michigan Society of Professional Engineers, Colonel James stated that the objects seen by the various airline crews were quite possibly space ships.

'I know they are not from here', said the missile expert, 'and they are not coming from Russia. We in this civilisation are not that advanced yet'.

If the crews and passengers really saw what was reported, Colonel James said, the objects 'would have to come from outer space—a civilisation decades ahead of ours'.

One engineer asked him about Captain Killian's report that his radar screen had not shown the UFOs. 'That civilisation quite possibly has licked that problem', Colonel James replied. 'It might use a special coating or a composite of certain materials which might prevent such a reflection'. Or force field? (author.)

In an interview with Lex Mebane of Civilian Saucer Intelligence of New York, Captain Killian spoke about the changing glow of the objects. Not only had their colour altered at times, from yellow to bluish white, but their intensity had varied from extreme brilliance to temporary fade-outs. He had wondered if the UFOs had been trying to signal, but he did not see any pattern or regularity.

Captain Killian also said that the UFOs' speed varied. At times they would pull ahead quickly, then apparently lag as if to let him catch up. These movements were easily observed, since the airliner was flying a constant 300-degree course. The captain said that some passengers asked him to fly closer to the objects, but he had to consider their safety, even if regulations had permitted this. Also, he added, he obviously did not have enough speed to catch up with the UFOs.

For three whole days the U.S. Air Force was silent about the six airliners' sightings of the three objects. Then on 28 February, the Air Technical Intelligence Center released its official comment. This was to the effect that all the various airline crews had been misled by the Belt of Orion! Glimpsed through broken clouds, Orion's stars had given an illusion of fast-moving objects, deceiving the airline pilots!

This 'explanation', of course, does not hold water. Captain Killian had seen both Orion and the UFOs simultaneously and in a clear sky. So had members of the other airline crews. However, the Air Force were able to produce an Air Force Transport crew that had been flying from Washington to Daytona that same night. They affirmed that they had seen Orion through broken clouds at 8,500 feet. In any case the Air Force statement directly contradicting the reports of the six experienced airline captains was a direct reflection on them and they did not like it.

The author personally knows of at least one well authenticated case where a UFO *was* detected by radar. The report may be classifiable and the facts can only be generally quoted. In 1964, in a remote part of England, a UFO was detected by a radar unit, travelling at nearly three thousand miles per hour. It was seen to stop instantly, where it hovered for three minutes, then rapidly shot off in the direction from whence it came. Its acceleration was said to have been prodigious, 'going up through 13,000 m.p.h., then completely off the clock'. I find it very difficult to accept this as a natural phenomenon, or do some kind of meteorites hover?

If radar does strange things in the presence of a strong G. field, then we might conclude that with our imaginary space ship would be associated other strange side effects. For instance we may find light also behaves erratically.

Dr Albert Einstein showed the world that a ray of light is bent in the presence of a gravitational field. We have seen that the G. field of our craft would have a source of focus as do all fields, we shall hear more about this as our story unfolds, and what is more, if such a field were highly concentrated, then certainly we might expect optical effects. For instance, we would not be wholly surprised to learn that some observers would see our ship, while others would not. Now before our sceptical reader raises the objection that we are again making the glove fit, let me hasten to assure him, that this is not so.

UFO files are brimming over with this singular curiosity among sightings, in fact so much so, that I repeat some of them here for his benefit at the risk of offending those who are already well acquainted with this vanishing phenomenon.

Here is a somewhat amusing version:

On one occasion when a queer formation of 'foo fighters' got on the tail of a U.S. night fighter of the 415th Squadron, the pilot had swung round his plane and headed for them at top speed! As he approached, the lights vanished into nothingness, as a squadron of 'Etherian' flying machines!

The pilot had reported: 'As I passed where they had been, I'll swear I felt the propeller backwash of invisible planes!'

Came the reply from derisive ground radar station: 'Are you fellows all plumb loco? Sure, you must be crazy! You're up there all alone!' [Note there was no radar reflection also.—author.]

The puzzled pilot flew on, and, glancing back, was now startled to see that the balls had reappeared about half a mile astern of his plane. Said he to himself: 'I'll show these spook planes a trick!'

The night was starry, but near the zenith was a bank of thick cumulus cloud. He headed his plane at top speed right into the dense mass. Then he throttled back, and glided down for about 1,800 feet. He turned the machine round and headed back for the cloud the way he had entered it, but on a much lower plane. Surely enough, the balls had been surprised; they emerged from the cloud ahead, but now on a course opposite to his own!

1 January, 1954, at 2.30 p.m., observers at Box Hill, Surrey and Melbourne, Victoria, Australia, saw something high in the sky—and not a balloon—shaped like a queer box which was turning over and over slowly. Suddenly something like a ball of vapour dived on the 'box' at terrific speed at an angle of 45 degrees and both objects vanished.

As yet the following case is not on record. Due to the fact that the author was able to investigate it personally and therefore vouch for its authenticity, it is quoted here as a comparison.

It was 3 a.m. one cold December morning in 1964, when Mrs Joan Pyner (a relative of the author's who does *not* believe in flying saucers) was awakened by 'a noise like a helicopter coming over the house'. She said the noise, which was pulsating, got slower and lower in pitch, as the thing came closer towards the meadow which adjoins the garden of the house, then finally stopped altogether. Mrs Pyner likened it to 'the rotating blades of a helicopter coming down, though I was puzzled by it not being as loud as I thought a helicopter would have been, that close, and it was close, I knew that'. Thinking that whatever it was might be in trouble, she got out of bed and went over to the window to investigate.

The grass in the meadow was quite visible, and there, only some 25 yards from the house, sat the 'thing'. It was black, 'I couldn't see the exact shape for certain, but it would be the size of an armchair', she said.

When I asked if in fact it might have been a cow, Mrs Pyner laughed outright, 'don't you think I can tell a cow even in that light?' she asked.

In order to obtain a better look, she had thrown up the sash window which made a loud squeaking noise. Whereupon the dark mass simply vanished. The witness had no idea, if in fact it moved at all. There was no noise, nothing, it simply ceased to be there.

Perhaps I should add that Mrs Pyner told me she had been annoyed when local people quoted her as having seen a flying saucer. 'I said nothing of the kind. I don't know what it was I saw, only that I can't find an explanation for it', she said.

A rocket engineer friend of the author's once pointed out that if we could build a gravitational space craft which radiated electro-magnetic waves including visible light, we might expect something like a red shift to occur as a result of the intense force field employed. In other words, the light emitting from such a craft would be continuously fighting against the intense local gravitational field. There is such an effect well known in astronomy as spectra displacement, which we shall examine later on. As a result, light would be slowed up or more correctly, we might expect an observer beneath the vehicle to receive frequencies at the red end of the visible spectrum. Briefly, saucers may appear orange or red at fairly close range. I do not have to say that this is frequently the case, for even those unacquainted with the subject, must have read of orange coloured UFOs seen in the sky at night, even if they accept them for being interplanetary or otherwise. Here is a typical example, in which the UFO appeared large enough for the witnesses to detect spin.

Mr and Mrs Stanley Cadwallender, Hoole Street, Walkley, Nr. Sheffield, were driving home late on Saturday night, 2 August, 1958. Suddenly they saw an object which 'spun like a burning top' across the sky towards Sheffield.

'It was about 10.20 p.m.', said Mr Cadwallender, 'when my wife saw the object. We were halfway between Ladybower and Riverlin Dams, driving at about 25 to 30 m.p.h.

'I stopped the car and we both got out to watch it spinning through the sky'.

He added: 'We followed it in its course for a full minute. It scared us stiff. The top part of the object glowed like a neon light, while the bottom of it looked like a ball of *orange* fire. It seemed to be about half the size of the sun'.

Radiation Effects

We have now reached the stage where we have considered most of the more obviously predictable effects which we might expect to be associated with G. field propulsion. Now for a few possible effects which we can only speculate on with any degree of certainty. Although from now on, the conclusions are somewhat more conjectural, research has indicated that if we could generate such a powerful gravitational field then it is believed, there would be released other radiations of various wavelengths accompanying it. This would probably include frequencies right through the spectrum, thereby giving both light and radio effects. If this is true, then we are in a position to predict some results of such radiation. For example we could expect similar results to those obtained by Tesla, when he oscillated electrical fields to an extremely high order. Radios would be interfered with if too close to the originating source. Again this is an almost common occurrence. Here is an interesting case.

At 7.17 a.m. on 31 May, 1957, a British airliner was flying over Kent, just south of Rochester, on its way to Holland, when it sighted a UFO. The object was seen by both the Captain of the aircraft and by his First Officer through different windscreens. Here is the personal account of this amazing sighting given to *Flying Saucer Review* by the Captain himself. (The names of both the Captain and the First Officer are being withheld upon request.)

'I was in command of a scheduled airline service from Croydon Airport to Holland. As we got to a position two nautical miles south of Rochester, my First Officer and myself became aware of a brilliant object bearing 110 degrees (T) from north and elevated about 10 degrees above the haze level. We were flying at 5,000ft above sea level, heading 082 degrees magnetic 074 degrees (T). The UFO was about two-thirds the size of a sixpence in the windscreen at first. It then appeared to come towards us. When it was about the size of a sixpence, the object became oval in shape and turned away. Then it became as before and reduced in size to about half the size of a sixpence.

'Then to our astonishment the UFO *disappeared* completely as we watched it. We did not see the UFO go, but became aware that we were looking at an empty sky.

'We were unable to contact 'London Radar' due to a complete radio failure in the aircraft, nor were we able to report to 'London Airways', nor to 'London Flight Information'.

'Radio failure, especially complete radio failure, is rare these days, and in our case was due to our circuit breakers not keeping 'in'. A radio circuit breaker 'breaks circuit' when the system is overloaded by an extra source of electrical or thermal energy. On this occasion we were not using all our equipment, so there was no cause for overloading. However, our radio equipment became fully serviceable after the UFO had gone, and all circuit breakers stayed 'in'.

'Is it too much to ask if the UFO was able, through overloading our electrical system, to prevent our reporting it or asking for radar confirmation?

'When we returned to the U.K. a similar report to the account I have given you was made to both the Ministry of Transport and Civil Aviation and to the Air Ministry'.

Close range saucers have also interfered with cars, domestic radios and T.V. sets, but perhaps one of the most amazing clue packed incidents is the following. Because the present author shares much sympathy and respect for Jules Lemaitre's introductory comments, I feel it is fitting to include the report in full.

'A Strange Story from Brazil', by Jules Lemaitre.

In the November-December issue of *Flying Saucer Review* I commented upon certain conclusions to be drawn from that remarkable book *Flying Saucers and the Straight Line Mystery.* Some readers have misunderstood my article, for they seem to have gained the impression that I had said that the saucers were hostile. On the whole I do not think so, but I do think that many of the incidents reported (and authenticated) tend to indicate that they are not necessarily friendly—which is not the same thing as being hostile. A gardener who steps back from weeding and crushes the life out of a beetle he has failed to observe, cannot be dubbed hostile to the beetle. At the moment of impact he would have been unaware of its existence. When he noticed what he had done he would probably be indifferent and would return to his work without another thought. He might even never notice the remains of the insect. Also, it is not true to say, as some correspondents seem to think, that I reject Adamski and others out of hand. Here the readers seem to have confused my views with those of Aimé Michele, as expressed in his epoch-making book.

In the September issue of *The A.P.R.O. Bulletin* published in New Mexico, there appears a sensational account by Dr Olavo T. Fontes of a terrifying incident in Brazil. Dr Fontes would seem to think, with many of the readers of my former article, that there is no midway between hostility and amity, for he heads his article 'Friends or Foes?' I would suggest that an indifference similar to that of the gardener in my analogy might better explain what happened in the story that follows:

'On 4 November, 1957, at 2 a.m., something sinister took place in the Brazilian Fortress Itaipu. This fortress belongs to the Brazilian Army and was built along the coast of Sao Paulo State, at Sao Vicente, near Santos.

'It was a moonless tropical night. Everything was quiet. The whole garrison was sleeping in peace. Two sentinels were on duty on top of the military fortifications. They were common soldiers—they did not know that saucers existed. They were performing a routine task, relaxed because there was no enemy to be feared. Then a new star suddenly burst into searing life among the others in the cloudless sky, over the Atlantic Ocean, near the horizon. The sentries watched the phenomenon. Their interest increased when they realised it was not a star, but a luminous flying object. It was coming towards the fortress. They thought at first that it was an aeroplane, but the speed was strange—too high. . . . There was no need to alert the garrison however. In fact, so tremendous was the object's speed that the two soldiers forgot their patrol just to observe it. It was approaching rapidly.

'In just a few seconds the UFO was flying over the fortress. Then it stopped abruptly in mid-air and drifted slowly down, its strong orange glow etching each man's shadow against the illuminated ground between the heavy cannon turrets. It hovered about 120 feet to 180 feet above the highest cannon turret and then it became motionless. The sentries were frozen to the ground, their eyes wide with surprise; the tommy guns hung limply from their hands like dead things. The unknown object was a large craft about the size of a big Douglas aeroplane, but round and shaped like a disc of some sort. It was encircled by an eerie orange glow. It had been silent when approaching, but now, at close range, the two sentries heard a distinct humming sound coming from it. Such a strange object hovered overhead and nothing happened for about one minute. Then came the nightmare. . . .

'The sentinels were startled, unable to think what to do about the UFO. But they felt no terror, no premonition, no hint of danger. Then something hot touched their faces (one of them thinks he heard a faint whining sound he could not identify at that same moment). In the darkness this would have been horrifying. But the UFO was bright and they could see that nothing had changed. Then came the heat. Suddenly an intolerable wave of heat struck the two soldiers.

'One of the sentries said later that, when the heat wave engulfed him, it was like a fire burning all over his clothes. The air seemed to be filled with the UFO's humming sound. Blind panic yammered at him. He staggered, dazed, heat waves filling the air around him. It was too hot. . . . He went stumbling and lurching, his whole conscious purpose that of escaping from that invisible fire burning him alive. He fought, and gasped and beat the air before him. He was suffocating. Then he blacked out and collapsed to the ground—unconscious.

'The other sentry got the horrible feeling that his clothes were on fire. A wave of heat suddenly enveloped him. Horror filled him and he lost his mind. He began to scream desperately, running and stumbling and crying from one side to another, like a trapped animal. He did not know what he was doing, but somehow he skidded into shelter, beneath the heavy cannons of the fortress. His cries were so loud that he awoke the whole garrison, starting an alarm all over the place.

'Inside the soldiers' living quarters everything was confusion. There was the sound of running footsteps everywhere, soldiers and officers trying to reach their battle stations, their eyes wide with shock. No one knew what could explain those horrible screams outside. Then just a few seconds later, the lights all over the fortress collapsed suddenly as well as the whole electric system that moved the turrets, heavy cannons and elevators. Even the ones supplied by the fortress's own generators. The intercommunications system was dead too. The strangest thing, however, was the behaviour of the alarms in the electric clocks, which had been set to ring at 5.00 a.m.—they all started to ring everywhere at 2.03 a.m.

'The fortress was dead, helpless. Inside it, confusion had changed to widespread panic, soldiers and officers running blindly from one corner to another along the dark corridors. There was fear on every face—fear of the unknown—hands nervously grasping the useless weapons. Then the lights came on again and every man ran outside to fight the unexpected enemy who surely was attacking the fortress. Some officers and soldiers came in time to see an orange light climbing up vertically and then moving away through the sky at high speed. One of the sentinels was on the ground, still unconscious. The other was hiding in a dark corner, mumbling and crying, entirely out of his mind. One of the officers who came first was a military doctor and, after a brief examination, he saw that both sentries were badly burned and ordered the men to take them to the infirmary immediately. They were put under medical care at once. It became clear that one of them was a severe case of heat syncope; he was still unconscious and showing evident signs of peripheral vascular failure. Besides this, both soldiers presented first and deep second-degree burns of more than 10% of body surface—mostly on areas that had been protected by clothes. The one that could talk was in deep nervous shock and many hours passed before he was able to tell the story.

'The nightmare had lasted three minutes . . .

'Next day the commander of the fortress, an army colonel, issued orders forbidding the whole garrison to tell anything about the incident to anyone—not even to their relatives. Intelligence officers came and took charge, working frantically to question and silence everyone with information pertaining to the matter. Soldiers and officers were instructed not to discuss the case. The fortress was placed in a state of martial law and a top-secret report was sent to the Q.G., at Rio or Sao Paulo. Days later, American officers from the U.S. Army Military Division arrived at the fortress together with officers from the Brazilian Air Force, to question the sentries and other witnesses involved. Afterwards a special plane was chartered to bring the two burned sentinels to Rio. It was an Air Force military aircraft. At Rio, they were put in the Army's Central Hospital (HCE), completely isolated from the world behind a tight security curtain. Two months later they were still there. I don't know where they are now.

'Three weeks after the incident, I was contacted by an officer from the Brazilian Army, a friend who knew about my interest in UFO research. He was at the fortress of Itaipu the night of the incident. He was one of those who questioned the two sentries. He told me the whole story as it was described above. His name was suppressed from this report in order to protect him. The reasons are obvious; he told something he should not tell. As a matter of fact, this officer has asked me to forget his name and he wasn't laughing. He was too frightened.

'I was aware, however, that the information was not enough despite the fact that it had come directly from one of the witnesses. The case was too important. On the other hand, to get more information through the security ring built by Army Intelligence would be an almost hopeless task. The only way was to attempt to break the secrecy around the two soldiers under treatment in the Army's Central Hospital. As a physician, I might perhaps contact some doctors from the hospital and even examine the two patients if possible. However, all my attempts failed. The only thing I was able to determine was the fact that the two soldiers from the fortress of Itaipu were really there for treatment for bad burns. Only that.

'The case remained in my files until two months ago, when the final proof that it was real was finally obtained. Three other officers from the Brazilian Army who had been at the fortress on the night of the UFO were fortunately localised and contacted. They told the same story. They confirmed the report transcribed above in every detail.'

There are many more cases to substantiate the authenticity of this affair, we now turn to France in the year 1954 for further confirmation.

On the evening of 20 October, M Jean Schoubrenner of Sarrebourg, was driving on the road between Schirmectond St-Quirin-en-Moselle, about half a mile from the village of Turquenstein, when he suddenly noticed a luminous body on the highway some distance ahead of him. Instinctively he slowed down, and when he was about *twenty yards* from it, he suddenly felt as if he had been paralysed, his hands froze to the steering wheel. At the same instant the engine stopped, but the momentum carried the car onwards. As it did so a sensation of increasing heat spread through M Schoubrenner's body. Then the object sped away and the symptoms left him.

That same afternoon 125 miles west of Turquenstein, a lumber dealer M Roger Réveillé, was walking near the Lusigny Forest, not far from a place called Troyes. It was raining hard as he made his way along a woodside road, when suddenly he heard a *loud rustling* sound as would be made by a flight of pigeons.

Looking up, he saw at little more than tree top height, an oval shaped body of about 20 feet across. At the same moment he felt a wave of intense heat. In a few seconds the thing disappeared upwards. In the woods the heat was producing a thick local fog and it was almost a quarter of an hour before the witness could approach the site. When he did so, he found, in spite of the heavy rain, the trees and ground were quite dry as if exposed to strong sunlight.

The very next night, on 21 October, a motorist and his three-year-old son were driving in the vicinity of La Rochelle, near Pouzou, when the child began to cry, at the same instant the driver felt an electric shock and increasing heat pass through his body. The engine stopped and the lights went out, revealing a glowing object on the road directly ahead, which until then had been invisible. Instantly the thing became brilliantly luminous, first bright red, then orange as it took to the air. Everything then returned to normal.

5 May, 1958, San Carlos, Uruguay. About 3.40 p.m., Carlos A. Rodriguez, an experienced and reputable pilot, was flying a Piper aircraft in the vicinity of Capitan Curbelo Naval Air Base, when he spotted a glowing object approaching his plane. The UFO stopped at an estimated 2,000 metres away and, according to the report, 'It rocked twice in a balancing motion'. Rodriguez said the object was shaped like a child's top, symmetrical above and below. As he closed to about 700 metres, he felt intense heat in the cockpit and was forced to open the windows and door of the plane. The UFO then accelerated rapidly eastward towards the sea, leaving a thin vapour trail.

In some of the cases quoted, some readers will no doubt have seen a vital clue, one of the more indicative signposts we have yet come across and for that reason it is perhaps fitting to close this chapter here while we take a rest to prepare ourselves for what is to follow.

We sit gazing back over the track for a while, then forward to where we must go. From our lofty vantage point the trail winds away into the distance, but from here we observe that the pathway divides into two for several miles, before it bends back and reunites as one. We cannot afford to miss any signposts, so tomorrow we must split up, one group will take the left fork and the other the right, we should have some interesting yarns to exchange when we meet at the distant junction.

11

Analysis of Technically Corroborative Evidence One

I WOULD remind the reader that in presenting these findings I have two main intentions in mind. That is to bring to the attention of the general public and sceptics of flying saucers, the fact that far from being a lot of nonsense, detailed descriptions of sightings are beginning to make good engineering sense. In other words, information, which in the majority of cases is reported by discerning folk from all walks of life, is building up to a pattern which is technically corroborative. So much so, that the sceptic is once again faced with the familiar alternatives. A fantastic chance favoured coincidence. A world wide conspiracy. Misinterpreted natural phenomena. Or he must admit to the fact that a good many UFO reports are genuine descriptions of some kind of extraterrestrial craft.

My second intention is of course to help throw a little light on the possible techniques employed by those who build the fantastic machines men call flying saucers.

The following chapters in particular are in the form of a rational appeal to technically qualified doubters, to sincerely judge the corroborative value of many sightings, albeit purely from a mechanical or engineering point of view. I would say to you, if some of the observations fit, and only just fit, that might be of little more than passing interest. But if all the observations made by ordinary folk fit completely the G. field drive idea, then either the case for it and the UFOs is established and we have a fantastic truth to discover, or chance has so favoured a coincidental arrangement of facts so utterly neatly, as to render this latter alternative untenable.

Before we proceed further however, the following summary of some of the more well known phenomena associated with UFOs may help to prepare the less informed reader for the evidence and conclusions arrived at throughout the remainder of this work. Far more profuse details of this information will of course be found and recognised in the files of most keen researchers in the subject and therefore only notes of a technical nature are listed. No doubt an even more informative list would be found at the Air Ministry.

Colour

1. UFOs usually exhibit variation of colour, from deep red through orange, on to brilliant white.
2. Colour changes usually with acceleration or deceleration, but not always.
3. Red or orange in majority of cases as seen from underside of vehicle.
4. On occasions, red and bluish flame round perimeter of circular UFOs is seen.
5. In daylight and low down, discs usually appear dull grey.
6. A ring of glowing vapour has been seen during take off.

Noise

1. In the majority of cases saucers are completely silent.
2. When comparatively close, noise has been heard which has varied from humming like an electric motor, humming like a swarm of bees, cracking like splintering wood or snapping in of circuit breakers etc.
3. When saucers do behave noisily, then they do so with a vengeance, the noise which has been described as a 'terrible roar', a 'thunderous deafening noise, louder than a low flying jet', could be attributed to the sound which is sometimes known as 'white noise'.
4. There is reason to believe that when several otherwise silent craft are in formation, there can be a heterodyne effect producing an audible sound.

Motion

1. When descending, saucers have been frequently observed to appear to fall in the manner of a falling leaf.
2. On other occasions they have been seen to behave similarly when ascending.
3. Their velocity varies considerably from 'apparent speed of a shooting star', 'faster than a jet', to a mere crawl through the sky, direction of the wind seems to be of no consequence.
4. Saucers are often observed to hover momentarily before moving off vertiginously or even apparently vanishing.
5. Rates of acceleration and deceleration can be enormous, the craft have been seen to execute right angled turns and stop in mid-air instantly.
6. Another puzzling but common behaviour among the lists of sightings is the saucer which gyrates or dances all over the sky, often in the manner of a yo-yo.
7. The strange craft have been seen to apparently roll along on edge 'like a huge wheel in the sky'.

8. In a high proportion of sightings parts of the craft seem to rotate or to appear to rotate at varying speeds, more often than not at a very high speed.
9. The cigar type UFO often behaves very strangely, from standing on end, which is more common, changing shape, elongating, splitting into several parts etc., then the customary vanishing trick, only to re-appear once again etc.

Associated Phenomena

1. The electrics of cars have been affected by UFOs. Head lamps have been frequently dimmed or extinguished completely by the near approach of a flying disc. Then resume normal functioning as soon as the UFO left the vicinity. Car batteries have been found to boil, ignition coils heat up etc.
2. Car engines have been stalled by the near approach of a saucer.
3. Craters have been left in the ground after take off, 'as if the ground had been sucked upwards'.
4. On the other hand, impressions have been left in the ground after take off, 'as if an enormous weight had stood there'.
5. Strange and unpleasant smells are sometimes associated with UFOs like ozone after a nearby lightning flash.
6. Uncanny invisible fields which 'seemed to press me back' from a grounded saucer.
7. Animals are frequently terrified and have departed most hurriedly from the scene of a grounded saucer, even after some time has elapsed since the event took place.
8. Evidence of intense heat, fusing, and scorching often mark the departure of a landed saucer.
9. Vegetation is frequently laid down and fanned outwards, 'like the petals of a flower', while on other occasions there has been evidence of a swirling or rotary motion, 'like the effect of a rotary scythe' or 'a huge catherine wheel'.
10. Large chunks of soil, on at least one occasion, up to a ton or so in weight have been hurled aside, 'as if by some huge mad thing'.
11. Debris from the sky has often accompanied a UFO in flight. This has varied from falling ice to large chunks of stone, bits of alloy, iron etc., but most frequent of all is the strange gelatinous stuff commonly called 'angel hair'.
12. Ferro-magnetic materials have been found to be magnetised by the near approach of a flying disc.
13. Radio and television sets have gone haywire when saucers have flown overhead. The electrics of at least one army garrison mysteriously shut down.

14. Paintwork on vehicles has been mysteriously changed to another colour or even returned to the original primer undercoat. Equally mysteriously the colour has returned after a short lapse of time.
15. People have sustained burns to the face and body when near to, or in some rare cases, touching a grounded UFO. Usually burns have been similar in nature to those caused by radiation.
16. Sometimes saucers leave a trail in the sky as do conventional aircraft, but not always. On occasions a 'short exhaust flame' is often seen trailing behind UFOs.
17. Strange craft have been observed entering or leaving water. And on these occasions it was said, 'the water boiled and frothed up' and again 'the water seemed to be pressed back away from the craft.'
18. Sometimes saucers appear on radar screens, sometimes they do not, even when there has been evidence of their presence.
19. Although saucers have been photographed when observed in the sky, on occasions a UFO has appeared on developed film even though the photographer has seen nothing.

These then are some of the more outstanding observations on UFO phenomena, there are many more, but the above may help to establish the general pattern.

In the previous chapter some of the more mechanical evidence for the existence of the flying saucer was presented as phenomena we would expect to accompany the operation of a hypothetical space ship of our own design. We saw how at low power we could expect air to be displaced by a moving craft, while at high power there would be an aerodynamic shielding or cushioning effect. We saw how on take off, craters might be formed and other ground or aerial effects experienced. Now in the following chapters I propose to analyse some of these and other predictions a little more technically, let us see if the pattern continues.

The Inverse Square Law

Before we proceed however I would ask of the more technically informed reader, a little patience while we discuss very briefly the fundamental working of the inverse square law for the benefit of those unacquainted with it. To those among my readers, I would only ask that you first get a grasp of this simple physical principle in order for us to continue this journey together. This much is needed to enable you to follow and appreciate the remaining startling facts in the remainder of this book. I would earnestly ask you not to miss the implications in it!

Without much doubt, one of the simplest ways to imagine the inverse square effect is the optical application, in which a lamp throws an image on to a screen as in Fig 30(a).

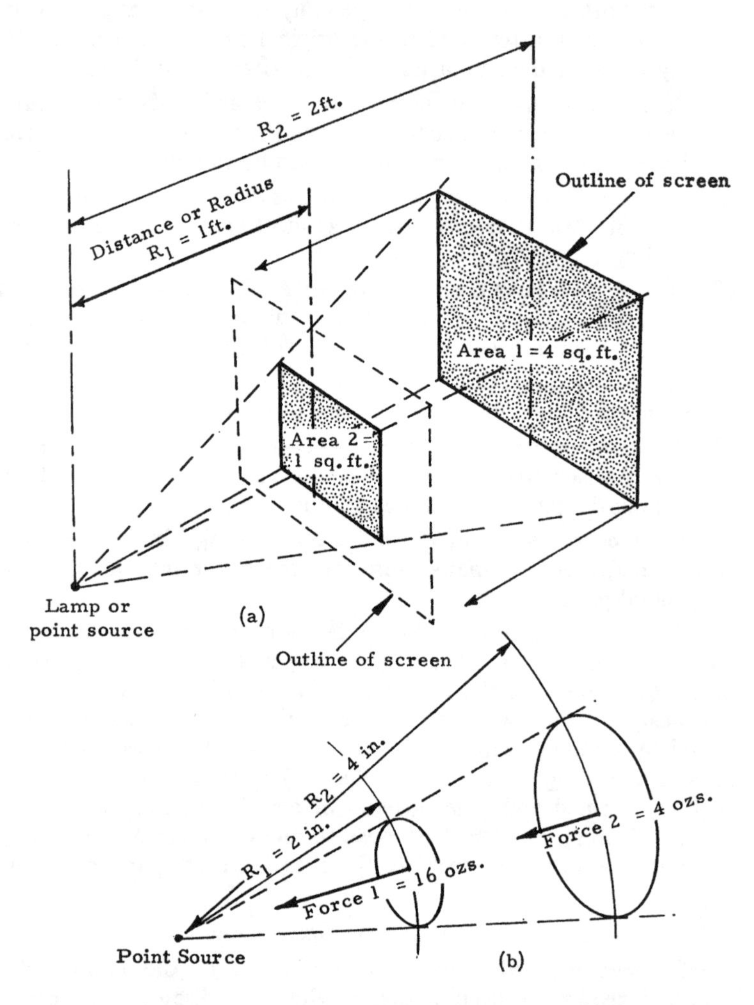

Fig 30. The inverse square law explained in terms of light.

First we consider the larger screen designated Area 1, the dotted lines portraying the rays of light from the lamp, which in this instance is placed say 2 feet from Area 1.

Now should we exactly halve this distance by moving the screen nearer to the light source, then obviously the rays of light intercept the screen at different positions producing the Area 2, which due to the inverse effect happens to be exactly one quarter of Area 1, *i.e.* 1sq.ft.

This can be quite easily found by the formula $A_2 = \dfrac{R_1^2\ A_1}{R_2^2}$ viz.

$A_1 = 4$ square feet.
$R_1 = 1$ foot.
$R_2 = 2$ feet.

$$\text{Hence } A_2 = \frac{1 \times 1 \times 4}{2 \times 2} = 1\text{sq.ft.}$$

Or if we know Area 2 and require to find what Area 1 would be if we doubled the distance, we get

$$A_1 = \frac{R_2^2\ A_2}{R_1^2} = \frac{2 \times 2 \times 1}{1 \times 1} = 4\text{sq.ft.}$$

As an example we have chosen the units of Area and distance. But the same rule holds if we use distance and *force*.

For example in Fig 30(b) we again have a radiating point source, but this time it is radiating not light but a *field* of force, it could be a magnetic field, an electro-static field, or a gravitational field, given the force at one distance, by applying the same formula, we can find the force at any other required distance.

Suppose in this example we have an electro-static field radiating from a source to which a body is attracted by a force of 16 ounces, when it is positioned only two inches from the source. What will be the force on the body if we doubled the distance to four inches?

Well by simply altering our units from Area to Force we get

$$F_2 = \frac{R_1^2\ F_1}{R_2^2} = \frac{2 \times 2 \times 16}{4 \times 4} = 4 \text{ ounces.}$$

Therefore the rule of the inverse square law: double the distance and we get one quarter of the force, or half the distance and the force is quadrupled.

Here we have halved the distance for convenience, the same will hold true for any other values we care to employ. Let us start by applying the rule generally to the conditions likely to be met in our space ship. Later we shall be able to calculate some of these conditions more fully.

Crew Protection in a G. Field Ship

To begin with we shall consider a G. field ship in the hovering condition Fig 31. In this we note that the focal point or generated point source is formed some distance (R) above the vehicle and we also assume the vehicle to experience a 1g acceleration towards that point to counteract the earth's field. Also we assume the field to follow the inverse square law, *i.e.* the strength of the field would vary proportionally

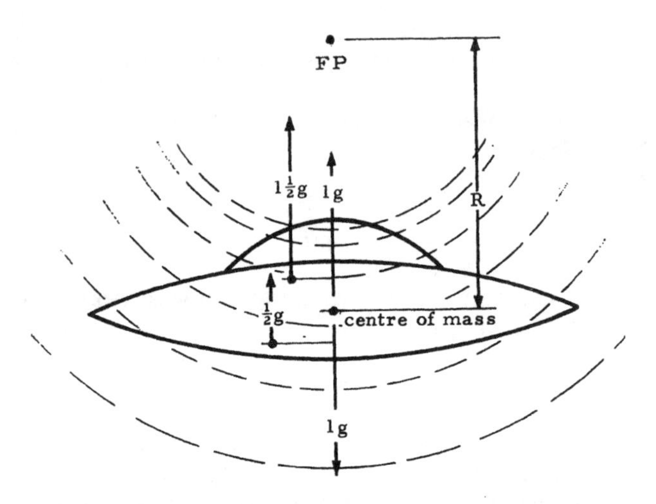

Fig 31. *G. differential set up in craft due to the inverse square law.*

to the square of the distance from the source, as we have seen from the formula $F_2 = \frac{R_1^2 F_1}{R_2^2}$

We shall assume also that the contra-gravitic acceleration of 1g is acting on the centre of mass of the vehicle. It follows that due to the inverse square law effect, different parts of the ship will experience more, or less than 1g, the magnitude of this differential being dependent on the distance of the point source from the vehicle. Throughout the rest of this work, the descriptive term 'point source' will be adopted, but it must be clearly understood here that theoretically a point source, as such, cannot exist. In actuality, the point source we are considering, may extend for some distance. Even so, if the focus of this were to be closely situated to the vehicle, the g. differential would be greater, so that two bodies free to move within the ship, one situated at the top and the other at the bottom, would be subject to different g effects. Remember the G. field does not affect the terrestrial gravitational field. The effect being similar to two bar magnets exerting a force on a piece of iron placed in the gap between like poles. Neither one pole nor the other destroys the opposing magnetic flux one bit, this is simply a tug of war using a field instead of a rope. Similarly, the body situated at the bottom of the vehicle is still being 'pulled' by the earth's field, but the opposing G. field might be exerting perhaps $\frac{1}{2}$g vertically at that point, therefore effectively the body now 'weighs' only half its original weight. On the other hand, above the

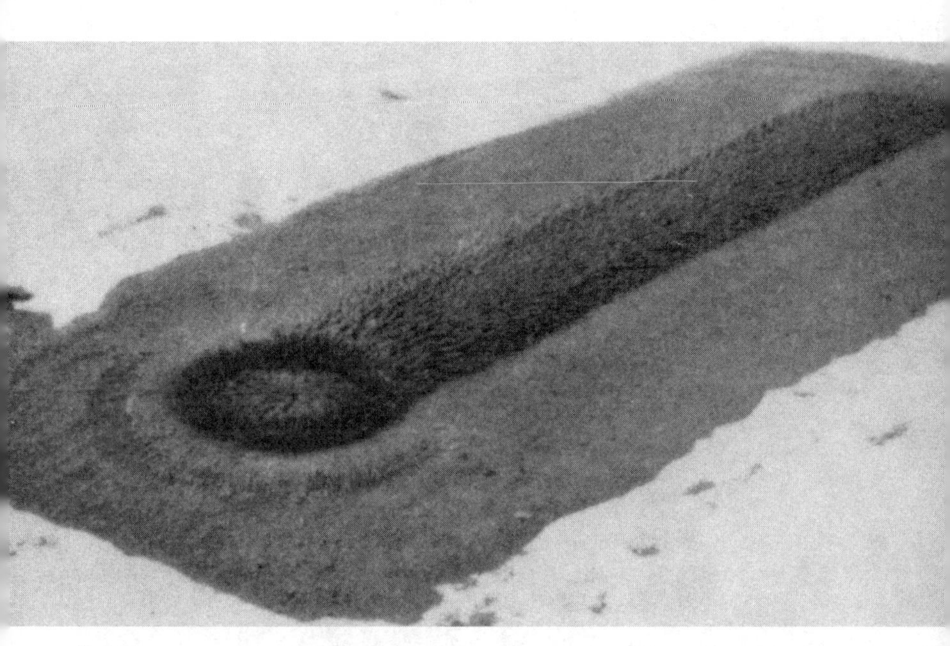

G field aerodynamic behaviour simulated by a powerful moving magnetic field in iron filings. In this case the point source would be situated a little in front and below the disc. Note the bow wave effect and the 'air-particles' following the point source.

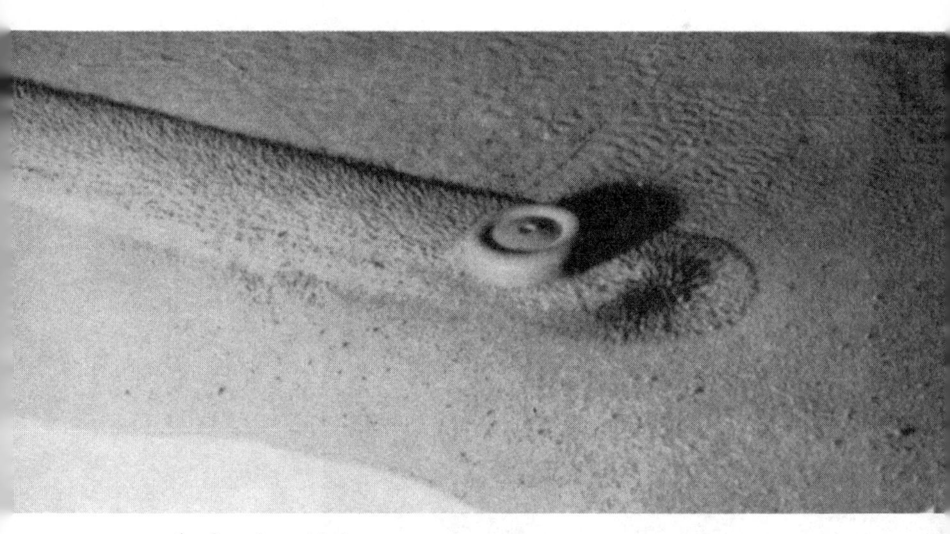

As above but with the point source situated more ahead of the disc.
Plate 16.

General view of the author's centrifuge showing test cell and camera. (*a*)

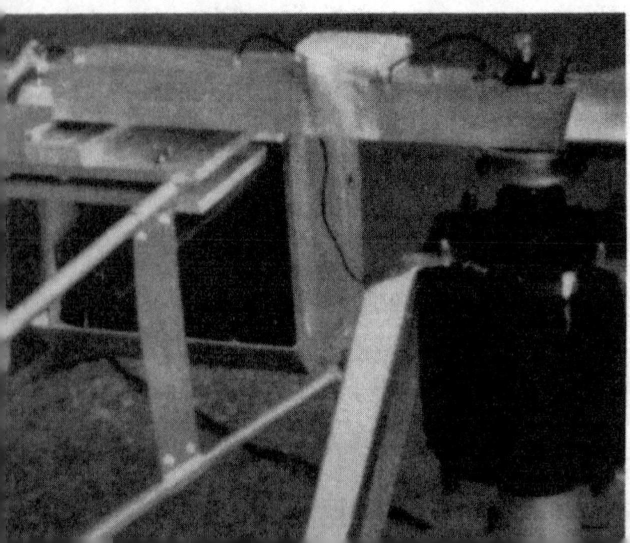

Detail of test cell and pivot mounting. (*b*)

Plate 17.

centre of mass of the vehicle, the second body may be subjected to say $1\frac{1}{2}$g's vertically and now 'weighs' nothing, but rather has a 'negative' or contra weight of $\frac{1}{2}$g—upward.

Now if we imagine the pilot to represent one of the two bodies, it becomes apparent that even in the hovering condition being considered, he is going to be subjected to some inconvenience by a loss of weight. To a lesser degree if we put him in the bottom of the ship, or more if he is placed above the centre of the ship's mass. In fact he might lose so much weight that he floats to the top of the canopy. A little thought will show that these conditions become more marked at high vehicular thrusts. From this it would seem obvious to place the crew below the ship's centre of mass, so that when the machine was subjected to a 2g field for instance, then due to the inverse square law effect, the crew might be subjected to only 1g giving them earth environment conditions. But this would be true only if the focal length of the field could be kept constant throughout the range of field intensities, 10g's on the ship, 9g's on the crew and so on. Due to the inverse square law and several other reasons however, a fixed focal length is not possible, therefore we have to consider an alternative solution. This was suggested in *Space, Gravity and the Flying Saucer*, but which we can now go into in greater detail. As we do so in the purely engineering sense, we are once again reminded of a blind genius in Germany. Therefore as a somewhat inadequate acknowledgment, from now on throughout the rest of this work I shall refer to my G. field generators as field inducers after Burkhard Heim.

Now let us continue by considering the sketches in Fig 32. In this we assume a spherical shape for the vehicle in order to clarify the issue somewhat.

From what has been said so far, it follows that the point source can be formed anywhere in space, whether this be outside the ship or even inside it, indeed it could be formed within the very heart of matter itself with all manner of interesting results, the effect being in proportion to the force of the field employed. But in this instance we assume a secondary point source to be created at the centre of the hollow sphere and because every part of the shell is acted upon uniformly, there can be no movement. Any free moving bodies placed within the sphere, will immediately gravitate toward the secondary point source at the centre, as in Fig 32(a). But should a partition or floor be placed at some suitable position away from the point source, then it becomes obvious that any body in that area will 'fall' towards the centre as before, until it is arrested by the floor. In this way the strength of the field can of course be governed to establish a 1g acceleration at the centre of mass of the moving body, which we can now visualise as being the pilot, Fig 32(b). In so far as the secondary point source *must* be created within

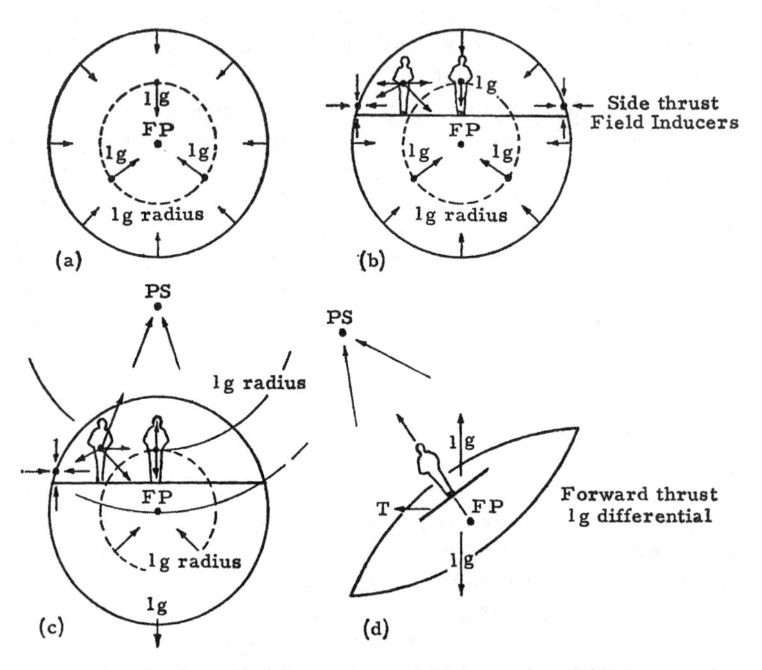

Fig 32. Ship/crew g. differential provided by secondary field inducers.

the craft and not too close to the bottom structure, and the crew's quarters must in turn be placed above it, this automatically positions the compartment fairly high in the ship. It will be apparent that the crew will also experience a side thrust component tending to pull them towards the centre of the craft. But suitably placed field inducers acting laterally as indicated in the diagram, would cancel this out quite easily. Later we shall see how the same result might be attained with the use of only one secondary field inducer, but to introduce this alternative solution at this stage would be tantamount to previewing the end of the plot in a detective story. Sufficient to say here that the discerning reader will observe a very important technical snag to this part of the theory, and we must for his sake halt in our tracks for a moment to examine it.

In several of the instances and calculated examples we shall see later, it will be seen that while we have managed to establish the required 1g differential between the centre of mass of the ship and the centre of mass of the crew member, nevertheless in some cases where high power is employed, his body would suffer extremes of g differential between feet and head. In Fig 32(c) for instance, it is obvious, without recourse to calculation, that the 1g differential will be acting approximately at the pilot's abdomen, while his feet, placed as they are much nearer to the secondary field inducer, will experience several g *downward*. On the other hand, his head is exposed to the more powerful regions of the primary G. field.

By careful selection of focal lengths and field power, this differential can be largely offset, or the crew could lay flat to minimise it, but this is not very practical. In addition, the side thrust field inducer ring could be moved up or down in relationship to the centre of mass of the crew in order to minimise the g loading, as indeed the whole crew's compartment could be adjustable in relationship to the centre of mass of the disc itself, bringing about the same result. The reader may immediately recognise this very manoeuvre as being described in some classical UFO sightings.

Now the problem of g differential on the human body is not new by any means, in fact space flight engineers have acquainted themselves with it for years, though certainly not in association with G. field propelled vehicles. It is now fairly well known that some scientists believe prolonged weightlessness in space might have harmful effects on the human body. Some are inclined to think that after months in space, astronauts may adapt themselves *too* well, making life on earth intolerable for them when they return. Recent reports concerning the physical state of astronauts Ed. White and Jim McDivitt for instance, revealed that there was evidence of slight bone hardening which has been attributed to prolonged weightlessness in their epic Gemini 4 flight.

In order to obviate this risk, scientists have long considered the possibility of producing an artificial gravity in an otherwise weightless orbiting space station, by rotating it about its central axis. Fashioned like a huge wheel, the vehicle would go on rotating perpetually at such speed as to produce a centrifugal acceleration equal to 1g on the crew housed within the rim, see Fig 33. In this, a hypothetical design of 212 feet in diameter is considered. From this it will be apparent that the diameter should be as large as possible from optical considerations as well as the issue raised here, for should there be no dividing walls within the tubular accommodation rim, one would get the impression of the *floor* rising up in front and behind. However, provision of dividing walls restricting the cubicle lengths, should obviate this effect, though if the floor was flat, one would get the impression of tilting over a little, while rollable objects 'dropped' onto the floor would tend to roll towards either of the end walls. But the whole point of this exercise is to illustrate the fact that the astronaut in such a centrifuge, would be exposed to bodily differential in exactly the same manner as the pilot in the G. field ship we have just examined, and some scientists are of the opinion that such variations may produce effects even worse than weightlessness.

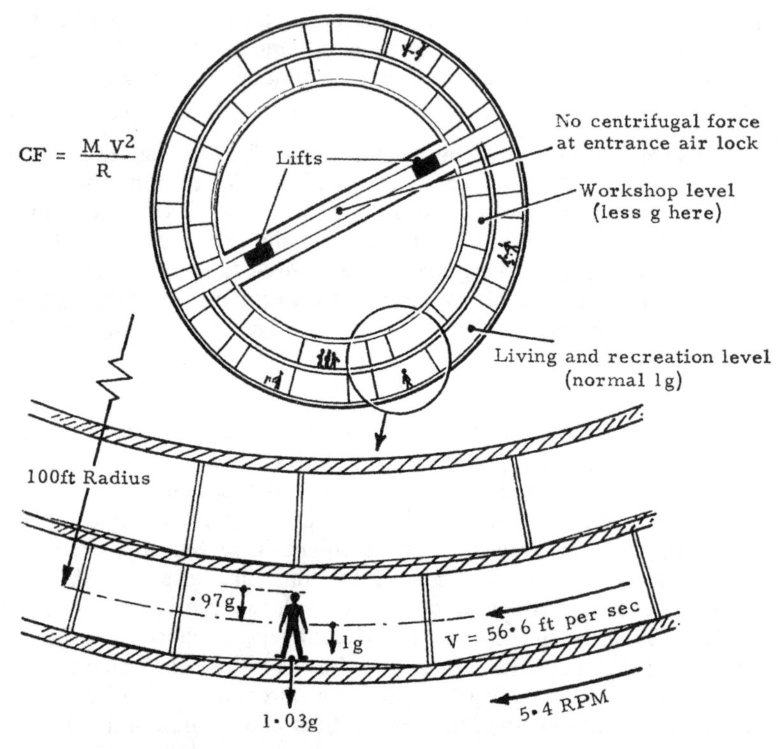

Fig 33. Differential centrifugal stress on the body in a rotating space station. Differential becomes greater as radius diminishes.

A brief consideration of the factors governing 'the centrifuge' effect will establish the relationship.

To begin with, centrifugal force is one of the only forces in nature which so utterly duplicates the gravitational field, for in it, all material bodies are acted upon, molecule by molecule and atom by atom, simultaneously. This means that if a 1lb weight were to be taken to a remote position in space and whirled on a string, it could be made to exert its original 'earthly weight' in centrifugal pull of 1lb on the string, depending on the length of the string and the speed of revolution. In such a condition we would say the mass was subjected to a centrifugal acceleration of 1g. If either the rotational speed or radius of gyration were changed we might get more or less than the original 1g acceleration.

Therefore knowing the diameter of the space station (212ft) we can find the rotational velocity required at the perimeter to give a 1g centrifugal acceleration, at say 100 feet radius. The following will help to compare the centrifugal and G. field effects.

$$\text{Now CF} = \frac{M V^2}{R} \text{ or } \frac{W V^2}{g\,Rcm}$$

Where CF = centrifugal force in lb acting at centre of mass of body.
W = weight of body in lb.
V = velocity of body in feet per sec.
g = acceleration due to gravity (32.2ft per sec per sec).
Rcm = radius of gyration at centre of mass of body (100ft).
ω = angular velocity in Radians per sec.
But CF also = Weight. Therefore

$$V = \sqrt{GR} = \sqrt{32.2 \times 100} = 56.6\text{ft per sec.}$$

As a matter of interest this would amount to a comparatively slow rotational speed, only some 5.4 revs per minute in fact. In other words, a body rotating on an arm of 100 feet radius and at a velocity of 56.6ft per sec, would experience a 1g centrifugal 'pull' comparable to the 1g 'pull' in our gravitational field. Therefore we can now say, at the astronaut's midriff, he would be subjected to an acceleration of 1g pushing him on the floor of the rotating space station, but what of the rest of his body?

Well, we know the angular velocity at the centre of mass is

$$\omega = \frac{V}{Rcm} = \frac{56.6}{100} = .566 \text{ Radians per sec,}$$

or centrifugal acceleration on astronaut at any station $= \dfrac{\omega^2 R}{g}$

Therefore assuming a 6ft astronaut, we get the radius to his head and feet as being Rah = 94ft and Raf = 103ft, respectively. From which we can now calculate the accelerations acting at these two extremes as being:—

Acceleration at astronaut's head Gah =

$$\frac{\omega^2\,Rah}{g} = \frac{.566 \times .566 \times 97}{32.2} = .97g.$$

and

Acceleration at astronaut's feet Gaf =

$$\frac{\omega^2\,Raf}{g} = \frac{.567 \times .567 \times 103}{32.2} = 1.03g.$$

So in the case of the space station rotating to furnish the crew with earth gravity environment, we find an exact counterpart to our own problem, *i.e.* a variation in 'g' forces over the body due to the relatively near focal point of the g field in our case, and the relatively short radius of the rotating space station in the other. Also it is obvious that one solution to the differential lies in making the radius in *both* cases longer. A more realistic example of course is the earth's gravitational field, which is so uniform as to render the body differential virtually undetectable, but it exists nevertheless. Later we shall examine a far more convenient method of obviating the difficulty of G differential and the exclusion of it here for the time being does not materially affect the conclusions to any great degree. So now let us continue with the spherical space ship in Fig 32.

Above the sphere we can place the primary G field point source tuned to 1g as in Fig 32(c) and the conditions for hovering flight are established, also we can revert to our disc shape in order to minimise the g differential on the ship.

By tilting the machine and increasing thrust, flight attitude is obtained, the performance depending on focal length of the field and power factor. If during the flight phase the secondary field is maintained at 1g differential, the crew of the ship will experience no accelerating forces nor sense of movement, only normal weight at right angles to the floor of the ship, entirely independent of the vessel's tilt angle relative to the surface of the earth. To them it would seem that it was the earth which was moving, while they were standing still, completely immobilised, Fig 32(d). It is interesting to note that some witnesses who have claimed to have ridden in a flying saucer have said just this. Also it will be obvious that the 1g differential might be adjusted to suit any planetary g value, hence agreeing nicely with popular science fiction stories. From this foregoing analysis, four very important factors emerge.

One, the position of the crew's compartment in the ship is established, for in order for the G field idea to work, the cabin would logically occupy the *geometric centre* of the craft coincident with the power source, and due to the necessity of the secondary point source, it would be more convenient for the crew's quarters to be placed *above* the centre of mass! I need hardly say that this is nearly always the claim of witnesses in the great majority of near proximity saucer sightings.

Two, there are on record testimonies of people who have sworn to have ridden in a flying saucer, note in all these cases witnesses claimed they experienced no sense of movement whatsoever. If they *are* hoaxers etc., strange that they should invent an important requirement which is inherent in this type of vehicle.

Three. At least one leading scientist whose mathematical genius places him above the normal category of scientists, is working on a so-called anti-gravity machine. I refer of course to the work of Burkhard Heim and may I remind the reader, he claims to be able to protect the crew of his space ship by 'the appropriate arrangement of field inducers'. We have seen that if for 'field inducers' we read 'G. field generators', there is no fundamental difference, though I would stress concerning this latter that I would not wish to be so presumptuous as to relate myself in any way with the work of this great mathematician. By training the author is an engineer, and it is solely from an aircraft engineer's point of view that I conceived the G. field theory. In this context I wish solely to stress the importance of the issues at stake here, that of completely independent technical corroboration.

Four. Last, but not least, we do not have to labour the imagination much to accept the fact, that if a device is designed to beam and create a focal point of energy, then that device will almost certainly be circular in plan form and part parabolic in cross section!

Remember my intention is not to make the facts fit, I shall be the first to acknowledge when they do not, but from our very first analysis of the fundamental requirements of the G. field ship, we have arrived at the very same shape described by so many impartial thousands upon thousands of the world's peoples, the disc.

Aerodynamics of the G. Field

While it is generally true to say that when the lift-propulsive field of a moving saucer is restricted to within the confines of the structure, the craft will cause a displacement of air, what is more correct, is that no matter how nullified by the inverse square law effect, the field will always extend outward into space, which is equally true of all fields. So we shall qualify this statement somewhat by saying, 'to any effective degree, restricted within the confines of the ship's structure'. For in this condition although the power of the field might cause the craft to move, only the innermost layers of adjacent air will receive a measureable thrust. In such event, the disc will be subject to some aerodynamic force, but at low speed this would not amount to much and present no more problems than it does to any moving vehicle, such as a train, bus or automobile. Certainly the effect would give no undue concern to the designers of the disc, for as we shall see, corrective trim or stability would be taken care of by other means.

But when the power of the field is intensified, then we have to cope with a far more complicated set of conditions, but the general idea can be conveyed in the following simple terms.

Fig 34(a) represents diagrammatically the side elevation of a disc which is tilted in the forward vectored thrust position. It will be seen that the vertical lift component equals 1g, compensating the weight of the craft, and sustaining it in level flight, while V is the forward velocity component. It follows therefore that the focus of the G. field will also receive this same component, as will the adjacent air molecules according to their juxtaposition. From this an approximate pattern of the moving air 'bubble' can be visualised, in fact it will in effect resemble an oblate spheroid trailing off into a tear drop Fig 34(b) depending on its velocity. The local changes in atmospheric density due to the field would also be accompanied by a change in refractive index in these

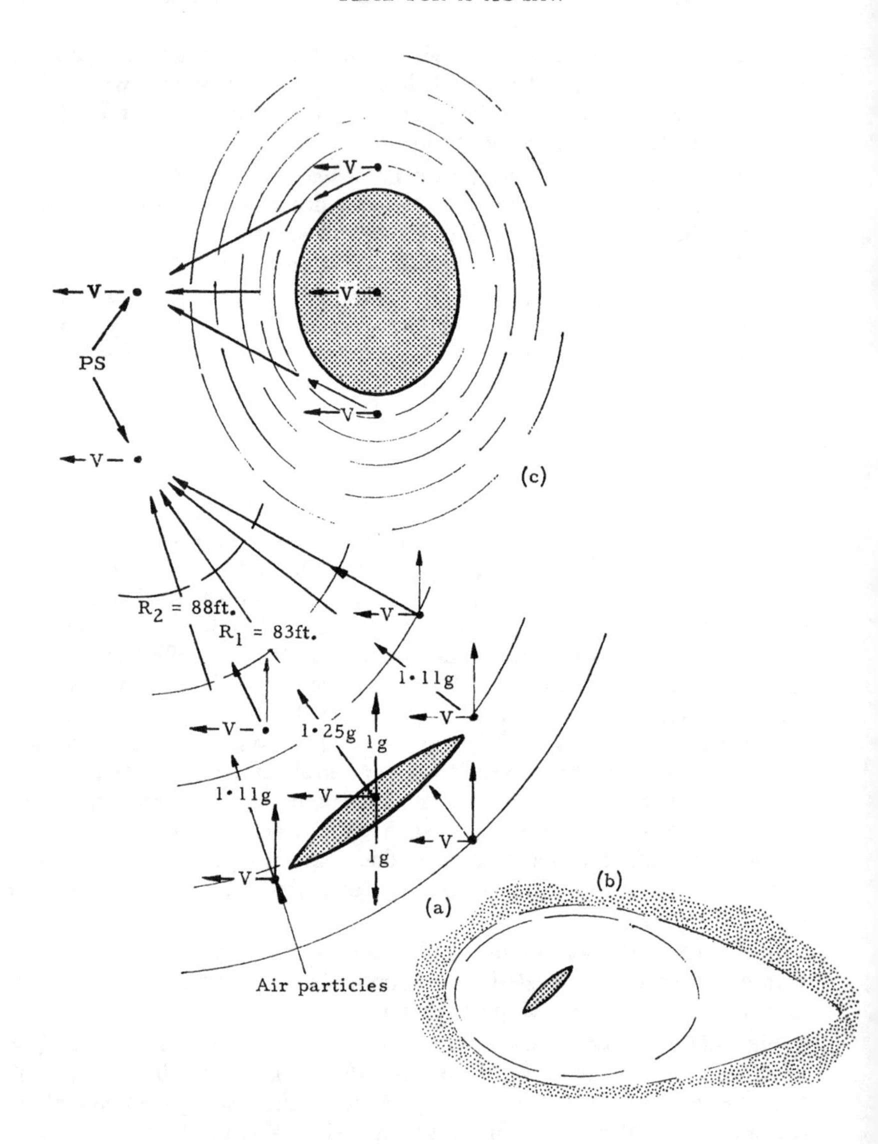

Fig 34. Atmospheric displacement in a moving G. field due to translational velocity and acceleration towards point source.

regions, which if illuminated, might conceivably define the shape of the moving belt of air. This has been reported often, here is an appropriate case:

On 10 August, 1944, Capt. Alvah M. Reida, Commander of 468 Bomber Group, 792nd Squadron based at Kharagapur, India, had an encounter with a UFO. In a later report to N.I.C.A.P.* he said:

'I was on a mission from Ceylon, bombing Palembang, Sumatra. The date was 10 August, 1944, time shortly after midnight. There were 50 planes on the strike going in on the target at about two or three minute intervals. My plane was the last one in on the target and the assignment was for us to bomb, then drop photo flash bombs, attached to parachutes, make a few runs over the target area, photographing damage from the preceding planes. The weather was broken clouds, with an overcast above us. Our altitude was 14,000 feet, indicated air speed about 210 m.p.h.

'While in the general target area we were exposed to sporadic flak fire, but immediately after leaving this area it ceased. At about 20 or 30 minutes later the right gunner and my co-pilot reported a strange object pacing us about 500 yards off the starboard wing. At that distance it appeared as a spherical object, probably five or six feet in diameter, of a very bright and intense red or orange in colour. It seemed to have a halo effect. Something like this:

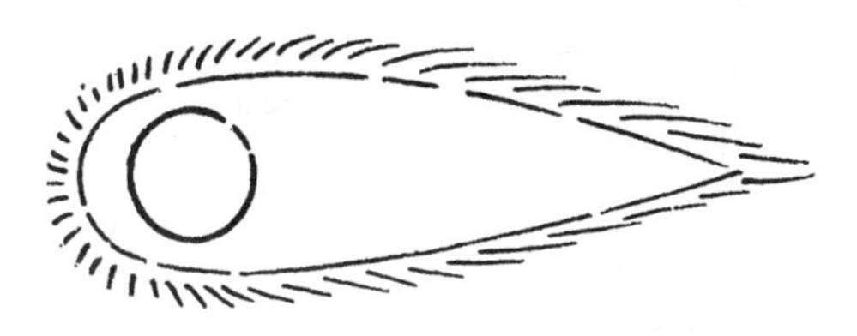

'My gunner reported it coming in from about five o'clock position at our level. It seemed to throb or vibrate constantly. Assuming it was some kind of radio controlled object sent to pace us, I went into evasive action, changing direction constantly as much as 90 degrees and altitude about 2,000 feet. It followed our every manoeuvre for about eight minutes, always holding a position of about 500 yards out and about two o'clock in relation to the plane. When it left, it made an abrupt 90 degree turn, up and accelerating rapidly, it disappeared in the overcast.'

Capt. Reida added, 'During the strike evaluation and interrogation following this mission, I made a detailed report to Intelligence, thinking it was some new type of radio controlled missile or weapon.'

* National Investigation Committee of Aerial Phenomena in U.S.A. (Has the greatest supporting membership in the world and no doubt is the best informed civil organisation for the study of UFOs.)

Continuing our aerodynamic analysis we see that Fig 34(c) represents a plan view on the disc shown here foreshortened as an ellipse while the concentric ellipsi represents various stations of the field. As the angle of tilt might in some conditions be very small, it will be more convenient in future illustrations to portray the configuration as concentric circles rather than ellipsi.

In Fig 35(a) the disc is shown with the force field intensity marked by the concentric circles, while V_1 is the velocity of the disc in magnitude and direction. V_2 and V_3 etc., over the radius R in Fig 35(b) represent

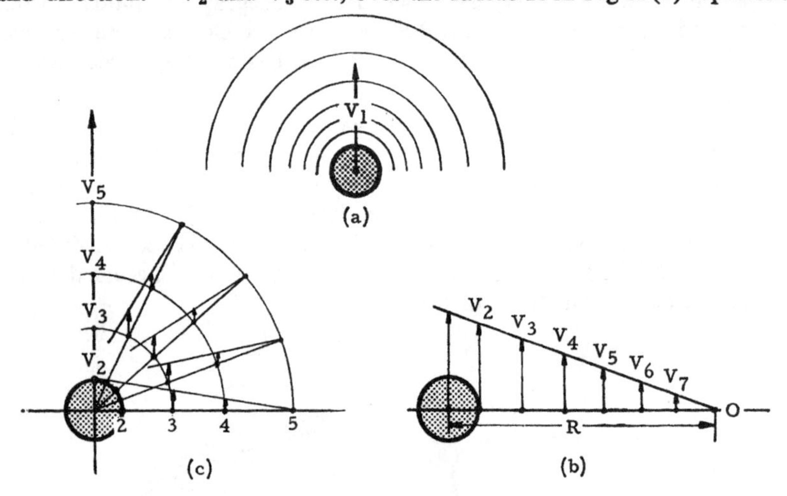

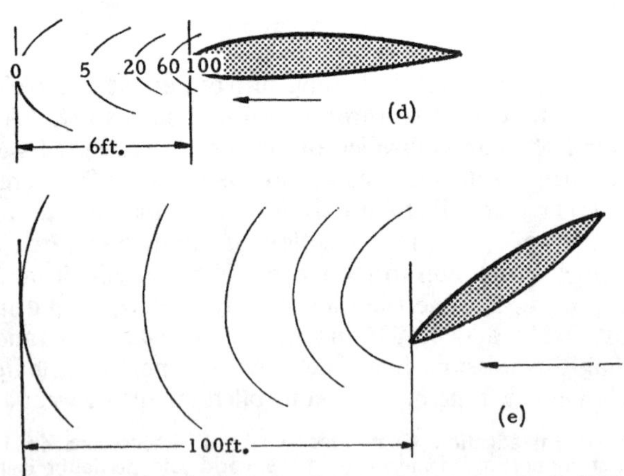

Fig 35. Atmospheric velocity gradient in the vicinity of a G. field propelled vehicle.

different velocities at these stations which diminish to zero at some hypothetical radius R from the ship. This then represents the velocity gradient imparted to the air surrounding the disc when viewed from above or nearly so.

Fig 35(c) is simply the development of the diagram showing similar gradients at different angles round the craft from which can be plotted the shape of the moving air belt.

From this it will be appreciated that immediately in front of the craft there is a *progressively* built up velocity, as indeed there is in front of the leading edge of conventional aircraft wings, Fig 35(d). The difference of course being, whereas with the aeroplane, this region effectively extends only a matter of inches, or at the most a few feet, it could extend a hundred feet or more in front of the G. field propelled vehicle, Fig 35(e).

As the surrounding air molecules take on the translational velocity of the point source, so they will also be accelerated towards the source by an amount proportional to their distance from it. Such a flow would be very complicated and beyond the scope of this book, but in general, the flow pattern would be similar to those shown in Plate 16. In this, iron filings substitute the air particles, while an electro-magnet simulates the G. field. As the oncoming point source moves across the 'atmosphere' of filings, so they are attracted back to meet it, then to follow as the point source passes. The analogy is useful as the magnetic field obeys the inverse square law, and when seen in motion, clearly shows the smooth shearing effect in the iron filing layers.

The heating and sonic cushioning effects can be visualized from this and quite obviously a more thorough investigation into these aspects alone could take up another volume. On the other hand, the falling leaf tendency which was also predicted in chapter 8 as an aerodynamic effect is comparatively simple to evaluate. We cannot be sure of course that such behaviour with some UFOs can be ascribed to aerodynamics, along with the other thoughts expressed in this book, we are merely trying to present facts which make sense when reviewed in an orderly manner, so long as observable phenomena continue to support the theory, there is no point in rejecting it.

As we saw earlier, if a saucer were to be hovering, then almost certainly the vertical lift component would have to be less than the normal weight. For if it were to be entirely weightless, then it would ascend a little due to the aerodynamic effect. So in order to hover we would expect the pilot to adjust the lift vector to a little under 1g in order to compensate this lift. Similarly we would expect such upcurrents of air induced by the field to cause the disc to rock slightly, in fact a slight wind would move it unless corrections were made, no doubt automatically.

Although we shall consider later other rather more scientific methods, it is none the less interesting to note in this present context, that should the pilot wish to descend, then he must reduce the lift until the machine began to fall very gently. Even so it would be subject to upsetting aerodynamic forces which again could be presumably corrected. Now if for some reason, perhaps even to relieve the monotony, for remember in such a ship the occupants would have no sense of movement, the pilot reduced the lift considerably, then the disc would fall, much the same as a parachute. If in addition the pilot momentarily shut everything off, then the craft, unstabilised, would almost certainly go into a series of stalls.

Fig 36 is almost self explanatory and it is interesting to find, in the cases I have come across where a UFO behaved in this way, they have been significantly bi-conic in section as shown in the diagram. This would be necessary if this explanation is the correct one. The reader can quite easily duplicate an experiment by forming a light weight disc out of two paper cones, holding it slightly out of level and dropping it from a height. There you will see the characteristic of the falling leaf.

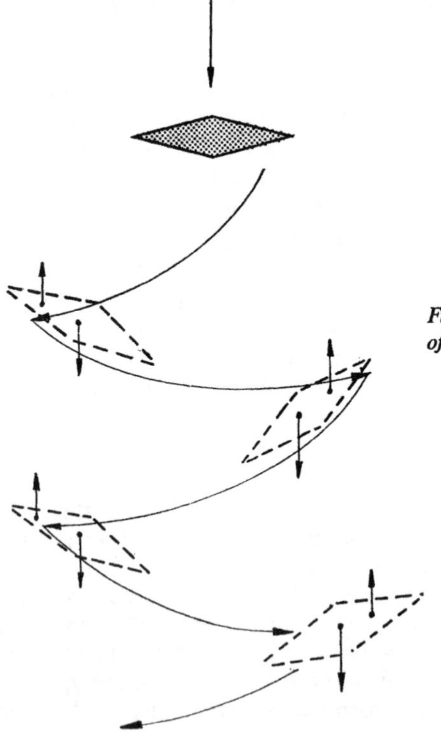

Fig 36. Aerodynamic pendulum effect of a descending bi-conic shaped disc.

In the sketch a cross section of a disc is shown, which due to its symmetrical shape can be regarded as presenting a leading edge whichever way it falls. If it slips to the left, then it will stall in just the same manner as a conventional wing, because a moment will be caused by the displaced centre of pressure about the centre of gravity. But having reached the stalled position, the disc will now dive to the right and the process is repeated. If in addition, a slight rotary motion is imparted, the disc will oscillate and gyrate on its way down, in fact just like 'a falling leaf'.

The extraordinary thing is, but quite easily understood if you think about it, should a craft be hovering and the pilot *increase* the lift vector to 1g or over, then the machine will begin to ascend, in other words it could be regarded as 'falling' upwards, and nearly all the aerodynamic factors we have been discussing above are reversed, *i.e.* 'falling' leaf upwards. We shall see later on there might be a more specific reason why saucers should so behave, but it is touching to think that it might be solely the personal whim of the pilot. The fact remains, when people have claimed to see saucers behaving like leaves, falling or otherwise, they are making at least one good piece of engineering sense.

9 February, 1957. At 1.00 a.m. in Philadelphia, Pennsylvania, Roger Standeven observed a white oval shaped UFO with a red light visible on top. The object would stop, 'fall like a leaf', speed up again, and repeat the sequence, gaining altitude each time it sped up.

One of the most incredible stories on record concerns the inverted falling leaf and I would be failing in my task if I omitted it simply because the witness claimed to have talked to the crew of a landed saucer. In fact, in this analysis I have deliberately included such cases where an untrained witness bears testimony to evidence of a technical nature.

For I regret to say that prejudice can be found even among researchers who appear not to recognise the same failing in themselves with which they often accredit others. No truer measure of this fact is evident than the consistency with which such cases as the following are left out of their serious study. In criticising such bias, I need no other substantiation than to remind the reader, that it was only after including *all* the reports of sightings, including the 'doubtful' ones, that Aimé Michele discovered the pattern which enabled him to develop his Othotenic theory. Such is the work of the true researcher, we would not be fair to this subject or ourselves to treat it otherwise.

In August 1953, Salvador Villanueva was 37 years old, he lived with his wife and seven children in Mexico City, and ran a one-car taxi service for a living. He never asked for or wanted the publicity and interrogation which became his lot after that fateful day when he was employed by a couple of Texan tourists, to drive them from Mexico City to the Texan border.

They had covered about 60 miles and just passed Caidad Valleys when the car broke down. Examination revealed that the car would go no further that night, and the Texan passengers engaged another car, and irately drove away without paying. Salvador tried to get help at the lonely spot but without success. By then it was beginning to rain, so he decided there was no alternative but to spend the night in the car and make arrangements for repairs in the morning.

He had crawled under the car to investigate the damage once more and then became aware that he had company. Right by his head were two pairs of feet encased in a substance 'like grey corduroy'. Salvador scrambled up to find two pleasant looking men, but they were no more than four feet six in height. This did not seem very unusual, because in Mexico there are many short people of that stature. He said they were clad from neck to toe in a one piece suit made of the same grey material, broken only by a wide perforated shiny belt. Round their necks were metal collars and on the backs of their necks were 'small shiny, black boxes'.

The men carried helmets under their arms similar to those worn by jet-pilots or football players, so he assumed them to be fliers who had landed nearby.

One of them smiled at him and asked if he was having trouble, Salvador said he was and the man smiling sympathetically, spoke of one or two casual matters. He asked Villanueva a little about himself, during which time the Mexican noticed that the man had a peculiar accent as though he seemed to be stringing his words together. During this time his companion said nothing, but smiled or made other expressions suggesting he understood. 'Doesn't your companion speak Mexican?' asked the driver, 'No, but he is able to understand you', came the reply.

Now it began to rain again so Salvador invited the men to shelter in the car with him and the conversation continued.

'Are you aviators?'

'Yes, we are.'

'Is your plane near here?'

'Not very far.'

'Where have you come from, if I might ask?'

'We have come from very far.'

By now the Mexican began to feel uneasy, and as the night wore on the feeling increased. Somewhere in the back of Salvador's mind he knew there was something different. The men knew far too much about too many places all over the world. Finally, as dawn was breaking, he asked the question which had crept into his mind.

'No,' came the answer, 'we are not of this planet. We come from one far distant, but we know much about your world.'

Feeling that this was some kind of joke, the driver did not believe the man right away, and anyway he didn't like the constant smiles of the silent one and more than once accused the men of pulling his leg. At dawn, Salvador was a very confused man, and was a little relieved when his companions said they must go. Then they asked him if he would like to see their machine, and still half expecting to see a conventional aircraft, he followed them. They led the way through the bushes, across swampy land for half a kilometre, and all the way the land became wetter and more boggy. Salvador found himself sinking up to his ankles, but he swears the men in front of him did not retain a mark from the muddy ground. Instead, he noticed, that as their grey clad feet touched muddy pools, the mud sprang away from them as if repelled by an invisible force. Please note, we shall hear of this same force later on. [author.] Turning to him, his strange companions smiled encouragement, he hesitated, then struggled on. The men's feet fascinated him, why did their perforated belts glow each time the mud sprang away? Suddenly they came to a clearing and there it stood. A great shining craft unlike anything the simple Mexican had ever seen. 'In form it had the shape of two huge soup plates joined at the rim. Above it was a shallow dome with portholes.' About thirty-five to forty feet across, the strange thing rested on three large metal spheres.

As they approached, Salvador heard a faint humming sound coming from within the machine and he noticed a portion of the lower part opened outwards forming steps, 'much in the manner of the rear entrance to a Martin 404 air liner'. The two men went up the short flight of steps, pausing at the top to turn and look at the Mexican. 'Would you care to come inside with us?' came the invitation. Salvador shook his head. His wife, his family, his job and all the things he understood and loved seemed suddenly very real. He turned and fled.

Hardly able to believe his senses he regained the road, stumbling and out of breath. Then he took a glance back the way he had come. There among the bushes something was happening, 'something glowing white rose slowly into view, hovering for a moment, then gaining speed it began a kind of *pendulum* motion, a backwards and forwards arcing movement, like a falling leaf going up instead of down', Salvador said. In this manner the disc had attained an altitude of several hundred feet, then glowing brighter, it shot up vertically with incredible speed. In seconds it was lost to sight and only a faint *swishing* sound to mark its passage.

Salvador Villanueva went back to get attention for his car, he had work to do. It was the next night by the time he arrived home. He told no-one of his experience, not even his wife, until she questioned him about his pale and strained appearance. Making sure the children were asleep he told her, faultering and hesitant, convinced she would think him a little deranged. She did believe him, but even so begged him not to tell anyone about his experience. And it was only when others had similar experiences that the Mexican finally told his story.

Now Desmond Leslie was the investigator of this incident, and he was able to pay Salvador a visit. He now takes up the story:

'When I visited Mexico in November 1955, I sought out the group of journalists and investigators who had been probing Villanueva thoroughly since the occurrence. They were most helpful and arranged a meeting with an interpreter at which I could interview him personally. A few days later I was loaned a car and Villanueva volunteered himself as my driver. During this time I was able to study him as a human being. I found him quiet, unassuming, well mannered and an excellent driver. The way he navigated the dangerous 11,000ft mountain passes by day and by night won my admiration. His judgement of speeds and distances was first class. On a trip to the Great Pyramid of Cholula—eight times greater in bulk than Cheops—he and his eldest son followed me through the six mile labyrinth of tiny tunnels and galleries honey-combing this ancient structure. We had a wonderful adventure which they enjoyed every bit as much as I did, and I found them intelligent and pleasant companions. At the end of the long drive he surprised and even embarrassed me by refusing to take any payment, not even a 'pourboire' or present for his 'senora'. He gave me every impression of being a trustworthy, reliable human being, the kind you would trust to take your jewellery to the bank or to look after your children if suddenly called away. I liked him very much, and I thoroughly believed his story.

'But I am not alone. All who have investigated him have come to the same conclusion. On one occasion Mr and Mrs Charles Reeves, ardent researchers from Ohio, took Villanueva with a group of investigators and journalists to find and establish the place of contact. They found the pull-in beside the road where he had parked the broken-down car, and after a little recollection Villanueva set up with his stick the line of direction where he remembered seeing the ship take off. The party followed this line until they came across a clearing where bushes and sticks had been broken down by some heavy object within a circle roughly 40-45ft in diameter. Later, one of the party secretly moved the stick about fifteen degrees and when they regained the road, asked Villanueva to re-confirm the direction. He studied the line of sight carefully and moved the stick back about fifteen degrees to its original position. He was quite certain, he said, that this was the true direction because he had noted the exact background in the distant landscape against which the saucer had first appeared on take off.

'Now such powers of observation may seem remarkable. But I had a chance to see them demonstrated for myself when we took the mountain roads. At times our wheels were but six inches from the edge.

But I never felt afraid for I sensed that here was a man who knew exactly what he was doing and whose sight and judgement were above average.

'Out on the road he related and re-enacted the story to his examiners without change or contradiction. The whole episode was still vivid in his memory. He knew what he had seen and heard; just that and no more. A practical working man, he had learned to use his eyes, and he was not in the habit of being deceived.

'When I showed him the photos of the Adamski saucer he said that though it was similar to his ship there were several major differences: for example, the double convex hull and the curved underside. He did not believe his visitors were Venusians. They were small and clad in this one piece garment covering the feet as well as the body. He had the impression from their talk, though they did not name any planet, that they had come from somewhere much further than Venus, maybe from a world beyond our vision entirely.'

Now I want to impress on the reader that I cannot explain how it is that beings similar to ourselves could originate from our solar system because anyone who knows anything at all about Astrophysics must sympathise with the findings of modern astronomers. Errors there may be but it's nonsense and wishful thinking to consider that they are *completely* wrong. On the other hand I cannot explain how beings like ourselves could be visiting us from another part of the galaxy. But I do want to keep to the facts and simply because I do not understand these things, I see no reason why I should dismiss claims like Villanueva's solely to suit the convenience of my limited intellectual powers.

Therefore I earnestly ask the reader to consider the technical clues among such sightings in order to evaluate them, rather than dismiss them untried as worthless nonsense. In the course of my investigation, I have come across more than one very responsible scientist, who though bewildered and reluctant, has finally had to admit, that there is something in stories such as Salvador Villanueva's.

Now let us turn again to the inverted falling leaf phenomena, but this time it was seen by a trained witness. I would advisedly add, why accept the one case and denounce the other? For the whole crux of the matter revolves round the same technical inclusion.

11 October, 1951. J. J. Kaliszewski (then Supervisor of Balloon Manufacture for General Mills Inc.) and Dick Reilly, were flying near Minneapolis, Minnesota, observing a balloon at 6.30 a.m. They noticed a bright glowing object overhead, moving at high speed from E to W. The UFO had a dark undersurface and a halo of light. Finally it slowed and started climbing 'in lazy circles'. Kaliszewski said, 'The pattern it made was like a falling leaf inverted.'

Once again I would stress that I am merely pointing out here, that such behaviour would be easily explained in aerodynamic terms, though

I would hesitate to assume such. In some cases, and so far I have not come across any, where discs might be undercambered, or concave in section, a different aerodynamic behaviour would occur, *i.e.* the disc on descending and unless stabilised, would immediately turn turtle, then continue the falling leaf pattern. There are several other aerodynamic conditions we have not yet dealt with, but involving as they do rotation, I have gone out of context so as to include these aspects in the appropriate section later on.

12

Analysis Two

Craters and Surface Proximity Effects

IN Chapter 8 we saw how we might predict crater phenomena to accompany a swift lift off. These ground effects will not always be the same, depending as they do on several factors, such as the nature and moisture content of the vegetable soil, type of craft used and length of focus of G field, etc.

In order to establish the effects negative g would have on plants and soil, the author built several centrifuges, one of which is shown in Plate 17(a) and (b). In operation the idea is based on the fact that if a sample of soil is inverted, then gravity pulls it and any vegetation, downwards, in exactly the same manner in which a saucer's G field might be pulling it upwards at 2g, in other words, 1g effective.

If the soil sample is now secured upside down in a box (with its anchorage resembling as near as possible the normal ground conditions) it can be installed in the centrifuge to impose even higher g levels. In order to be able to take photographs at certain g, an accelerometer was installed and wired up to trigger off a flash light. The four shots in Plate 18 show confirmatory results to the G field theory as far as craters are concerned. Plate 18(a) shows meadow grass and buttercups at normal earth gravity, while Plate 18(b) shows immediate response to negative or upward 1g (effective), Plate 18(c) shows the effect at 3 negative g, while an instant later the sample is torn apart and hurled 'upwards', the jagged remains testifying to the effectiveness of the anchorage, Plate 18(d). The effect occurred at just over 3g!

Now let us consider a grounded machine resting on legs with the centre of mass some five feet above ground level as in Fig 37(a). The focal length of the point source R_1 is 15ft, making the distance from the point source to ground level R_2, 20ft.

In order for the machine to take off gently, the field intensity must of course be generated slowly, until at radius R_1, passing through the centre of mass of the vehicle, the G field produces a vertical component equal to the normal weight, at which point the craft will be weightless and rising due to aerodynamic effects. As before, the upward acceleration

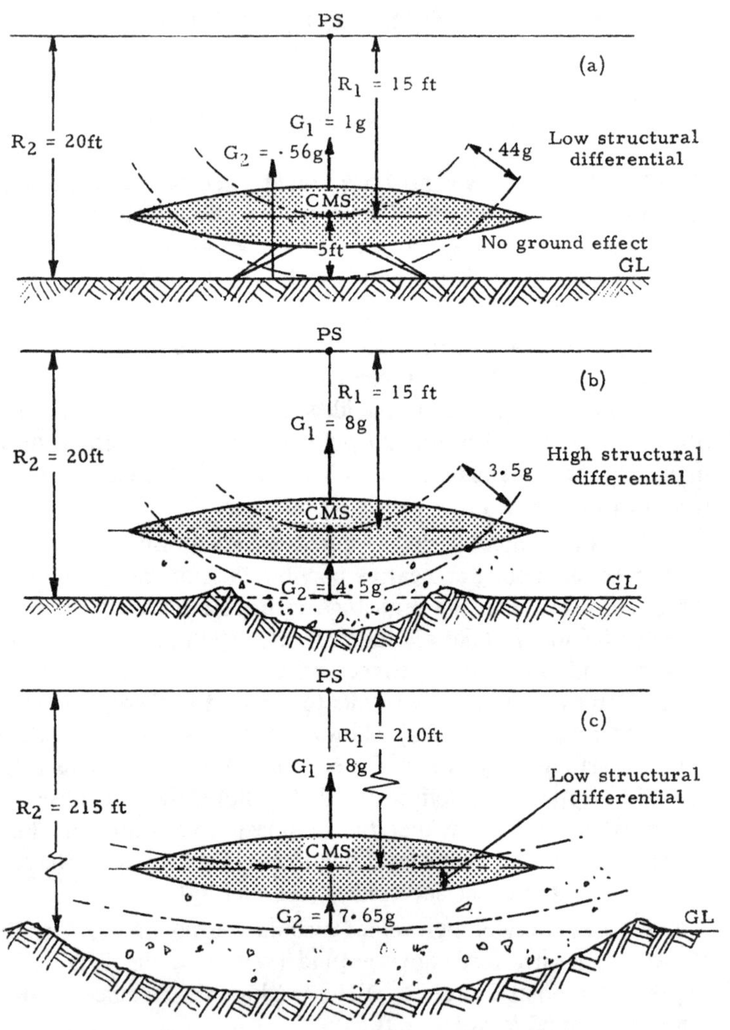

Fig 37. Structural differential and ground effects due to varying focal lengths and field strengths.

'exerted' on the ground immediately beneath the disc can be shown by the inverse square law as:—

$$G_2 = \frac{R_1^2\ G_1}{R_2^2}$$ and substituting we have

$$G_2 = \frac{15 \times 15 \times 1}{20 \times 20} = .56g.$$

Or in other words the soil loses approximately half its weight at the surface, and it is extremely unlikely that any effects could be discerned. But it would be a different case entirely if the craft employing the same focal length were to be subjected to a G field of say 8g as in Fig 37(b), in which case the vertical acceleration would be

$$G_2 = \frac{15 \times 15 \times 8}{20 \times 20} = 4.5g.$$

Which means 4.5 times contra weight, less 1 terrestrial weight. So that the soil would be subjected to a total lifting force of 3½ times its own weight. In which case some of it would almost certainly find itself wafting away spaceward, following the craft which caused it. But not for long, for remember the machine itself is accelerating due to 8 minus 1g. So there exists a total g differential of 3½ between the machine and the particles. Therefore, although the debris would 'fall' upwards initially, the craft would rapidly leave it behind, when it would finally fall back to earth.

Due to the comparative nearness of the point source to the craft at high field intensity, a fairly large g differential will be set up, and we have seen one way to avoid this is to increase the focal length of the field. Let us see what happens in an extreme case if we do. As in Fig 37(c), suppose the focal length R_1 is increased to 210ft while making R_2 215ft, then we get:—

$$G_2 = \frac{210 \times 210 \times 8}{215 \times 215} = 7.65g$$ on the particles and

the original 8g on the ship, which means that there now exists a total differential of only .35g between the ship and the particles of soil, and a much less differential within the structure of the vehicle itself, to the order of .08g in fact. Certainly an easily tolerable one from an engineering point of view, allowing of course that the machine is built of light materials.

Now the picture is somewhat changed, for the soil particles travelling at extremely high initial velocity would be separated from the machine by only .35g. Even so, before very long they would be out of the effective zone of the G field and fall back to earth again. But this part of the picture is not yet completed, there is still one other factor to consider. It is the crew g differential component we discussed earlier. Fig 38 will make this clear. In this we note that the radius from the primary F.P. to the centre of mass of the pilot R_3 is 207ft which gives him a vertical g component of 8.2g due to his being nearer to the point source.

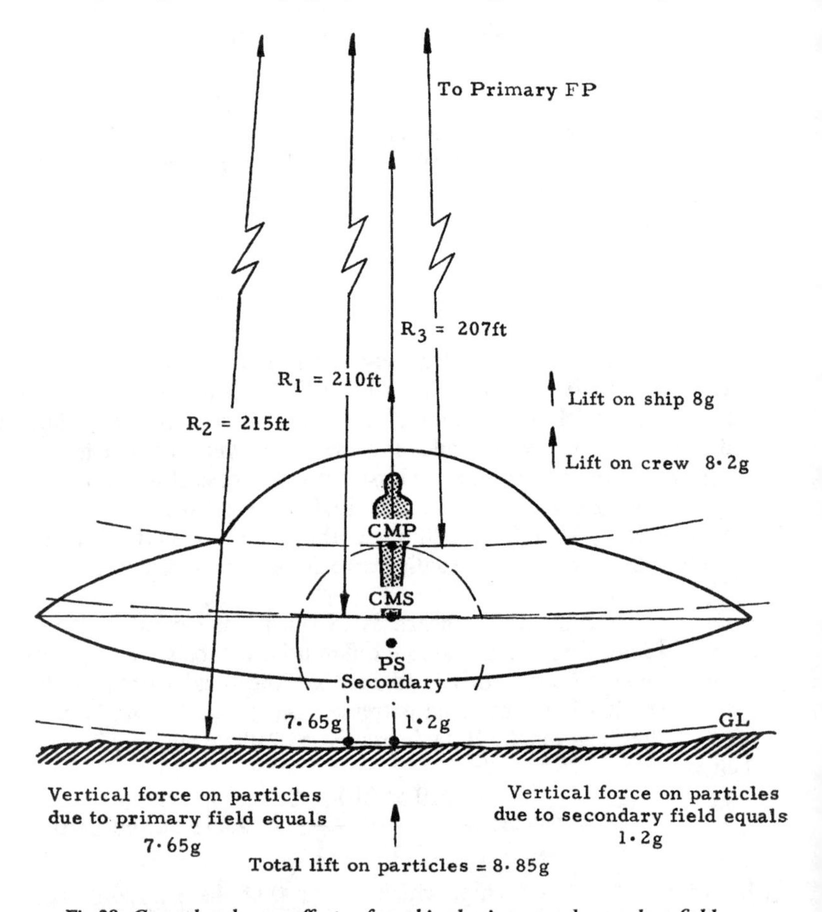

Fig 38. *Ground and crew effects of combined primary and secondary fields.*

The resulting differential will be counteracted by the small internal G. field within the ship. But now it becomes obvious that the upward vertical component of this will also act on the particles beneath the ship, for they must be considered as free moving bodies in the same sense as the pilot was. Therefore the 1.2g downward acceleration required by the pilot will also be experienced by the particles vertically, towards the secondary point source, therefore they will be subjected to a total vertical acceleration of not 7.65 but 8.5g, that is *more* than the ship itself. Therefore the particles would overtake the ship and appear to 'cling' to its undersurface.

From this it will be apparent that so long as the secondary field inducer is in operation, then any loose material beneath the ship must experience a greater thrust than the craft itself, even though the G field focal length is kept small in proportion on take off. The only way to avoid the inconvenience of airborne material, is a *slow* lift off at just 1g or over. In addition there is another factor to be borne in mind when considering the alternatives. Should a long focal length be employed then this represents a proportionally larger area of disturbance, atmospheric and locally around the ship. Outside the atmosphere, the longer focus field might be preferable. But there are conditions which we shall review later, which may change this.

I have tried to show how a device employing the G field principle might cause craters. You, the reader, have been patient over this more tiresome climb, you have earned this little respite; in many respects craters are both varied and weird, so now let us continue by having a look at some of them.

October 1954. That historic year of UFO activity over France when this interesting landing case was reported from Rovigo, in Italy, not far from the mouth of the river Po. At Po di Gnocca several Italian peasants saw a circular object floating slowly along, which then landed noiselessly on the ground where it remained motionless for a few minutes, before finally taking off, straight up. Where it had landed, the witnesses saw a crater almost twenty feet in diameter. As at Poncey on 4 October, the earth had been torn out as if by powerful suction and scattered around the edges of the hole. In addition six poplar trees standing nearby had been carbonised. An investigation was at once conducted by the military authorities.

12 October, 1962. The *Portsmouth Evening News* reports that more holes in the ground have been causing speculation in a number of places in Kent, Hampshire and Berkshire. These holes have varied from six inches to three feet in diameter. The holes were at first thought to be due to bombs and this naturally caused alarm and despondency. The Royal Engineers' Bomb Disposal Unit at Horsham received 30 to 40 calls a day reporting the sudden appearance of the mysterious holes.

July of 1963 saw quite a spate of craters in the British Isles, which kept flying saucer researchers and army bomb disposal units very busy. Although by no means the most spectacular, the Charlton crater received unusual publicity, covered as it was for over a week in most of the national dailies, and the television programmes. So much so in fact, that there is no point in my giving prominence to it here. Sufficient to deal with the main facts.

July 1963, at Manor Farm, Charlton, near Shaftesbury, Dorset, Reg. Alexander found an eight foot diameter crater, one foot deep. In the centre of the crater was a three feet deep hole. Radiating out from the crater were four slots, three of them in the barley crop, Plate 20. Note, over a circular patch of some 12ft radius, *the ground was denuded of potato crops*, there was much speculation as to where and why they had gone. Army bomb disposal units failed to find a suspected bomb, but they did find the source of their instrument readings, a small piece of material which was identified as a meteorite by Patrick Moore—then finally and very quietly, settled as natural iron stone. That was the end of the Charlton crater. Now the author has no wish to interpret anything significant or sinister into the fact that at that time, craters were appearing up and down the country in pairs and at least two of these so enormous as to make the Charlton crater look like a child's seaside paddling pool compared to a fair sized gravel pit! But, was it just coincidence that at that same time the Charlton crater should have been plugged so much? I shall leave the decision to the reader. Perhaps I should add, that with typical one sidedness, the daily Press jumped eagerly at the opportunity when a would-be hoaxer came along to confess, for which pains he received about £50 from a prominent newspaper and several columns of nonsense. It seems that in order to start a hoax concerning interplanetary visitors, he and two student friends went over the country making holes in the ground. Later when questioned by representatives of the British Unidentified Flying Object Research Organisation, who *were* equipped with the facts, the would-be hoaxer broke down and admitted the whole story was an attempt to cash in on the phenomena. This counter claim was never given any further copy; the public still remain ignorant of the facts.

Here is an approximate time-table of the appearances of the craters; note there were coincident and independent UFO reports at the time.

28 June, 1963. The *Oldham Chronicle* of 29 June reported that a group of Royton people recently saw a 'flying saucer'. It was oval-shaped and hovered silently 150 feet above their heads for 15 minutes, then it glided over a hospital towards nearby playing fields. It was a very bright object with something spinning above it.

7 July, 1963. Two giant sized craters found on the slopes of Meldon Hill, Dufton Fell, Westmorland. Mr W. Richardson, who owns the land, thinks there may be a connection between the craters and the disquieting fact that 40 sheep are missing from his flock. Nearly at the top of the hill was a huge circular depression about 200ft in diameter, '2 to 3 feet in depth, from which all the topsoil, peat and heather had disappeared, leaving only the clay subsoil, looking as if it had been scraped clean by a bulldozer'. To the right of this crater facing uphill and 150 feet away, lower down, there was a second crater,

slightly smaller. The lower ends of the craters terminated into 40ft wide ravines, which ran down the hill and joined to form a gigantic 'Y'. On each side the soil was piled up to a height of 10-12 feet, 'as though shoved to the sides by an enormous force'. Giant clods, as big as a car, with the peat and vegetation still intact, were thrown about as if by some mad thing. Investigators claimed that it looked as though torrents of water had gouged out the soil.

10 July. Two craters several yards across were discovered in a field owned by farmer Brown at Middle Moneynut, Scotland.

11 July. Six people claimed to have seen a flying saucer high in the sky over Edinburgh. They watched it for ten minutes. It was grey-white on top and dark underneath, shaped like two saucers one on top of the other, with a hump on top. No sound, and the object appeared to gyrate.

13 July. Whitley Bay, Northumberland. A Mr Walker saw a large white shining disc flying over his house. There were 'strong sound vibrations' and witness saw a smaller object orbiting around the first.

20 July. Flamborough Head, Yorks. 10 feet diameter crater in 20 acre field, surrounding grass dead and blackened, soil scattered over 20 feet. Explosive expert called in could not explain it. Hastily suggested, yes you've guessed it, a lightning bolt 'which ignited a small pocket of gas'. One comment stated, 'The thought that Flamborough finds at once most appealing and appalling is that the answer lies not in the soil—but in space'.

27 July. Flamborough Head, Yorks. A shining object which coast-guards described as 'balloon shaped and almost transparent' had holiday makers mystified for hours. It moved *against* the wind.

27 July. West Cumberland. A tumbler shaped object was seen in the night sky. It was glowing brightly in the northern part of the sky and was under observation for nearly three hours.

27 July. Two craters at Sanquhar, Dumfries, discovered by well known sheep breeder, Mr Gavin Hendrie, on the hillside at Tower Farm. About 6ft in diameter and about twelve feet apart, the two craters were linked by a furrow a foot deep. Visited by an official from Eskalemuir Observatory, who after a minute inspection declared the craters had been caused by a 'powerful lightning stroke' which went into the ground converting moisture into steam, then 'burst the ground upwards'. An eye witness said, '*You could lift the grass up like a skin*'. (Note: Author.)

30 July. Mr James McGill, holidaying at Luce Bay saw an unusual stationary object in the sky east of the Isle of Man. Three lights appeared to emerge from the main one, then return.

7 October. Report in '*Hastings Evening Argus*'. 20ft deep hole found

in barley field at Fittleworth, West Sussex. Bomb disposal experts found—nothing.

16 January, 1964. Another mystery hole on Puckwell Farm, Niton, I.o.W. Farmer Ray Peach had to stop ploughing his barley field when it was discovered. A bomb disposal unit dug a 10ft square by 20ft deep hole before giving up. No bomb was found.

12 February. Two men almost fell down another mysterious hole which suddenly appeared at Slackstead, Nr. Winchester. Investigators were told it was a 'clay' hole dug out by Ancient Britons! Then the bomb disposal unit said they thought an old bomb had caused it.

There was another hole in a barley field at Leicester in the Midlands, a crater in Sussex and so on, usually the public accept the explanation 'old bomb'. But no-one seems to ask the question, 'If this is the correct explanation why have they all appeared now?' To which I might echo another, 'Why so often a barley field?'

Now let us investigate more closely some of these g effects in terms of the statements made by those who are fortunate, or unfortunate as the case may be, to claim contact, or near contact with visitors from outer space.

From time to time we hear of a case where a witness has claimed to have actually stood near to a hovering saucer, with no apparent ill effect, and the objection may be raised, if saucers are said to take aloft all kinds of loose objects, and even the ground beneath them on occasions, surely such claims are technically inconsistent? Indeed this may not be so, such accounts may be quite compatible with the theory. Let us consider a hypothetical case.

Fig 39 shows a machine hovering at 20ft in an extreme case, immediately over a man. While the point source is 7ft above the centre of mass of the machine R_1, the distance from the P.S. to the centre of the mass of the man is 27ft, R_2. Assuming the lift at R_1 to be due to 1g acceleration and applying the inverse square law, we can write:—

$$Gm_1 = \frac{R_1^2 \; Gs}{R_2^2}$$

Where Gm = contra-gravitational acceleration of main field at C.M. of man.

Gs = contra-gravitational acceleration of main field at C.M. of machine.

R_1 = distance from P.S. to C.M. of machine in feet.

R_2 = distance from P.S. to C.M. of man in feet.

and substituting we have $Gm_1 = \frac{7 \times 7 \times 1}{27 \times 27} = .067g.$

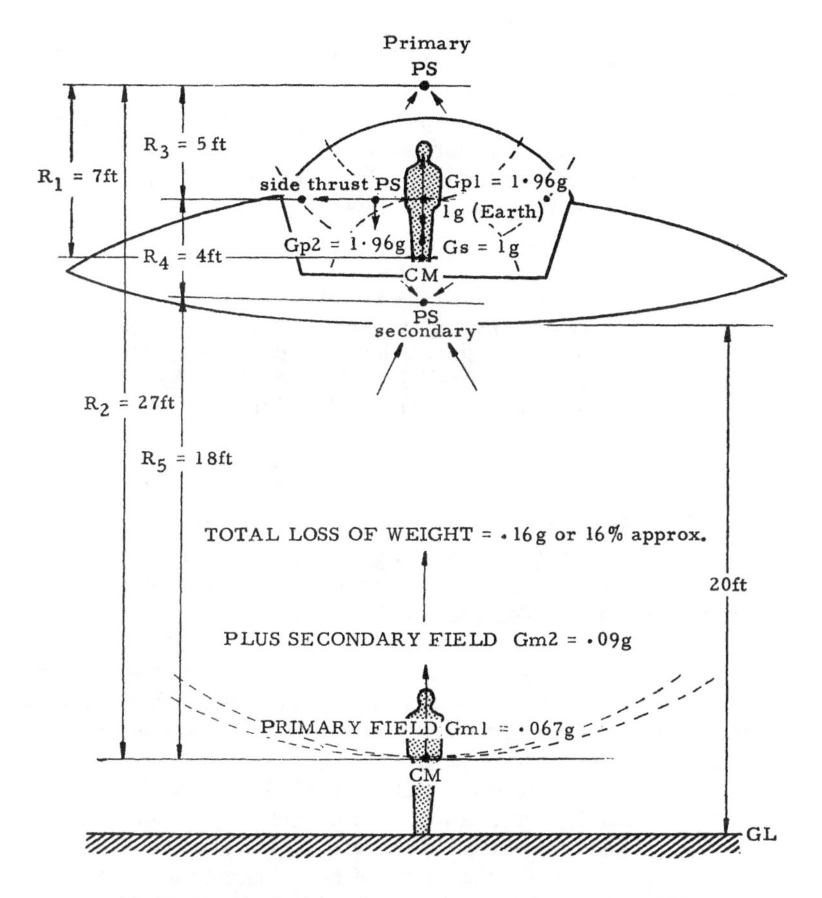

Fig 39. Combined effects of saucer primary and secondary fields immediately above a witness.

The centre of mass of the pilot is situated some 5ft below the FP (R_3), from whence we get

$$Gp_1 = \frac{R_1^2 \ Gs}{R_2^3}$$

Where Gp_1 = contra-gravitational acceleration of main field at C.M. of pilot.

and R_3 = distance from P.S. to C.M. of pilot in feet.

substituting again we get $Gp_1 = \dfrac{7 \times 7 \times 1}{5 \times 5} = 1.96g.$

In other words the pilot will be experiencing a vertical acceleration of 1.96g, from which we must subtract 1g due to the earth's field, leaving .96g contra-acceleration on the pilot.

Now the secondary P.S. is placed 4ft below the C.M. of the pilot (R_4) and in order to give him normal 'weight' of 1g, a total downward effect of 1.96g must be exerted, Gp_2. From which we derive the secondary field effect on the man standing below the craft and we have:

$$Gm = \frac{R_4^2 \ Gp_2}{R_5^2}$$

Where Gm_2 = contra-gravitational acceleration at C.M. of man.
Gp_2 = secondary gravitational acceleration at C.M. of pilot.
R_4 = distance from P.S. to C.M. of pilot in feet.
R_5 = distance from P.S. to C.M. of man in feet.

and substituting we get $Gm_2 = \dfrac{4 \times 4 \times 1.96}{18 \times 18} = .09.$

So we have contra-gravitational effect on the man due to the primary field=.067g, and contra-gravitational effect on the man due to the secondary field=.09g. Therefore the combined effect on him due to the primary and secondary fields is .16g.

In other words the man would experience a weight loss of 16%, which, assuming he weighs about 170lb, represents a loss of approximately 27lb. Now, as I pointed out, this is an extreme case with the man placed immediately beneath the point source. Standing to one side he would, in all probability be unaware of any change.

In chapter 9 I correlated the G field theory to Senhor Lustosa's sighting at Saude, north eastern Brazil, where a giant disc hovered over the sea. I have reconstructed this case to scale in Fig 40 and assumed hypothetical values for the length of the 'hanging thongs' and the focal length of the P.S. I have also assumed the machine was hovering at 1g and there would be local uprising currents of air.

Now the water immediately beneath the craft would experience a slight decrease in weight and would tend to be 'pushed up' by the weight of the surrounding mass of water, see experiment, Plate 22. This together with a local decrease in atmospheric pressure beneath the craft would combine to produce the effect described by the witnesses. Note also in this case the crew of the machine seemed to be having trouble of some kind. This may have caused an excessive amount of radiation, which together with high frequency eddy currents set up in the salt water, may have violently agitated the swelling mass, causing it to 'appear to boil'. Plate 23 shows a similar effect using iron filings in a high frequency, alternating, electro-magnetic field.

But the witnesses claimed to have seen appendages in the form of 'thongs' which they described as hanging down. UFO researchers will recognise this latter effect as being in the 'jelly fish' UFO category, in which witnesses have often described ribbon like appendages apparently hanging beneath the saucer, and the reader may ask 'If the machine was in the hovering or 'weightless' state, how could the thongs hang loosely downward, surely this is an inconsistency?' Indeed not, as a brief examination will show. First of all notice the witnesses mentioned the 'porthole windows' placed on the middle band, therefore in this instance the centre of mass of the crew would coincide with the centre of mass of the ship, that is if they were occupying the centre of the machine. But if the occupants were near the 'windows', if indeed they were windows, for this exercise we shall give this the benefit of the doubt, then they occupy a region which will experience *less* than 1g vertical force, as is required. So assuming R_1, R_2 and R_3 to equal 25, 37 and 49 metres

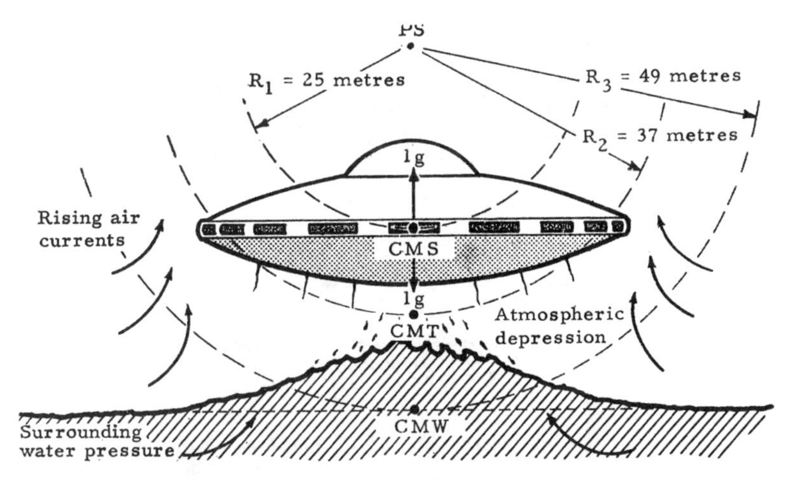

Fig 40. G field analysis of the Brazilian sighting at Saude in which the water would be 'sucked up' by the combined aerodynamic and G field effects. Note, the hanging 'thongs' are not inconsistent with the theory.

respectively as in Fig 40, then the acceleration acting on the centre of mass of the centre thong and therefore the crew situated at R_2 will be

$$GT = \frac{25 \times 25 \times 1}{37 \times 37} = .46g.$$

Therefore for every pound weight of the thongs, they will lose approximately one half. Which means they will still have positive weight and if they *were* pliable, would certainly hang downwards. In addition, there is good reason to believe that the hulls of the discs are highly electrostatically charged, which at close quarters would also thrust the thongs away from the underside. Note this static charge is of another function and bears no relationship to the much pursued idea that saucers are electro-statically repelled from the earth.

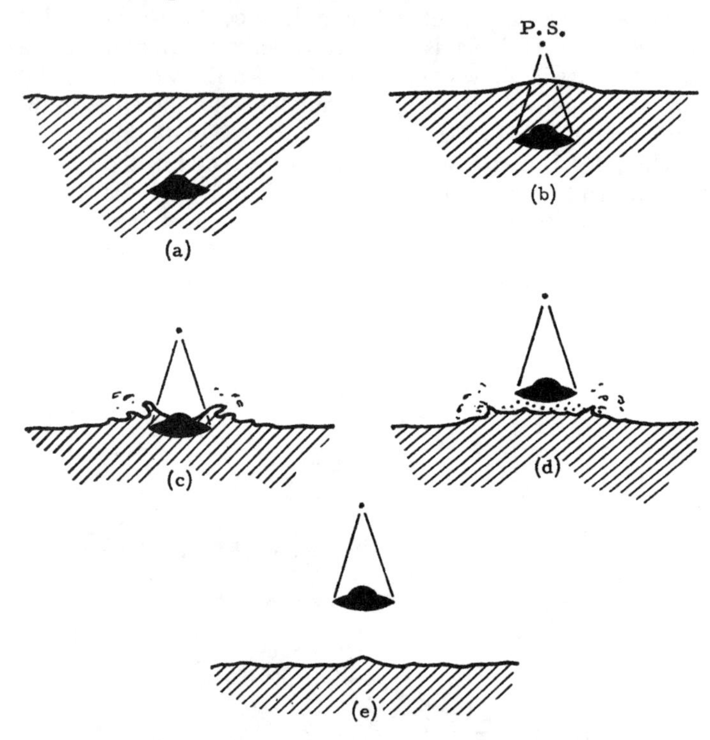

Fig 41. *Sequence sketches of G field craft leaving water.*

Now the G field effect on the surface of the water beneath the ship can be shown as

$$G_w = \frac{25 \times 25 \times 1}{49 \times 49} = .26g.$$

Which again means every pound of water will lose about a quarter of its weight. Over a large expanse of water this would amount to quite a large figure, certainly large enough to produce the effect mentioned by the witness of the Saude case.

In this respect we should also consider the bizarre case which occurred at Leba on the Baltic. Note, in describing this incident in chapter 8 Mr Kawecki said, 'About 300 metres from the shore, the surface was rising in one spot. It looked like a round hill'. Fig 41(a) and (b) At this stage we can imagine the point source projected above the waves by the submerged machine 'pulling' the waters up to meet it as the craft rose; it would be natural for the witness to interpret this as being 'pushed up from beneath'. Then splashes of water gushed from the top, lessening G effect as the point source rose higher, and displacement, Fig 41(c), 'and like fountain jets fell into the hole left in the waves'. Fig 41(d) and (e) Note once again, both these cases could be predicted by the G field theory and much information could be gained from other reports like them and perhaps even more would be available if there were more UFO researchers. In the Leba case for instance, the G field point source theory would predict that the water should be *heaped up* higher towards the hovering craft, as with the Saude case. Closer technical questioning of the witnesses involved in this incident might have revealed if this was so. This is where researchers can help to fill in the technical jig-saw, it makes fascinating searching.

Debris from the Skies

'What goes up must come down,' would be a handy maxim to introduce this aspect of the G field principle, and quite obviously, if ever we do succeed in building machines operated by such means, then no doubt someone will have to introduce a residential by-law which in effect says, 'All lift off refuse, to be dumped into parking orbit before touch down.'

The late Charles Fort and Dr M. K. Jessup have already made a more than adequate contribution concerning mystery falls from the sky, and I reproduce a few cases here only as typical examples. As to whether such phenomena can be definitely attributed to the theory being developed in these pages, must rest of course, mainly on speculation, for there may be all kinds of causes. But I cannot help feeling it explains some of it. My readers will be the best judges.

Earlier, I included the domestic cat as being taken aloft by the force field of a departing space craft. Now the well intended jibe has gone a little sour on me, for turning to my file of falling matter, ironically, the first case I should pull out was the following.

The Star, London, made a front page news story on 21 October, 1957 of a rabbit which had apparently dropped from the sky. First there was an aerial *explosion* over Dublin, Ireland. Soon after the explosion, heard over a wide area, Mrs Sally Moran found a rabbit near the front step of her home in Avoca Avenue, Blackrock, Dublin. It was still alive but badly injured in the legs. The amazing thing was that it was quite dry, although there had been heavy rain shortly before it was discovered.

Although regretfully sad, if associated with UFOs, I like to think that here was an accident. In writing this, I am of course aware of the more disquieting facts about falling debris. The acres of flesh-like material, blood-like stuff by the ton, dead birds, frogs by the thousand, fish and the rest which have been so adequately dealt with in other works. Such phenomena, although unpleasant, may not necessarily be sinister in origin and I propose to deal here with that part of it which might have some bearing on our case. There have been falls of various substances from clear skies almost since the days of recorded history. This whole book could easily be filled with accounts of such phenomena and I would emphasise, an investigation into UFOs is not complete without a perusal of the books of Charles Fort. There are accounts entered there, so numerous, so prodigious, as to leave the reader in not very much doubt that very strange things go on in the skies above us. Natural stones, stones with hieroglyphics, shaped stones, pig iron (manufactured), copper, sheets of beef . . . ice, lots and lots of ice, not just the common or garden hailstones mark you, but chunks of the stuff, and it still continues to fall. Ice from passing aircraft perhaps? Sometimes, but let us go back and listen to a few.

12 May, 1811. Derbyshire, England. Lumps of ice a foot in circumference.

1828. Candeish, India. A mass about a cubic yard.

15 June, 1829. Cazorta, Spain. Block weighing four and a half pounds.

22 May, 1851. Bangalore, India. Ice, size of pumpkins.

August 1857. Found in meadow after a storm by a Mr Warner of Cricklewood. 'Pure ice' weighing 25lb.

16 March, 1860. Upper Wasdale. Blocks so large that they looked like a flock of sheep, fell in a snowstorm.

11 July, 1864. Pontiac, Canada. Storm in which not hail fell, but ice half an inch to over two inches in diameter. But the most extraordinary thing about this was—a respectable farmer picked up a piece with a little green frog embedded in the middle.

June 1881. Iowa, U.S.A. Mass 21in. in circumference fell with hail.

August 1882. Near Solina, Kansas. Mass about 80lb, shopkeeper packed it in sawdust.

12 July, 1883. Chicago, U.S.A. Size of brick weighing 2lb.

Chunks of ice, a cubic yard of ice falling from the skies in the days before the advent of the aeroplane, were almost commonplace occurrences, unexplained by weather experts. Chunks of ice in which were embedded fish of all varieties and frogs—lots and lots of frogs!

But nothing has changed, we still have inexplicable falls of ice today, but of course we do have aircraft. And aircraft do sometimes collect ice on wing leading edges, tail surfaces etc. In fact, such a phenomenon was a hazard at one time, but it is very rare today. Just imagine the airline companies accepting such a risk to town folks if in fact falling ice formations were a common occurrence. Nevertheless ice continues to fall.

In April 1961 a Mr Charles Boseley, who lives at Windsor, declared that a chunk of ice 'the size of a football' crashed on to his roof. He

(*a*) (*b*)

Meadow grass in centrifuge test cell shows immediate response to applied negative g.

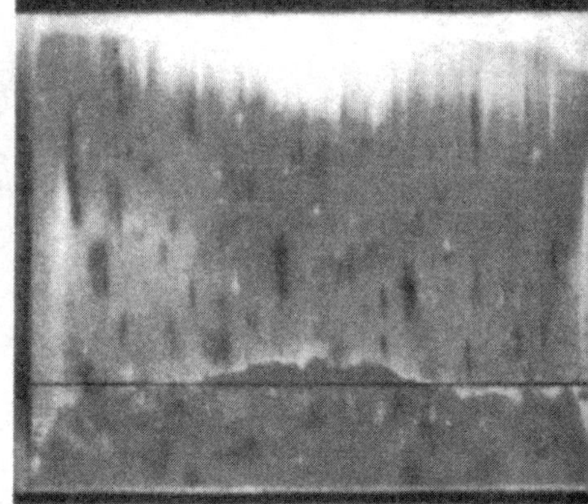

(*c*) *Plate* 18. (*d*)

A comparison of (*b*) *and* (*c*) *shows the moment in time just before a 'crater is formed'. The lineal velocity of the cell while these pictures were taken was in the neighbourhood of 20ft per sec.*

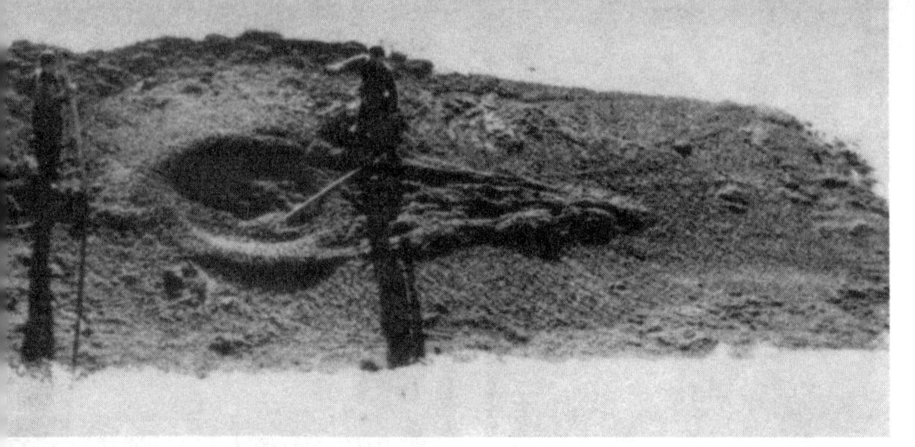

Plate 19. Tapering crater in iron filings left by an electro-magnet 'taking off forwards and upwards' as did departing UFO in the Poncey-Sur-L'Ignon case.

'Crater' left in iron filings by a vertically ascending electro-magnet.

Plate 20. Charlton crater. Note the central hole, main crater dotted and two of three indentations in the barley. Note also the ground was completely denuded of

said 'someone might have been killed'. We can understand the anxiety which prompted him to write to London Airport Commandant, John Warcup, to enquire, somewhat testily, if a plane had dropped it. Not that it will be of any consolation to Mr Boseley to learn that he is not alone in his dilemma, for there are others.

Mr J. E. Southern, writing from the White Cottage, Colston Rd., East Sheen, to *The Times* on 22 January, 1962, complained that there was no fund from which he could be compensated for the damage resulting from ice falling on to his property. He claimed that the ice had become detached from an airliner flying overhead. Mr Southern suggested that the Ministry of Aviation should set up a fund to indemnify householders and others who had suffered similar damage.

Although Mr Southern stated categorically that the ice had fallen from an airliner, we wonder whether he could, in fact, support this assertion. It sounds, indeed, as though he had placed too ready a reliance on this explanation for his misfortune. The ice weighed about 14lb.

Now it may be asked, is it possible for a large mass of ice to fall from a great height, at great speeds, without being vaporised through aerodynamic friction? The answer is yes, for it has been established quantitively, that a solid ball of ice, which might have entered the atmosphere at a speed of 6.9 miles per second (24,840 miles per hour), can slow down to terminal speeds without completely ablating, the terminal velocity of a falling body being that speed where the aerodynamic drag equals the downward force due to gravity. When these values become equal, the body then falls at a steady speed, which only varies a little as the plunging mass enters more dense layers of the atmosphere.

The steeper its angle of entry, so the greater are the drag stresses, but as with re-entering space-shot capsules, the smaller is the net heat influx. For example, it has been shown that a ball of ice about 2ft in diameter, entering the atmosphere at a tangential angle of 6 degrees would lose only half its radius on the way down. By the time it had descended to an altitude of 100,000ft, the speed would have decreased to something like 1,700 miles per hour, when the frictional heating would be negligible. Again a sphere of ice some 5ft in diameter, while falling under the same conditions, would lose about 8in, and the ablation would have ceased at an altitude of approximately 50,000ft, where at that point in its downward plunge, the ice would be travelling at 2,000 miles per hour.

From this it will be seen that chunks of ice, perhaps dropped from a UFO flying at great height and great speed, would still survive the journey down to earth, without being dissipated into steam and water. Though what would be left after impact, is anybody's guess!

What *did* happen to the potato crop on Mr Roy Blanchard's field at Charlton, what caused the crater there? Was it a lightning flash that denuded the ground, leaving not a trace, or were those plants hurled sky high along with the soil that nurtured them, by a force field propelling a space craft? The same kind of field we would expect to 'suck up water' as witnesses have vouched for.

Falling debris, including thousands of birds of all kinds of world wide species, have been neatly, too neatly, dropped in well defined shapes on the ground. Tons and tons of water and thousands of white frogs, frogs which have appeared to have been bred devoid of sunlight, have hurtled out of an otherwise clear sky. Such phenomena as this may have its explanation in other possibilities that some writers have suggested, viz. huge tanks being opened up somewhere aloft, for some unknown reason, by intelligences equally unknown to us.

Be this as it may, I feel equally certain that if a space craft powered by a gravitational field took aloft with it inadvertently-gathered large masses of water, then sooner or later it would be desirable to jettison this unwanted cargo and further, if this was done at very high altitude, then an equally large mass of ice would find its way earthward. In all probability, a pilot of a disc would do this at some convenient position, perhaps over the sea, but this might not always be possible. There would come the time when the other alternative might be unavoidable. Granted there might be others who don't care anyway, but ice still falls, as do the other various kinds of debris. But there is just one more technically corroborative point with which to close this section and I have included two of many appropriate sightings to substantiate it.

In Chapter 12 we saw how we might expect a gravitational device on occasions to take aloft with it all kinds of materials and we have correlated this to falling debris from the skies. We also saw how the glowing spheres which drifted outward from the discs photographed by Alex Birch, finally disintegrated and drifted down as they left the supporting regions of the craft's field.

Now the reader will appreciate, that in order to discharge unwanted debris or waste products from such a craft, be it hovering or in flight, then either the pilot must shut off the power (which admittedly doesn't seem very practical) or the unwanted cargo must be ejected out of the supporting field by force, presumably by an explosive of some kind.

Here then are the sightings which UFO students will know are typical of many.

21 June, 1947. Mr H. A. Dahl, the Captain and crew of U.S. Coastguards were patrolling the southern end of Puget Sound, Washington. They had put their launch into an eastern bay off a very thinly populated island, Maury Island, about three miles from the mainland.

Also on board were two members of the crew and Dahl's fifteen year old son with his dog. Mr Dahl had looked up from the wheel and was startled to see 'six very large doughnut-shaped machines' almost directly overhead. The strange craft were stationary and silent.

At first he thought they were a kind of balloon until the others began to circle round one machine which began descending rapidly as if in trouble. Keeping about 200 feet above it the others followed it down until it came to rest about 500 feet above the watching crew.

The things were silent and reflected light from their metallic surfaces. All of them had what appeared to be large portholes spaced round the hull.

Later Mr Dahl said 'Fearing that the central and lowermost machine was going to crash in the bay, we pulled our boat over to the beach and got out our harbour patrol camera. I took four photos of these balloons, as I still thought they were. All the time, the five were circling round the one which was stationary. Five minutes passed, and then one of the circling machines detached itself from the formation and came right down to the stationary one. It seemed to touch it, and stayed motionless for about four minutes. Then we heard a *dull thud* and the central craft spewed out what looked like thousands of newspapers from the inside of its centre! But these falling fragments turned out to be a white type of very light metal that fluttered to earth, and also fell into the bay. The machine then seemed to hail on us in the bay, and over the beach, black and darker types of metals, which hit the beach and the bay. All these latter fragments seemed molten. Steam rose when they hit the water. We ran for shelter under a cliff and got behind logs. My son's arm was hit by a falling fragment of metal, and our dog was killed. Then the rain of metal stopped. The strange craft silently lifted and went westward toward the Pacific. All the time, the centre one remained in the formation. We found the fallen metal too hot to touch, for some time. But when it cooled, we loaded a large number of pieces into our launch.'

We cannot assume with any certainty of course that the dull report was in fact evidence of a means to eject the falling material, but there is good reason to accept this as such when there are other cases like it, the following is another example. In this case UFO students will recognise the falling material as what is commonly called 'Angel's hair' frequently seen falling from the sky after UFO activity and about which we shall hear more later on.

13 October, 1957. Graulhet (Tarn) 35 miles east north east. of Toulouse. There were many witnesses to this report, the following is an extract. M Carcenac, a tanner of Graulhet. He said, 'At 4.30 p.m. I noticed at a high altitude toward the north west, moving southward at full speed, a white object which seemed to have a curious shape. I first thought it must be a jet plane of an unfamiliar type. Not making out any vapour trail, I went and got my opera glasses. I could then see very

distinctly a sort of huge, flexible, soft disc, white, which was swaying as it moved along at tremendous speed'.

'I had been following the bizarre craft for several seconds when it exploded in full flight. At the same time a circular object, very much smaller and silvery, seemed to spurt out from the mass and continued straight toward the south, where it soon disappeared, while the burst fragments of the soft disc scattered out through the sky in a multitude of shapeless fragments which began to fall gently like shreds of cloth or paper.'

Everyone who had seen the explosion rushed toward the place above which it had happened; they were able to watch the debris settle on the ground, sometimes it caught on trees or telegraph wires. Many witnesses picked up fragments of the material which resembled 'silvery filaments clinging together like cobwebs' which wilted away when handled. Samples were taken to the police, and a chemist in Graulhet tried to analyse it, but without success. As always, in the heat, the strange material evaporated without leaving a trace. When brought near a flame, it 'disappeared almost instantaneously and produced neither fire nor smoke'.

13

Analysis Three

Radar and Optical Effects of the G Field

SEVERAL cases where UFOs have been observed visually, yet failed to be monitored on radar screens have already been quoted, and we saw how on other occasions the objects *have* been detected by radar. In fact there have been so many unexplained targets located by radar that they have a special name allocated to them by radar crews. Known as ' Radar Angels ', some of these targets have been satisfactorily explained, many have not.

During the 1952 wave of UFO sightings in the United States, many radar reports were made by ground and air crews, to which the reader is referred. Some of the effects were straightforward, others were more difficult to explain. One of the first of more than a dozen simultaneous radar-visual sightings occurred at midnight, 19 June, at Goose Bay AFB, Labrador. A red light, which changed to white, hovered for a little, oscillated and sped away. The strange thing about it was, when the light oscillated, the image on the radar scope *flared up*, returned to its original size and disappeared.

July 1952, Washington, D.C. CAA officials asked an Eastern Airlines pilot to check a radar target, he reported back that there was nothing to be seen. The radar operators confirmed that the UFO vanished off the screen the moment the aircraft entered the area, 'then came back *behind* him'.

December 1956, The Far East. The report states: Two USAF jet pilots were practising ground radar positional intercepts on each other in the vicinity of . . . , when during one run, the intercepting pilot picked up a large unexplained radar blip, at about 20 miles range. 'The pilot called the Ground Control Intercept site to ask if they had a target which would correspond to the unidentified blip. After receiving an answer in the negative, he asked for and received permission to determine the nature of the source of the radar return.'

The pilot then closed in at over 700 m.p.h. and at eight miles range, a round-shaped object became visible exactly where radar had placed it. He estimated that the UFO was as large as a B-29 bomber, or 'the size of a lead pencil eraser if placed against the windscreen'. Experts used figures supplied by North American Aviation to compute the UFO to be about 350ft. The pilot was able to get a radar 'lock-on' (a device which automatically guides an aircraft towards the source of the echo). But as he closed in 'his radar was *suddenly jammed* by a strong interference'. He then used anti-jam procedure and switched frequency. This eliminated the mysterious interference for ten seconds, then it began again.

Even so, the emanating pulses were not strong enough to break the radar lock-on and the jet held its course. The pilot closed within five nautical miles of the UFO, but thereafter could close no further. As it banked, the pilot thought the UFO had a circular shape and was golden in colour. Radar speed check indicated a departure speed of up to 2,000 m.p.h. An official Air Force report states that the pilot had many hours of flying time, was conscientious and had 'reported the incident in a straightforward, slightly embarrassed manner, saying that he would doubt the possibility of such an occurrence if it hadn't actually happened to him'.

Now radio jamming is a comparatively easy thing to understand, but having talked to radar experts on the problem, it seems that radar jamming is not. And, of course, any radar engineer could take himself right to the top of government-sponsored posts if he had the remotest idea of what lay behind such phenomena.

A radar pulse is an electro-magnetic wave and it is logical to predict that UFOs might play similar tricks with light itself, in fact we saw earlier that some sightings bore evidence of this. There are two main effects which a powerful gravitational field may have on light. These are one, the direct gravitational effect and two, the refractive effect produced in the atmosphere by the field. Let us examine them in this order.

To begin with, it is known that a light ray is made up of photons which have a rest mass of zero. But at the velocity of light they have a measurable mass and should therefore react to a gravitational field. This was predicted quantitively by Einstein and was established during the 1919 Solar Eclipse, though not all results are satisfactory as measurements are very difficult to take.

Fig 42 shows the effect more graphically, where the light from a distant star passing close to the sun is deflected or bent by its intense gravitational field.

For the sun the angle of deflection $= \dfrac{1.7 \text{ secs. of arc}}{\Delta}$ where Δ is the number of sun's radii that the ray of light is situated from the sun's centre. Therefore we can say that, while an intense gravitational point source, produced by the craft, might *deflect light*, it would not appear to be of any significant degree.

Now it follows that if a ray of light *is* deflected by a strong gravitational field, then in the case of a luminous body such as the sun, light must do work against the gravitational field in order to escape from it. Indeed, this was proved to be so. The effect is explained in the following general terms. A quantum of radiation of a particular wave-length emitted outwards at the sun's surface starts off with a certain amount of energy, part of which it gives up in escaping from the gravitational field of the sun. In consequence, according to physical principles, the wave-length is somewhat increased. Taking all such quanta into account there is a

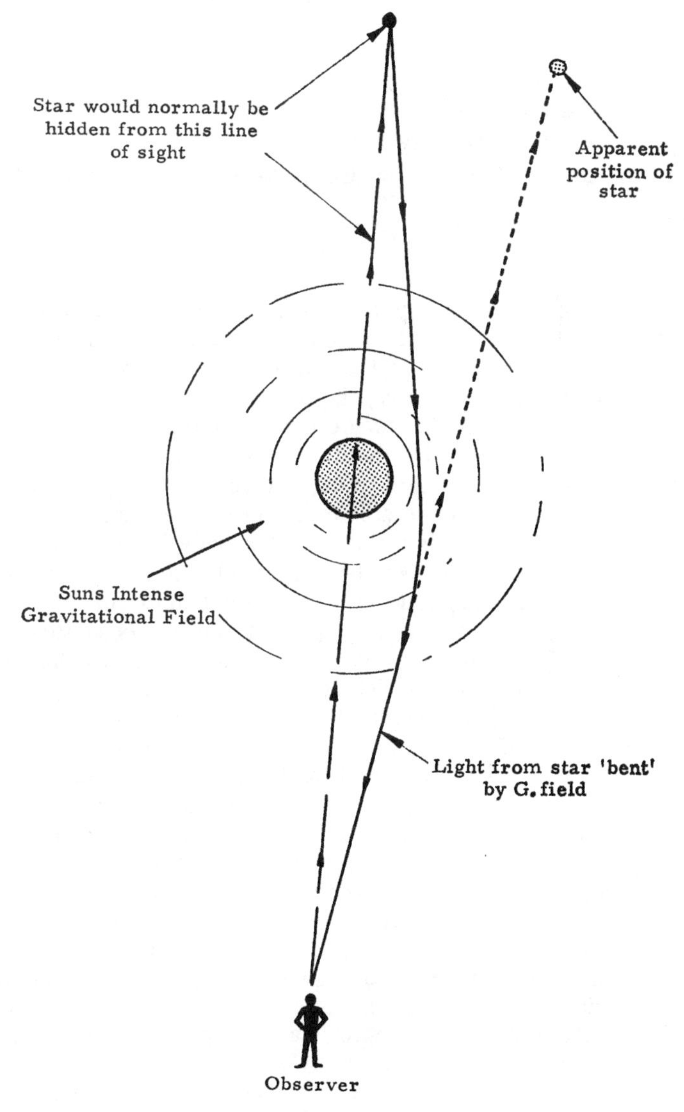

Fig 42. Light is attracted towards a strong gravitational field.

displacement of all parts of the spectrum, including absorption lines towards the red end. The magnitude of this displacement varies directly with the sun's mass and inversely at the sun's radius, and is given by

$$\text{Frequency displacement} = \frac{Vo - V}{V} = \frac{KM}{C^2R}$$

Where K=Newton's gravitational constant.
M=Mass of star.
R=Radius of star.
Vo − V=Frequency of change.
V=Frequency.
C=Velocity of light.

In the case of the sun the displacement towards the red end of the spectrum would amount to $\frac{2}{10^6}\lambda$

Where λ = Wavelength.

Because of the relatively large size of the sun the displacement is hardly detectable. On the other hand, the Companion of Sirius with practically the same mass as the sun, but with a radius only one thirty-sixth part of the solar radius, the displacement of any line is 36 times larger than for the sun. This was observed by Adams in 1924. Note: This 'red shift' must not be confused with the velocity of recession of the remote galaxies which is said to account for the theory of the expanding universe, it is due simply to gravitational displacement.

Now, according to this reckoning, if a body has a great mass and a small radius, then there will be a marked shift of spectral omissions towards the red end. Later we shall see that during the process of generating a G field, accompanying radiations will be emitted, including visible light, which therefore will also be subject to this shift, but if the spectrum given out by the field were continuous, it would appear the same no matter what strength the field may be. For all the radiation from it would suffer the same displacement, i.e. the invisible ultra-violet would move into the violet, and at the other end of the spectrum, red would move into the infra-red and therefore become invisible. In a word, if there were no distinguishing marks in the spectrum, such a shift might be impossible to detect; the operative word here being 'if'.

So much for the direct gravitational effects on light in the presence of the craft, but we shall now see there might be an even greater contributing cause of some optical phenomena, due also to the craft's G field, and even an elementary understanding of simple optics will explain it.

In the first place it should be understood that not all of the light from a given source is completely reflected at an angle, according to the incident rays as in Fig 43(a), as shown by the fact that we can still see the illuminated surface of most materials when viewed from position 1. This is due to the scattering of light by the minute ridges and indentations of the illuminated surface shown enlarged in Fig 43(b). This scattering decreases as the surface becomes more polished, as with a mirror.

From a myriad of reflected rays of light radiating as they are in all conceivable directions and due also to the natural cone of vision, the eye is automatically furnished with a multiplicity of rays perfectly focused to it as in Fig 43(c), all other rays go on past. But to every ray there are countless other rays running exactly parallel to it; in fact, exactly like the creative rays in the *Unity of Creation Theory*. Now a ray of light is said to be refracted when leaving a medium of a particular density and

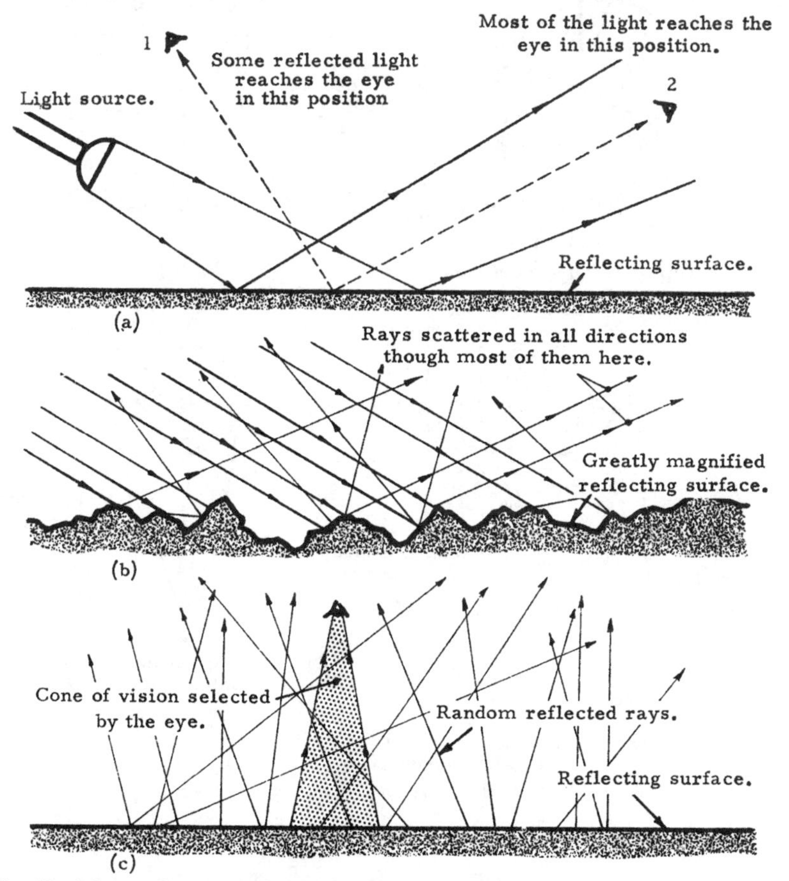

Fig 43. Most reflecting surfaces are rough in texture, a truly perfectly polished surface would reflect light in one direction only.

entering another of different density at an angle to what is known as the normal line. This is shown in Fig 44(a) and is represented by a line drawn at right angles to the point of entry of the incident ray of light. The rule in optics governing this is, 'When a ray of light leaving a less dense medium (say, air) and, at an angle to the normal, enters a more dense medium (glass), it tends to move *towards* the normal line. But when a ray leaves a denser medium for a less dense medium, and at an angle to *its* normal line, then it tends to move *away* from it'. This holds true for a prism, lenses, water, air, etc.

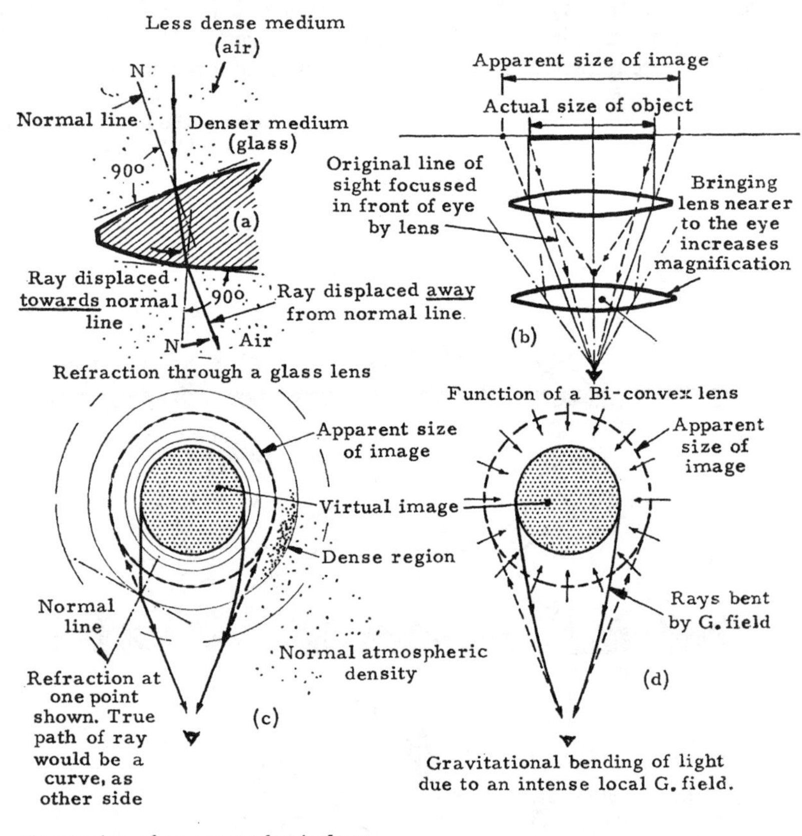

Fig 44. Some possible optical effects caused by a powerful G. field.

Therefore if we interrupt the cone of vision with a bi-convex lens, as in Fig 44(b), we find that the original rays are brought to a focus in front of the eye and therefore never reach it. But we might now find that the parallel rays leaving the object are brought to a focus by the lens and directed to the eye instead. Continuing this line of sight, or cone of vision, our eyes tell us that the object is now much larger. Bring the lens nearer to the eye still and the magnification is increased further. A spherical gold fish bowl is such a lens of poor efficiency, but the fish within it are convincingly enlarged nevertheless. In a similar manner, an intense gravitational field point source would, locally, condense the air, producing an atmospheric lens as it were, which, augmented to some small degree by the field effect, would conceivably produce optical phenomena in the vicinity of a saucer, Fig 44(c) and (d). But, as we have already seen, such an increase in atmospheric density below the dew point would almost certainly produce fog. Therefore we can say, in the vicinity of a G field type craft, we could expect to find fog and other optical effects.

Many and varied might be the visual effects of the field of such craft, but of these, one in particular stands out specially as a possible means of UFO detection, it involves the use of plane polarised light.

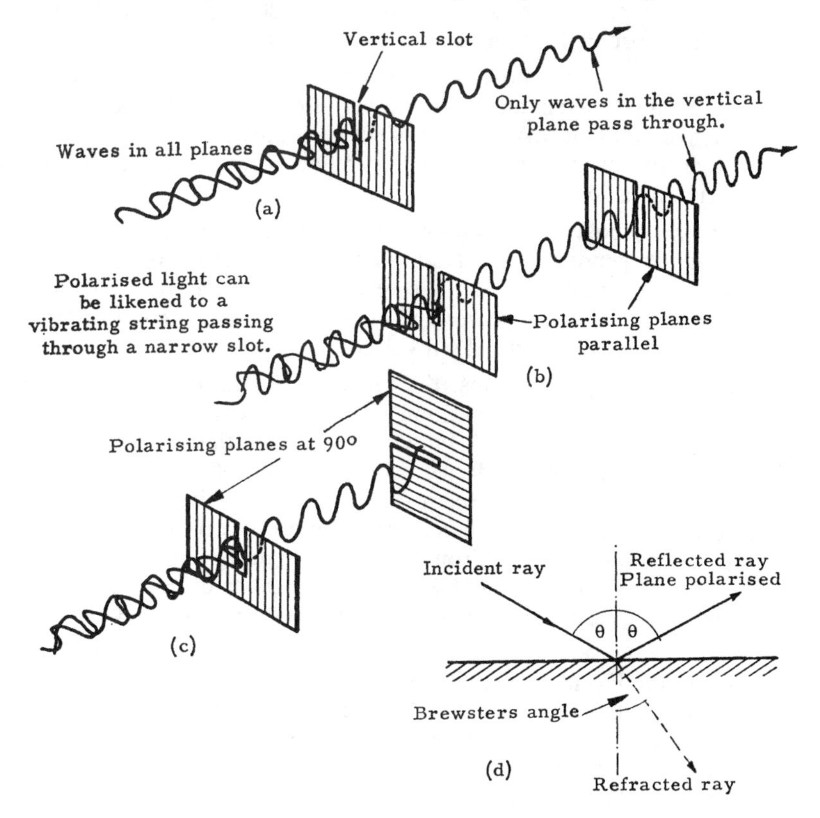

Fig 45. Polarisation of light.

A beam of ordinary light is said to have transverse wave motion in all planes, while a polarised beam of light is said to be oscillating in only one plane. If a fixed string along which waves were passing was led through a narrow vertical gap in a piece of card, the only motion which the string could have in the gap would be vertical, so that no matter in what direction or plane the string was vibrating *before* the card, on the other side it must necessarily be polarised in a vertical plane. Fig 45(a). In exactly the same manner, light which is normally oscillating in all planes, is plane polarised by entering a sheet of polaroid material. So that any light waves leaving the substance are vibrating in the same plane. If two sheets of the polaroid are placed one in front of the other with their polarising planes parallel, then it follows that a beam of plane polarised light leaving the first screen will be uninterrupted by the second screen, as in Fig 45(b), but the light would be stopped by the second screen if this should be rotated ninety degrees, as in Fig 45(c). In exactly the same manner, in fact, as the string vibrations would be stopped if these were led into two gratings orientated at right angles to each other.

Now light is also polarised when reflected from different substances at a certain critical angle known as Brewster's angle, so that if the refractive index of, say, a piece of glass is known, it can be so arranged for light to fall upon it at Brewster's angle, when the reflected beam will be plane polarised. Fig 45(d).

Similarly, polarisation occurs when light is scattered from tiny dust particles suspended in the atmosphere, or even from the air and water molecules themselves. For this reason, light received from the sky is partially polarised, the degree and direction of the polarisation depending largely on the part of the sky involved.

Under certain conditions, the plane of polarisation of a light beam can be rotated by the application of magnetic and electric fields. Three well-known effects are the Faraday effect and the Kerr magneto-optic effect, in which light reflected from the pole of an electro-magnet is polarised, and the Kerr electro-optic effect in which light traversing a dielectric medium, which is saturated by an electric field, has its plane of polarisation rotated. From which it is logical to assume that in some circumstances polarised light would have its plane of polarisation rotated by the strong associating magnetic and electric fields of a powerful G field.

A more convenient way to visualise this might be to consider for a moment the Faraday effect illustrated in Fig 46(a), in which the plane of a beam of polarised light is rotated when passed through a block of 'heavy glass' situated between the poles of an electro-magnet. Which means that if the beam was fairly wide, most of the light would be free of the magnetic rotation and could therefore pass unrestricted through a second screen, whose plane was set parallel to the first. But the portion of the light beam closest to the pole pieces would be rotated and therefore could not pass through the second filter, where it would show up as a shadow. If instead of one electro-magnet there are perhaps eight arranged in a circle each fitted with 'heavy glass' through which the light is passed, then there will be eight accompanying shadows also arranged in a circle, Fig 46(b). From which it will be seen that if all the spaces between the magnets are filled with more magnets, so as to form something like a solenoid or ring magnet, then the resulting shadow would take on near-ring form, as in Fig 46(c). Note, this effect is obtained when the line of sight is coaxial through the 'core' of the magnets.

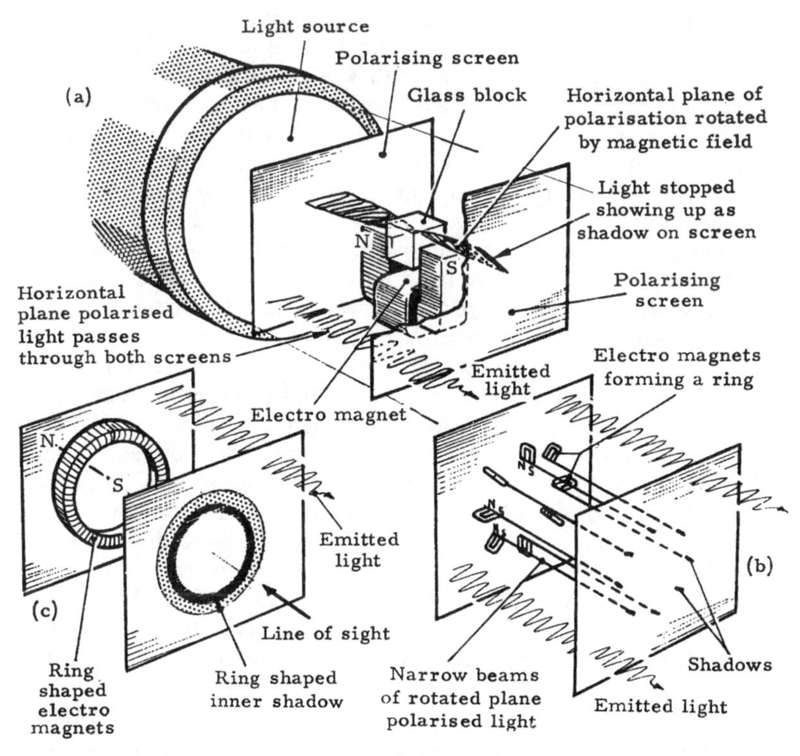

Fig 46. Illumination of a Magnetic Field due to rotation of the plane of polarisation.

Now this is very interesting, for if some parts of the sky are plane polarised and we are right in our assumption that intense electro-magnetic and electric fields accompany a saucer's G field, then we could logically assume also that certain areas of polarised sky light passing close to or through such a field would be rotated when viewed from below the UFO, and furthermore, should UFO spotters be wearing Polaroid screen type sun glasses, then they might just be lucky enough to see such an effect. Would the sceptic admit to the technically corroborative value of such a case, I wonder? You would say he would have to? Let us see.

Mr W. A. Webb, author of *Mars, the New Frontier*, is a research chemist of repute and considerable achievement. Speaking of a UFO sighting on 5 May, 1953 he said:

'It was a clear sunny morning, between 9.45 and 10 a.m. The author was standing in a field near the Vacuum Cooling Company plant, not far from Spain Flying Field, and about a mile north of the Yuma Air Force Fighter Base. His attention was drawn by the buzzing of jet fighters taking off in quick succession, passing directly overhead travelling northward. As he scanned the northern sky, his attention became fixed upon what first appeared to be a small white cloud, the only one in the sky at that time. He was wearing Polaroid glasses having a greenish tint, and as was his custom when studying clouds, he took the glasses off and put them on at intervals to compare the effect with and without Polaroid. The object was approximately oblong with the long axis in a horizontal plane. It floated at an elevation of about forty-five degrees. During the course of about five minutes the object travelled approximately thirty degrees towards the east. Then it appeared abruptly to turn and travel northward; at the same time its oblong shape changed to circular section. As a circular object it rapidly became smaller, as if receding. While receding, the object did not noticeably lose any of its brightness. In about thirty seconds of this its diameter became too small for him to hold in his vision. During the first period Mr Webb had not noticed a change in the oblong nor in the field of view about it, as a result of putting on and taking off his Polaroid glasses. But during the second period several uniformly spaced concentric circles appeared around the now circular object. The circles were distinct dark bands which enveloped the silvery disc. The largest of these circles was, perhaps, six times the diameter of the central disc. When Mr Webb removed his polarising glasses, the silvery disc remained, but the concentric rings vanished. When the glasses were put on again, the rings reappeared. He repeated this several times, each time with the same result. The rings, with glasses on, faded to invisibility before the disc became too small to see.'

Notice the witness did not see the ring effect when the object was oblong, that is when viewed sideways or on edge, but then he says: 'Then it appeared abruptly to turn and travel northward, at the same time its oblong shape changed to circular section', tilt for forward flight *away* from the witness, for he goes on to say, 'As a circular object it rapidly became smaller *as if receding*'. Therefore he was viewing the object very nearly co-axial with the propulsive field, in fact in the very attitude we would expect a polarisation effect to be seen, for he says, 'But during the second period several uniformly spaced concentric circles appeared around the *now circular* object'. The fact that the witness saw several bands is of course extremely interesting to the researcher, but the effect is no doubt similar.

Now did you notice that the witness was viewing the *northern* sky at the time? In the morning, oh, and did I tell you what particular part of the sky light is plane polarised . . . I did not? Well, it happens to be at right angles to the sun; in the morning that would be due south . . . or *north*!

The pattern continues. Here and there, taken from a host of seemingly inconsistent reports, comes more information and slowly the bits of the jig-saw puzzle continue to fall into place, but no ordinary jig-saw puzzle is this, but one complicated and disguised by many different aspects, which certainly would never fit so completely by mere chance alone. For us the game gets more exciting, let us look round for a few more clues. Well, we have discussed several UFO optical effects, and there are others. Let us see if some of these can also be explained by the G field theory. Can we, for instance, reconcile the mother of pearl cloud-like saucer with the blindingly flashing type, the shiny metallic disc, and the dark silhouette UFOs that have all been seen under the same daylight conditions? And further, can we relate such presentations with the red, orange, green and other coloured objects seen at night? I would stress that I am concerned here solely with the optical effect, not so much the initial cause, which we will analyse later on. Yes, I think we can explain these effects as being in part what we might expect of such devices, and far from being inconsistencies in the flying saucer evidence, these contributions help to consolidate it. Let us examine some of these visual effects in the broad terms of the G field idea.

First the 'mother of pearl', or milky-white, UFO seen by day. As we have seen, it is logical to accept that this effect is caused by the 'private cloud' produced by the field, a collection of 'angel hair', or both, acting like a ball of opaque glass which reflects the sun's rays. If the atmospheric conditions were not suitable for the formation of either of these phenomena, then we would see the disc itself. If the field intensity were low, then the polished surface of the machine would be seen, as with conventional aircraft. But should the field intensity be very high, then the combined atmospheric, gravitational and electro-dynamic effects of the field would produce dazzling displays of optical phenomena. To such an extent that the associated craft would appear to 'flash and pulsate'. This fact was brought home dramatically to the author when at the time of writing I had cause to call my young son David to tidy up his room. Traipsing through our little wood on that bright spring morning, I was met by a blinding flash of light from several hundred yards away in the adjacent corn field. Truly this seemed to be more brilliant and more blinding than the sun itself. In fact, the source was only a two-foot mirror David had found and was using as a 'searchlight'. I could not help but imagine a far more pronounced effect created by a device several hundred square feet in area, at about half-a-mile range.

To elaborate a little further, the reflective power of a surface is purely a function of its smoothness and, as we have seen, even a polished piece of glass has minute irregularities when microscopically examined. When we polish the surface still more, we reduce the irregularities further, but there is a limit to the degree of perfection in smoothness which can be mechanically obtained. But what if the rays of light round a body could be brought to a focus and reflected by an intense *field*? Such a surface might represent a mirror of infinite reflecting power, indeed beyond this stage, the very body which was producing the effect might even appear to vanish.

On the other hand, should the machine be flying fairly low and at low field intensity, and at the correct angle with the right light conditions, we would see just a shiny metal surface or even a dull grey shape, again as in the case of a conventional aircraft. Even so, saucers do glow and probably there is an emission of light at all times, depending on the power used, even when they do appear dull. There is a perfectly simple explanation which I would like to illustrate. We all know that a poker when heated to a bright cherry red will look quite dull when seen in an outdoors light. Now obviously this is not simply because it is chilled, but rather a function of the ambient light, and I should like to describe an experiment which was made to illustrate this point.

An electric bowl heater was hung up to simulate a saucer emitting a red light. This was photographed outdoors at night, suspended in the air as shown in Plate 24(a). The same heater, still switched on, was photographed by day, at the same angle, and, as will be seen, it simply shows a dark silhouette against the sky, Plate 24(b). The same can be done with a white light, which might be dazzling to the eyes in darkness, but appears considerably subdued when seen against a bright sky.

Should a craft be moving with high field intensity, then it may well generate the private cloud in addition to the emission of yellow or white light, which reflecting on the mantle, may again amplify the general effect to one of dazzling brilliance. In addition, the prevailing conditions of the sky itself contribute to the visual effect of some UFOs, see Fig 47, while it is well known that the eye is not equally sensitive to different wave lengths in the visible spectrum. For instance, under identical conditions of brightness, a yellow-green light radiating at 5,550 Angstroms in daylight would appear brightest to most people. But at dusk or poor light the eye becomes more sensitive to green or blue-green in the neighbourhood of 5,050 Angstroms. All these factors, and many more, help to produce the diversity of witnesses' descriptions, as to the varieties of shapes and colours of UFOs.

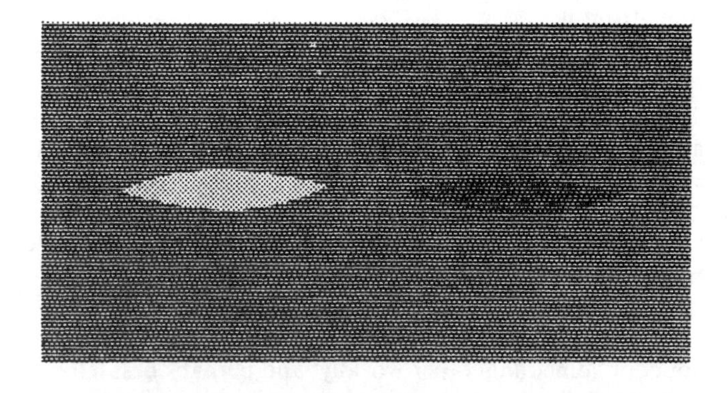

Fig 47. In addition to field effects, various brilliances or silhouettes are exaggerated by prevailing atmospheric conditions.

Electro-magnetic Radiation Effects

Paramount among the annals of flying saucer reports, perhaps more than anything else is evidence of electro-magnetic radiation. By this I mean radiations of varying frequencies, both visible and either end of the spectrum.

In this section we continue the analysis and evaluate corresponding sightings by correlating some of these effects.

To begin with, let us take another look at the extraordinary affair of the Brazilian Fortress, Itaipu. You will remember the object appeared at 2 a.m. in the morning of 4 November, 1957 as a bright star in an otherwise clear sky over the Atlantic Ocean. The approach speed was 'so tremendous that the two soldiers forgot their patrol just to observe it'.

Then the UFO had stopped abruptly and drifted slowly down, emitting a strong orange glow. It then hovered motionless at approximately 120-180ft above the cannon turrets. It had appeared silent at first, but now at closer range, the watching men could hear a distinct humming noise coming from it.

The thing had remained thus for about one minute, and, quote, 'Then came the nightmare. . . .'

The sentries felt something hot touch their faces and one thought he heard a whining sound. Next came the intolerable wave of heat 'like a fire all over his clothes'. Shortly after the ensuing commotion, the entire electric system collapsed, including: (a) The Fortress lighting system. (b) Heavy equipment power supply. (c) Independent generating sets. (d) Intercommunications system. The electric alarm clocks, which had been set to ring at 5 a.m., now began ringing; the time was 2.03 a.m.

Now, to the layman, these effects may sound strange, bizarre and chaotic, or something 'other worldly' and perhaps best left alone. But to the engineer, the events in this affair make sense, for we can reasonably assume that either an immensely powerful radio wave of very high frequency was superimposed on the electrical circuits, causing the alarm clock motors, which operate at mains frequency, to speed up, in which case we can calculate the imposed frequency, or, and more probable, the alarm mechanism relays were closed by a powerful superimposed induced magnetic field in the coils. Whichever way you look at it, this affair really did happen and once more there is some pretty convincing evidence of a kind scientists should consider.

Now this is not an isolated case where UFOs have interfered with mains electricity supply, there are others, the following is typical. 14 July, 1959, Salisbury, N.C., United States. A loud oscillating noise was heard when a circular UFO was observed. Television sets were blacked out and house lights reported out of action.

But one of the most thought provoking and one of the best authenticated, yet least known, of such incidents took place on 26 June, 1965. On that evening a round luminous object was observed for several minutes over Washington, D.C. It was described as 'like a yellow grapefruit', which left behind a luminous trail. The object, moving in a rather erratic fashion, approached Washington National Airport and hovered over it, then oscillated and departed at some speed. As the disc approached, the *ceiling lights* of the airport went out; when the thing departed, they returned to normal operation. Searchlights were trained on the UFO, but when they caught it in the beam, the searchlight went out!

Here is another case where electric clocks failed. 5 November, 1957, Philadelphia, Penna. A milkman reported a flaming disc. Apartment lights went dead and electric clocks stopped.

Orogrande, New Mexico, 4 November, 1957. James Stokes, a long-term Navy veteran and a high-altitude missile engineer at Hollman AFS, was driving southwards from Alamogordo to El Paso (100 miles south-

south-east of the A-bomb site), he was eight miles beyond the town of Orogrande when his car radio began to fade, then died, and his motor quit. People in others cars ahead of him, apparently similarly afflicted, were getting out and pointing north-east. Stokes saw a 'mother of pearl coloured', egg-shaped object approaching rapidly in a shallow dive. It did a sharp right turn, then 'made a pass' at the highway to the north, disappearing in that direction. It reappeared two or three minutes later, crossing the highway again, swerved north-west, and went off towards the San Andres mountains. All the cars could now be started. Stokes estimated the object's speed at almost 2,000 m.p.h. He felt a heat wave as it passed, and found he had a slight 'sunburn'.

Immediately after Stokes reported the experience, the Air Force appeared to accept it; but ten days later they called his story a hoax.

6 November, 1957. Mr Jacobsen and three friends at a hunting lodge on Lake Baskatong, about 100 miles north of Ottawa, were listening to a battery-powered radio, when one of the party, returning from the out-house at about 9 p.m., excitedly reported the presence of a flying saucer. The others laughed—until they saw it themselves.

To the east of south a brilliantly luminous yellowish-white sphere, dazzling to the eye as a gas mantle, hung a few hundred feet over the summit of a hill two miles away. The sky was completely overcast; the object was beneath the cloud deck. From top and bottom of the sphere there fanned out conical beams of light, illuminating the trees on the hilltop and the undersurface of the clouds. The thing hung motionless without change in appearance.

Then the hunters discovered that there was no longer any reception on the radio. One of them, a professional electronics worker, had a portable short-wave radio; this, too, was found to be useless; not even the Government time signal, which invariably came in clearly, could be picked up. But, at one short-wave frequency, a *very strong signal was received.* It was a rapidly modulated single tone, somewhat like Morse code in effect —but it was not Morse, which would have been recognised immediately by the two radio hams present. The signal was so strong that it 'blocked up' the receiver—an effect well known to those who live near radio transmitting stations.

After about fifteen minutes the object began to rise and move slowly off to the south entering the clouds, by 9.30 it was out of sight. The radios then proved to work normally.

6 November, 1957. This time in the Eglinton-Dufferin (Western) section of Toronto, Ontario. Six persons had seen a yellowish white light in the eastern sky (over Lake Ontario), travelling from south to north. It was no meteor, for it travelled so slowly that there was time for Mr S. Beaumont to see it after being called outside by his neighbours, but just before they called him, he had noticed the sound on his television had

suddenly been overriden by 'an interfering noise that sounded like a car engine going very fast'.

12 November, 1957. Over Hazleton, Pennsylvania. 'Something resembling a triangle, with a light at each corner', hovered at three different times during the evening of 12 November. On 15 November, near the same city, a 'flame-coloured star' travelling rapidly south, in a rainy and foggy sky, was seen by two truck drivers. Shortly afterwards a Quakake farmer telephoned the police to report an object 'about 35ft long, spinning around and around', hovering in the air near Greenacre. At least ten other people reported other objects near Hazleton on the 14th and 15th, including a city resident who insisted that her TV reception was impaired on the night of the 14th by a helicopter-like object which emitted a 'velvety sound'.

And on 14 November, at Tamaroa, in Southern Illinois. Mrs John Riead, wife of Tamaroa's Justice of the Peace, said she heard a 'spluttering noise—like someone pulling into the driveway'. Then there appeared above the trees a bright, moon-shaped object which had 'a sort of tail or ray extending down from it as it moved along'. The object was accompanied by five or six loud booms and three brilliant flashes. Then the lights in her home went out. Electric power in a four-mile area between Tamaroa and Dubois was interrupted for about ten minutes. Service was restored when workmen closed an open circuit breaker; investigating crews could find no apparent cause.

Mr Richard Taylor said he counted the lights going off and on five times in his office.

For effects on car headlamps there is an abundance. Space restrains us to the use of a condensed few.

11 October, 1957. Fronfrède (Loire). M Baptiste Jourdy, aged 30, milk truck driver, crossing mountains south of St-Etienne. Suddenly his engine died and headlights went out. He stopped instinctively and got out to inspect the ignition.

Overhead, flying under the clouds and at right angles to the road, was a glowing multi-coloured object. It crossed the road in front of him and went into the distance at high speed; he watched for a minute or two, then, recovering from his amazement, he saw that his headlights were shining again. He climbed back into the truck and tried the starter; the engine turned over as usual.

Same time, 150 miles further north of Clamecy (Nièvre). M Henri Gallois and Louis Vigneron, grain merchants of Clamecy, on their way to the fair at Corbigny. 'We had not gone far from Clamecy', M Gallois said, 'when suddenly, near Sassier, I felt something like an electric shock all through my body, and so did M Vigneron. At the same time the motor stalled and the headlights went out. Paralysed, unable to move, we could only sit there wondering what had happened, and watching. Then we saw that about fifty yards away from us, in the meadow next to the road, there was a round object or machine, and alongside it we could see very clearly three small figures, with quick, lively movements and gestures. But soon the figures seemed to disappear into the object, which then flew off very rapidly. Almost at once our headlights went on again we could move and start the car'.

That same evening Mmes Julia Juste, Maria Barbereau and Marion Le Tanneur, all of Jarnac, were driving along Route D-14. 'We were coming back from Bordeaux', they said, 'and as we were about a mile from Châteauneuf, at about 10 o'clock, two luminous globes appeared in the sky ahead of us, at a low altitude. The car stalled and the headlights went out. We left the car and stood beside the road for almost five minutes, watching the two globes.

'One was much smaller than the other, and at first they moved slowly, apparently following the same direction as the road. Then they stopped, moved back and forth to right and left several times, and stopped again. The larger one became a brilliant white with a reddish halo. Finally both of them went straight down, disappearing from our view in the valley of the Charente River. We got the impression that they landed somewhere.

'We re-entered the car, which now ran properly, and started for Jarnac again'.

30 January, 1958. Near Lima, Peru. A truck, bus and car passengers felt shock and motors of all three vehicles failed, as UFO descended and hovered.

22 October, 1958. Cumberland, Maryland. Car headlights, engine and radio failed, as UFO hovered low over roadside.

26 October, 1958. Baltimore, Maryland. A UFO was observed hovering over bridge, motor and headlights failed, two passengers felt heat.

3 November, 1957. Levelland, U.S.A. Witness after witness telephoned the police to report an egg-shaped thing 200ft long and brilliantly luminous, which caused car and truck motors to stall and headlights to go out. After a few such calls, police cars and firemen were on the roads looking for the object; county sheriff Weir Clem saw a streak of neon red light crossing the highway less than a quarter of a mile ahead, that 'lit up the whole pavement in front of us for about two seconds'. The effects on the cars seemed to be connected with the light given off by the object; one witness said that the object's light was going off and on, and every time it came on, his own lights and motor would go off.

November 1957. Whitharral, Texas. Mr F. Williams said he encountered a disc-shaped object on the road at 12.15 a.m. His car headlamps and motor were affected. The light from the UFO pulsated steadily on and off. Each time it came on Mr Williams said the car headlamps went out. Finally the disc rose swiftly 'with a noise like thunder' and disappeared. The car then functioned normally.

This seems to indicate that the effect on car electrical systems is incidental to the normal operation of the UFO, rather than deliberate intervention.

The following nearer-to-home incident occurred to Mr Ronald Wildman when delivering a new Vauxhall car in the early hours of 9 February, 1962, and is well known to UFO students. He tells us:

'I left home at 3 a.m. to proceed to Swansea with a new estate car from the factory. I had driven through Dunstable and was approaching the crossroads at the end of the deserted Ivinghoe road at Aston Clinton, the time now being 3.30 a.m. approximately. Then I saw something—it was oval-shaped and white with black marks at regular intervals round it, which could have been portholes or air vents. It was about twenty to thirty feet above the ground and at least forty feet across—which, in my estimation, was fantastic.

'As soon as I came within twenty yards of it, the power of my car changed, dropped right down to twenty miles per hour. I changed down into second and put my foot flat on the accelerator—nothing happened. I had my headlights full on and although the engine lost revs, the lights did not fade. The object, which was silent, kept ahead of me by approximately twenty feet for 200 yards, then started to come lower—it continued like this till it came to the end of the stretch—then a white haze appeared around it, like a halo round the moon. It veered off to the right at a terrific speed and vanished; as it did so it brushed particles of frost from the tree tops on to my windscreen'.

If cars, why not aircraft? The answer may be that usually UFOs do not fly too close to airborne machines, or when they do, they pass each other very quickly, in which case one would expect some interference on the ignition of smaller aircraft, be it only a momentary one. Indeed, this proves to be the case.

14 August, 1957. Nr Joinville, Brazil. A Varid Airlines pilot observed domed disc. It affected the aircraft engines.

31 August, 1958. La Verde, Argentina. A light Piper Aircraft mysteriously increased engine revolutions 'abnormally' when a UFO was sighted. Engine returned to normal when UFO made off.

The reader will have noticed that although the witnesses on several occasions were immobilised by 'an electric shock', the saucer occupants were seemingly unaffected. This implies, either their metabolism is different and unaffected by the field, or what seems more likely, they are protected from it. Has this any relationship to the glowing suits and belts sometimes described?

In order to investigate the possible electrical effects UFOs may have on car ignition systems, Mr Thomas Thomson and Mr Alan Watts, members of the London Unidentified Flying Objects Research Association, installed a coil of 616 turns over the ignition coil of a Ford Cortina.

With the car engine running at high throttle and a d.c. of 6.5 amps in the coil, no noticeable effects were noticed. A reversed field again produced a negative result. But there was a noticeable falling off of engine power when an a.c. of 8 amps at 50 c/s was supplied to the coil.

Mr Watts said, the highest available current was 11.5 amps, amounting to a peak field of 700 gauss, but the higher field produced no greatly different effect than had 500 gauss. The general effect was a falling off in engine note, but there was no hunting or intermittent roughness, such as often described in sightings. He went on to say: 'This experiment only shows the feasibility of a magnetic field producing the observed effects when cars become slaves to saucers—it is not conclusive proof that the effect *is* magnetic in origin. I personally think, however, that it is the most likely explanation and it is interesting to accept its magnetic origin as a basis for argument'.

Mr Watts quoted in support of the magnetic field explanation, the case of an acquaintance whose car was very nearly struck by lightning. The engine had stopped momentarily 'in synchronisation with the stroke'. The simplest and therefore most likely explanation is that the very high current density in the lightning stroke produced enough magnetic field to temporarily saturate the coil, and the engine, deprived of spark, died in sympathy. Mr Watts analysed the case of Mr Wildman in terms of the E.M. effect. Later we shall review this same case from a possible alternative point of view. Mr Watts' findings are as follows:

'Using the inverse square law and assuming a field (B_c) of 500 gauss at the car when the saucer is 20ft away, then we can find the field B_s say 1ft, outside the saucer.

$$\frac{B_s}{B_c} = \frac{20^2}{1}$$

Whence $B_s = 400 \times 500$ gauss $= 200{,}000$ gauss.'

An idea of the gauss strength of magnetic fields might be gained from the fact that the Earth's field intensity is about half a gauss, while one hundred gauss would seriously disturb the working of a watch. One thousand gauss would stop it instantly, even if the watch were of the anti-magnetic variety. At fifty thousand gauss, *i.e.* fifty kilogauss, the very structure of the magnet which produced it, starts to collapse. Even so, by employing super-conducting magnets, megagauss over fractional time limits are now possible in the laboratory.

Now we have already seen several E.M. cases which involved car electrical systems and physiological effects, and the table in Fig 48, showing some typical examples, may help to analyse these effects further, particularly in terms of distance from the UFO. The fact that most of these cases are taken from the epic year of 1954 in France is additionally helpful in so far that it is reasonable to assume similar, if not identical, cause of the phenomena in each case. In ten out of the eleven instances

PLACE.	DATE.	EFFECTS.	APPROX. DISTANCE FROM UFO.
Seine-Infèrieure France.	October 1954	'Electric shock', car engine & headlamps went out same time.	300 feet.
Pouzou France.	October 1954	'Electric shock', increasing heat, motor died, headlights went out.	300 feet.
Wyoming U.S.A.	October 1954	Engine of car stalling.	249 feet.
Louisiana U.S.A.	November 1957	Engine spluttered & stopped.	210 feet.
Puy-de-Dôme France.	October 1954	Paralysis. Truck engine gave trouble, slowed down to 20mph. Go no faster.	180 feet.
Clamency (Nièvre) France.	October 1954	'Electric shock', car engine & headlamps went out same time.	150 feet.
Saône-et-Loire France.	October 1954	Motorcycle engine stalled.	150 feet.
Illinois U.S.A.	August 1963	Stalled car's engine.	99 feet.
Lusigny Forest France.	October 1954	Intense heat, ground & trees dry in spite of heavy rain.	78 feet.
Buckinghamshire England.	February 1962	Engine lost power.	60 feet.
Turquenstein France.	October 1954	Paralysis, sensation of heat & engine stalled.	60 feet.

Fig 48. Table of some typical UFO effects on cars and people at varying distances.

quoted, vehicle electrical systems were affected, either headlamps, ignition or both. At least six witnesses experienced physiological effects, electric shock or paralysis, which amounts to the same thing, *i.e.* muscular contractions, and three witnesses experienced heat or increasing heat as the distance from the UFO diminished.

It will be seen from this that many of these sightings involve distances greater than the 20ft datum used by Mr Watts, therefore in order to investigate the problem at the extremity of action it might be more helpful to use the distance at which Mr Wildman *first* detected a loss of power, *i.e.* 60ft, for remember he said: 'As soon as I came within twenty yards of it, the power of my car changed, dropped right down to twenty miles per hour'. Now this gives a gauss strength of 1,800,000 at the perimeter of the saucer, which by present-day standards is not formidable.

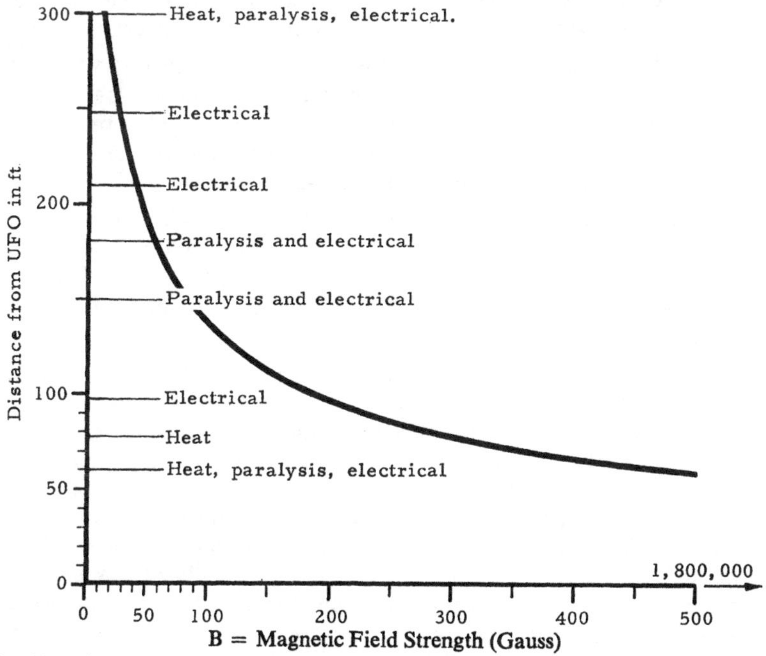

Fig 49. Electro-magnetic curve based on Mr Wildman's sighting compared with some other observed physical effects.

We must, of course, allow for errors of judgment where distances are concerned, but generally I think it is safe to assume 80% accuracy on the witnesses' part.

Notwithstanding power requirements as seen from our own technological standpoint, it is well to follow Mr Watts' praiseworthy initiative and examine similar evidence. Based on the above value, the curve shown in Fig 49 reveals that the spectrum of effects is manifest right through the range of distances, *i.e.* heat, paralysis and electrical disturbances were present at 300ft as they were at 60ft range. And further, E.M. induction heating in organic substances usually indicates high-frequency radiation; therefore the above experiment may well have yielded different results had a much higher frequency been used. In fact, the whole of the E.M. UFO evidence does point rather strongly to the existence of an oscillating field of extremely high frequency. Something of the order, in fact, that Nikola Tesla was working on in the later years of his life. Indeed, it is perhaps true to say that the whole of the UFO phenomena is more in keeping with some of the E.M. effects that this great genius produced, than anything else. As this story unfolds we shall examine further evidence of another kind which indicates the close relationship between E.M. radiation and gravitation. But in the present context we could wish for no more striking example with which to substantiate the corroborative value of the foregoing than to quote the following interesting case.

On 14 November, 1954, near Forli in Italy, two tractors were proceeding along the road side by side. When a red luminous disc flew over, one tractor broke down, the other continued to operate. The machine which broke down had a conventional internal combustion engine with electrical ignition. The other one had a diesel engine.

From this it would seem we should be able to observe other electromagnetic effects in the vicinity of some UFOs and many readers will know we do not have to look far for such evidence. Again I have intentionally found cases where beings were claimed to have been seen and others where they were not.

This incident occurred in the eastern part of Santa Fé, New Mexico, on 6 November, 1957. It happened to Joe Martinez, a taxicab driver, and one of his drivers, Albert Gallegos. Martinez later told reporters: 'This thing came right at us. It was so huge and so bright that it lit up the inside of the car. It was not round, but more of an egg-shape. When it got right over the car, moving rather slowly, we heard a distinct humming sound. Then it pulled up and shot away toward the south-east. Martinez was asked whether his engine had stalled. It had; 'but I may have killed it myself. But the clock in the car stopped too, and so did my *wristwatch*. That's how I know it was right after midnight. Tell me one thing', he concluded, 'am I going crazy, seeing something like that'?

The following interesting account comes from Marble Creek, California. We have already seen how we might fit a leg type undercarriage to a disc, should we build one, but we also discussed the merits of the single cylinder type. In this particular case, not only is there evidence of electro-magnetics, but also the witnesses describe both types of undercarriage. If this story was a hoax, why should the perpetrators needlessly complicate it?

Two miners, John Black and John Van Allen were working in a

titanium ore mine at Marble Creek on 20 May and again on 20 June, 1953 when they saw 'a strange silvery object composed of two large discs of metal about 12ft diameter and 7ft thick' land on a sand bar about 100ft from them. The object was cambered and 'on its crown was a plastic-looking dome. A being like a man, 4ft tall and broad shouldered, descended on a rope ladder from the machine, which rested on four retractable metal legs. The being wore a long kind of coat reaching to below the knees and a hood thrown back, which revealed hair on his head. He also wore a kind of gabardine trousers, tight at the ankles. He then took a gleaming metal container into which he drew water, went back to the machine, from which something took it from him.'

Disturbed by the watching miners, the machine had departed in a flash over the tree tops. Miner Black found his compass *spinning wildly*.

It is interesting to remember that John Worrell Keely produced just this effect in his laboratory, and note, among other things, he claimed to have solved the problem of gravitation. At one experiment involving varying chords of mass, an author of philosophical works was present, whose theories had not been in unison with those of Mr Keely on that particular subject. He sat for some time, after a demonstration, with his eyes fixed to the floor, 'wearing as serious an expression of countenance as if he were looking at the grave of his most cherished views'. The first remark that he made was, 'What would Jules Verne say if he were here?' 'The *rotation* of the needle of a *compass* at 120 revolutions per second, the compass being placed on a glass slab and connected with the transmitter by only a thin wire, had the same effect upon all the scientists present, one of awe; so completely were they transfixed and unable to form a conjecture as to the mysterious influences from any known law of science.'

When describing the saucer, miner Black said: 'The disc resembled two convex soup plates fastened together. It travelled soundlessly, slipping sideways between trees of the wood until altitude was gained.'

At the second sighting Black said: 'The second little man wore forest green trousers and black shoes; he had a green cap, black hair, good looks, fair skin and walked stiffly; it also seemed as if he had not been much in the sunlight. This disc rested on a projecting cylinder. When he raced for the machine, the little man put his foot on to a step and climbed into the saucer through the bottom. He went in as far as his knees, then raised his legs. Then the cylinder came up and the disc hung in the air for a few seconds and went off at an angle of 45 degrees with a *hiss*.' Black waved his hands as the disc went off and 'it seemed to wobble in reply'.

Incidentally there are other cases where saucer occupants have acknowledged friendly signals from earth people, several of them involving pilots. Here is one.

14 November, 1955, San Bernadino Mts., California. A pilot saw a ball of white light approach his plane. He blinked his landing lights, whereupon the object blinked in seeming response, reversed course and sped away.

Now let us go back to south-eastern France where again, in 1957, we find another E.M. clue. It was 14 April when Mme Garcia and Mme Rami of Vins (Var) were taking a stroll not far from the château of that village, when 'a deafening metallic racket' made them look back. About 30 to 40 yards away, flying slowly only a few feet above the road, was a small top-shaped metal thing, hemispherical on top and conical underneath.

This lower cone was made up of a sheaf of multi-coloured luminous rods which were in rapid movement. But the noise was coming from a metal road sign, which was vibrating violently, 'as if there was some kind of resonance set between it and the glowing rods'.

On a nearby hill, a M Boglio heard the noise and the screams of alarm, and thinking it was an accident, ran to the crossroads, where he saw the strange sight, which he later described in exactly similar detail.

Now the object which had come down to ground had given a hop, or flew over, another road sign, which also started vibrating, 'producing the same metallic uproar'. The thing pursued its otherwise silent flight for about 200 yards, before it again descended; the witness thinks it actually touched ground, where it remained motionless for a few seconds. Then, finally, it hopped into the air again and disappeared at moderate speed. Subsequent investigation revealed that the two road signs were abnormally magnetised; at a distance of two inches they deviated a compass needle some 15 degrees, while a third sign produced no deviation at all. It was also discovered that an iron irrigation pipe near to the spot where the object had presumably alighted, was similarly magnetised.

Scorch marks and signs of heating on grass and other vegetation is a common occurrence, and on at least one occasion, where the grass and brush were not scorched, though pressed down flat in a twenty-foot circle, the roots six inches below ground level *were* scorched to a cinder.

Engineers may recognise this as a kind of induction heating, but who among us will dare to credit this as a hoax perpetrated by the young native lads who claimed to have seen a 'great white metal eagle' land at the spot?

But we are not finished yet, for the scapegoats of the sceptic, the would-be 'hoaxers, misinterpreters of natural phenomena', etc., have still much information to offer us, and the pattern still holds good. Let us take a look at the evidence for other radiation phenomena we might expect.

On 6 November, 1957 at a place near Merion, Indiana, Rene Gilhan a 33-year-old iron worker, was at his farm home when he noticed a UFO from which a light was beamed. The disc made a whirring noise when it left. Gilhan's wife, children and father-in-law, also saw the craft. The day after Gilhan stared up at the UFO, his face began to itch and redden; that night the 'sunburn' was so bad he had to seek medical attention. The physician who examined him, Dr Joseph Drake, found that Gilhan's condition was not a rash such as might be caused by poison ivy or allergy, but a real burn, 'similar to the burns that are inflicted on the face and eyes when working near an arc welder without a face mask'. But Gilhan said he had not been near a welder for three weeks.

Three hundred miles to the north-north-east that same night, an even stranger incident took place. At about 11.30 p.m. Olden Moore, a 28-year-old plasterer was driving home from Painesville in north-eastern Ohio, just west of Montville when he saw a bright object in the sky.

'It stopped when it got to the centre of my side of the windshield, and then it split into two pieces. One part went straight upward. The part that remained seemed brighter than ever and kept getting bigger. When it got to be about the size of a sheet of paper, I pulled the car into a side road and got out. It seemed to be headed straight at the car. I had no idea what it was. The colour changed, as it approached, from bright white to a green haze, and then to blue-green as it stopped about 200ft above the field. I didn't hear any sound until it started to settle slowly to the ground. Then I noticed a whirring sound, something like an electric motor, only deeper.'

The UFO landed about 500ft from him. 'I stood by the car watching the thing for some fifteen minutes before I decided to walk toward it. No, I wasn't afraid at the time. The moonlight made it possible to distinguish the object itself from the haze. When the haze was dim, whatever the thing was made of looked the same as those mirrored sun-glasses—the kind where the outside of the glasses looks like mirrors, and you can't see the wearer's eyes. I didn't see any windows. Up until Wednesday night I figured, as most people do, that flying saucer sighters belonged in the booby hatch. Now, I do not doubt them at all.'

Moore approached only halfway to the saucer, then 'stopped and thought about getting witnesses, and returned to the car'. Finding no one on the road, he drove five miles to his home and returned with his wife, but the object was gone when they got back. Mrs Moore notified County Sheriff Louis Robusky the next morning and Robusky immediately questioned her husband about the incident.

County Civilian Defense Director Kenneth Locke was also notified, and searched the scene early that afternoon. Locke found two holes three feet deep. 'They were not like post holes,' an observer said, 'they were perfect'.

Mr Locke returned to headquarters for a Geiger counter. 'When we returned', he said, 'we got a reading of about 150 microroentgens in the centre of an area about 50ft in diameter, tapering to 20 or 30 microroentgens at the perimeter'. A few hours later, 'the meter showed only 20-25 microroentgens in the centre and no reading at all in the perimeter. This indicated that the activity was not caused by minerals in the ground. We concluded that something must have been here'. Later, Mr Locke expressed his conclusions even more specifically: 'A foreign object landed in that field,' he said.

Do little boys tell made-up stories about flying saucers? No doubt many have, but when they do, does their story produce physical results which make scientific sense? Perhaps. I leave this one for the reader to judge.

10 April, 1950. *Los Angeles Times*. Amarillo, Texas, 9 April (UP). David Lightfoot, 12, sighted what was at first thought to be a balloon, but turned out to be an object about the same circumference as an automobile tyre, and about 18in thick. It was rounded on the bottom with a top resembling a flat plate. He barely touched it to discover that 'it was slick like a snake and hot'. It was blue-grey in colour and had no opening other than the divided section. There was some release of gas or spray when the object took off, which turned his arms and face bright red, causing weals. A young boy of nine confirmed this story.

Radiation could do strange things. It might burn the skin; it might affect paintwork. If the following was a hoax, it must be one of the most elaborate and expensive on record.

In 1956, Mr Trygve Jansen was returning from Oslo, by car, to his home at Ski. He was accompanied by a lady, Mrs Buflot, a neighbour to whom he was giving a lift. Mr Jansen had travelled this road daily for several years. It was at Gjersjoen bridge that they first observed the phenomenon. An object came with great speed from behind a small hill, made a swing out over the nearby lake, and back to the road.

Jansen was keeping his eyes on the road and did not look closely at the object. He thought at first that it might be caused by some kind of light-reflexes, or perhaps by a large bird, but after a time, he became aware that the light seemed to follow the car. It circled again and again, and occasionally made great side sweeps.

Then, just after having passed the lake, when they were on a level stretch of the road, the UFO flew in front of the car and promptly stopped above the centre of the road. Mr Jansen felt as if he was compelled to stop the car, and finally did so when the object started coming down towards it.

When it was straight in front of the car, the object stopped again and stood completely still. The witnesses had a distinct feeling of being scrutinised. Suddenly the object took off straight upwards and disappeared with great speed behind the car.

While the object was hovering in front of the car, both occupants felt a prickly sensation on their faces; as if they were exposed to a strong beam of some kind. Mr Jansen's watch, which had kept perfect time for years, *stopped* at that moment.

Later, the watch had to undergo a very expensive repair, and the watchmaker said it had been exposed to a strong magnetic current.

When Mr Jansen arrived at his home, his wife excitedly came out to ask him if he had bought a new car.

'No', said Mr Jansen, 'why do you think so?'

Puzzled, his wife pointed to the vehicle.

The car, which had until the incident been a dull beige colour, was now shiny, and the colour more nearly green.

There were many witnesses to this change of colour, and what is more they all saw it before Mr Jansen or Mrs Buflot had said a word about their adventure. The next day the car had resumed its normal colour!

That evening the Jansens had a party, but Mr Jansen could not manage to swallow the least bit of food. He wasn't exactly ill, but he felt unwell. His skin was still prickly, as if he had been sitting too long before an ultra-violet lamp. He did not feel normal again before the next evening. Mrs Buflot had exactly the same experience.

Both witnesses had ample opportunity to study the saucer closely, and they say that it looked like a shining disc with wings. The disc seemed to rotate, and on the top it had what they described as a kind of cockpit. The light that emanated from the saucer was quite strong, and of a greenish-white colour. It often seemed to come in waves and at times lit up the whole forest.

There are other cases where cars have changed their colour after having a near encounter with a saucer and at least one occasion when a ship was similarly treated, these of course are the cases which the scientific sceptics will ignore in argument, or if they do acknowledge them, how will they explain such incidents away? Meteorite activity, perhaps?

November 1957. This sighting from the Irish Sea involved the trawler *Ella Hewett*, which was overflown by a very bright object. The incident was reported by many eye-witnesses ashore, but the shock was yet to come, for on the following morning the bridge of the vessel appeared to have been stripped of its paint, leaving only the red lead primer visible. The day after that the bridge of the *Ella Hewett* was mysteriously resplendent in its original white. In this modern day and age, scientists are making the most fantastic discoveries; in fact, even as I pen these words, such phenomena as the above may already be known.

More recently, two workers at Bell Telephone Laboratories—Drs Y. H. Pao and P. M. Reutzepis—have polymerised a sample of distilled styrene monomer by irradiation with light from a pulsed ruby laser. The simultaneous absorption of two photons by the monomer molecule was found to have caused this reaction. There is a similar phenomenon taking place in the skies, the answer to which may also already be in the laboratory.

Angel Hair

One of the strangest and most interesting phenomenon which is linked with UFO activity is the gossamer-like substance which both in recent years and in recorded history, has dropped inexplicably from the sky. The ancients called it 'Angel Hair' or 'Threads of the Virgin' and records are full of incidents concerning it which have taken place the world over. Its cause may be extremely complicated and beyond our understanding, or something comparatively simple. It might be, for example, a by-product of extremely high-frequency vibrations in the air and dust particles surrounding the discs, producing a kind of aerodynamic candy floss, though the answer is more likely to be found in mass/energy conversion processes. Be this as it may, the substance is frequently associated with airborne cobwebs, the following case is interesting, bearing as it does professional opinion.

In 1957, a trained biologist witnessed a fall of 'angel hair'. He said: 'Several years ago, I would estimate close to the summer of 1957, two others and myself witnessed a phenomenon that could best be described as 'a sky full of cobwebs' off the Florida coast a short distance south of Miami. At that time I held the position of curator of the Miami Seaquarium, and I was taking part in a specimen-collecting trip aboard the Seaquarium vessel *Sea Horse*, which was skippered by collections director Capt. W. B. Gray and his assistant, Emil Hanson.

'We were travelling northward after a successful day's collecting, somewhere between Soldiers Key and Key Biscayne and approximately three miles off the Florida mainland. The sky was clear on this particular day and little or no wind was blowing. For a period of two hours or more we observed occasional strands of what appeared to be very fine cobwebs up to two or more feet in length, drifting down from the sky and occasionally catching in the rigging of our craft. On being questioned by the others as to what might be the nature of these webs, I explained that an oft-repeated statement in natural history books is that very young spiders on hatching will frequently pay out long strands of silk from their spinnerets until the wind catches them and they eventually become airborne, sometimes being transported many miles and even, as I seemed to recall, far out to sea on occasion.

'At the time I assumed that some phenomenon of temperature or timing had resulted in the mass hatching and exodus of a certain type of spider somewhere on the mainland, and that furthermore, these webs must be fragments of the original strands which in themselves may have been of considerable length. Spiders can and do at times produce vast lengths, in proportion to their size, of web material at little expense to their own metabolism, and I visualised the little spiderlets, wherever they might be, continuing to emit their silken trails during their airborne journey as the wind broke and blew the first ones away. Although we

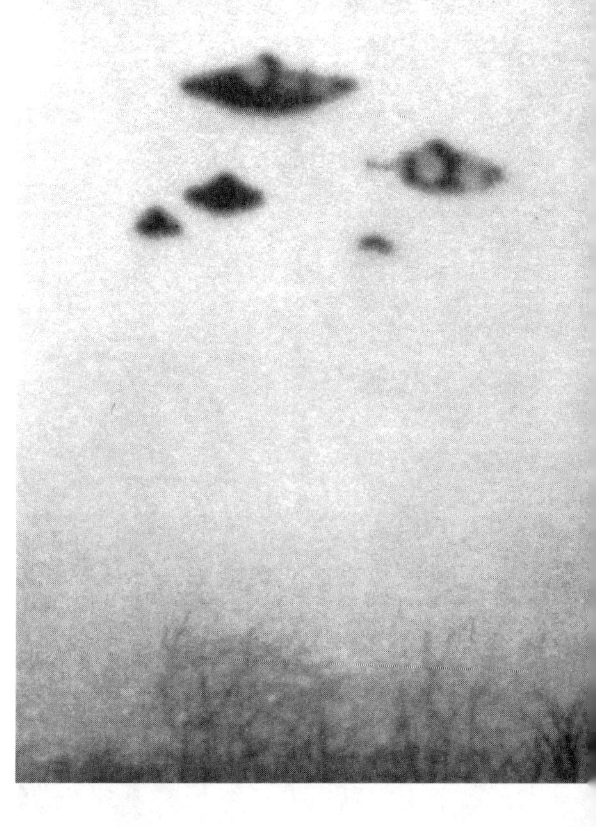

Plate 21. The Alex Birch photograph. The Air Ministry said 'Whatever this was a photograph of, it was not a fake'. Note the blobs of 'light'.

Plate 22. Water humped up due solely to an aerodynamic depression. The effect would be even more marked if the weight of the water near the depression was slightly less than normal.

Plate 23. Agitated iron filings in a rapidly alternating electro-magnetic field.

Plate 24. Electric bowl heater photographed by day with current on. Reflections on the underside are not caused by the heated element. (*b*)

The heater photographed from the same distance at night. (*a*) captured a number of these strands on our fingertips, no spiders were to be seen despite the likelihood that a certain percentage of them would still have spiders attached.

'With the intention of examining the strands under my laboratory microscope when we reached the Seaquarium, I carefully placed several of them inside a mason jar, allowing them to cling to the inside of the glass before I capped it. Under high power I had hoped to see the tiny adhesive droplets that adorn most but not all spider webs, and were these present, there would be little doubt of their true nature. However, when I uncapped the jar later in my office, no trace of the web material could be found.

'This phenomenon is to me still unexplained, and I have seen nothing comparable to it before or since. I will mention by way of information that I have always been interested in the biology of spiders and their webs, particularly the giant orb-weaver Nephilia, whose bright golden web is a fairly common sight through the Everglades. Strong enough to support small pebbles, this web has actually been woven into cloth by natives of the tropics.

'From the foregoing, I would say that it is possible that the strands we saw were something other than spider web, and I have no explanation for the apparent disappearance of the collected material in the mason jar.

Craig Phillips,
U.S. Fish & Wildlife Service,
Department of the Interior'.

17 October, 1952 saw one of the classic examples of UFOs and angel hair. It occurred at Oloron in France. The weather that day was splendid. The sky was blue and cloudless. At about 12.50 p.m. M Yves Prigent, the headmaster at the lycée there, was about to sit down to his midday meal. With him were his wife and their three children. M Prigent had just called his son Jean, who was looking out of the window, to come to the table, when the boy yelled out: 'Come quick, Papa, there's something fantastic!' M Prigent takes up the story from here. He said: 'Away to the north, showing clearly against the blue sky, a fleecy cloud of curious shape was floating along; above it a long, narrow cylinder, apparently tilted at an angle of 45 degrees, was slowly heading due south-west. I estimated its altitude at 2,000 to 3,000 metres. The object was whitish in colour, not luminous, and its shape was quite distinct. Puffs of white smoke were escaping from its top side. Some distance ahead of this cylindrical affair about thirty other objects were travelling on the same course. To the naked eye they seemed to be shapeless smoke balls, but field glasses revealed that there was a red ball in the centre encircled by a kind of yellowish ring *at a very considerable angle to it*, an angle so great that it hid nearly the whole of the lower part of the sphere in the centre, leaving the upper part visible.

'These 'saucers' were travelling in pairs in short, swift zigzags. When two saucers moved apart, they seemed to be connected by a whitish trail, like an electric arc. All these strange objects left very long trails which disintegrated and drifted slowly downwards. For many hours afterwards trees, telephone wires and the roofs of houses could be seen festooned with the remains.'

This amazing story of the 'Threads of the Virgin' scattered over Oloron by UFOs caused quite a stir. The threads resembled wool or nylon, which when rolled up quickly became gelatinous and vaporised into nothing. Scores of witnesses collected specimens and were able to observe this sublimation for themselves. A gymnastics teacher at the local secondary school picked up a substantial skein on the sports ground and excited the staff by lighting the threads, which 'flared up like cellophane'. Fascinated, M Poulet, the science master, watched the complete vaporisation of one thread, about 12 metres long, which he had twined around a stick. He had no time to analyse the strange substance.

In September 1955, at Edmore, Michigan, U.S.A., angel hair fell slowly to earth in clumps, 'some as large as big platters', a UFO accompanied the phenomenon.

Another interesting case is described in *Natural History*, September 1951, in which a formation of 'shiny silver-white balls' over California were seen to disintegrate with a prolific fall of the strange substance.

It is interesting to note that such work as that done by John Worrell Keely, De la Warr and others sometimes produced very unusual atmospheric phenomena. In fact, some readers will know that spider-like cobweb effects can be generated by extremely high-voltage experiments; indeed, it may well be that the eminent Nikola Tesla was producing effects which might well explain some of the electro-magnetic phenomena we have just been reviewing. Some of this great man's own remarks are well worth recording.

'I have produced electrical discharges the actual path of which, from end to end, was probably more than one hundred feet long: but it would not be difficult to reach lengths one hundred times as great.

'I have produced electrical movements occurring at the rate of approximately 100,000 horse-power, but rates of 1, 5 or 10,000,000 horse-power are easily practicable.

'Instead of sending sound-vibrations towards a distant wall, I have sent electrical vibrations toward the remote boundaries of the earth, and instead of the wall, the earth has replied. In place of an echo, I have obtained a stationary electrical wave, a wave reflected from afar.

'My measurements and calculations have shown that it is perfectly practicable to produce on our globe, by the use of these principles, an electric movement of such magnitude that, without the slightest doubt, its effect will be perceptible on some of our nearer planets, as Venus and

Mars. In fact, that we can produce a distinct effect on one of these planets in this novel manner, namely by disturbing the electrical conditions of the earth is beyond any doubt.

'We are whirling through endless space with an inconceivable speed, all around us everything is spinning, everything is moving, everywhere is energy. There must be some way of availing ourselves of this energy more directly. Then, with the light obtained from the medium, with the power derived from it, with every form of energy obtained without effort, from the store forever inexhaustible, humanity will advance with giant strides. The mere contemplation of these magnificent possibilities expands our minds, strengthens our hopes and fills our hearts with supreme delight.'

Nikola Tesla, 1895.

As we echo these words, somewhere in the back of our minds a familiar pattern takes shape, for there can be little doubt that Tesla was referring to the Ether. In order to proceed further, let us sum up the more important evidence of UFO electro-magnetic effects, which from the foregoing are:

(a) Failure of mains appliances and lighting.

(b) Failure of ignition in ground vehicle and aircraft engines.

(c) Television and radio interference.

(d) Heating of organic and inorganic substances.

(e) Physiological effects, temporary paralysis.

Now obviously we must first examine the evidence in terms of known phenomena before we venture into other domains. In which case there can be little doubt that practically all the effects described can be attributed to an alternating E.M. field which is generated in the vicinity of some UFOs.

Some researchers are inclined to think that the evidence suggests a very low frequency in the E.M. spectrum, on the grounds that mains supply of 50 c/s and spark plug frequency of car engines at 150 c/s would have to be subjected to a superimposed induced E.M. field of similar frequency to be affected. While other researchers are inclined to think that the effects might be caused by very high-frequency modulations. But an E.M. field which might produce some of these effects at the distances quoted, would require the expenditure of enormous amounts of energy. Even if we presuppose that where house and street lighting has been blotted out, the UFO responsible was hovering immediately over, or in the near vicinity of a take-off point or distribution transformer. No, we have to look for something else! Incidentally, this alone would offer an interesting line of research to those so minded, to establish how many, if any at all, of the cases bearing evidence of E.M. effects did occur near to electrical distribution points.

Now all through this work so far I have tried to establish the case for the existence of the UFOs by the corroborative value of some sightings, in addition to bringing attention to the evidence for the gravitational spaceship interpretation. More recently, in the previous pages, we have examined only a minute fraction of the evidence of E.M. accompaniment with the craft. Nevertheless, the evidence quoted is sufficient to go by. In the preceding chapters I have tried to emphasise the correlation between magnetic, electrical and gravitational fields and I know I am in good company when I restate I believe they are different aspects of one and the same thing. Therefore if a device employs a system which creates a kind of gravitational field peculiar to itself, then it is to be expected that the release of such field energy will manifest electro-magnetic waves of all frequencies right through the spectrum. We would *expect* radio waves, heat and local manifestations of electro-magnetic phenomena to occur.

14

Analysis Four

Weight of a Saucer

IN compiling this evidence, I am of course aware that there are some UFOlogists to whom the gravitational space ship idea would be unacceptable, due to their belief that such craft are interplane or 'etheric' in origin; therefore, it is argued, the earth's gravity does not affect them. Now I have every sympathy with other schools of thought, but I think it wise to first examine phenomena when we can and not to make it unnecessarily complicated. If a grounded craft leaves impressions in the soil, it is obvious to consider first that this was caused by weight, and therefore the machine is functioning very much in our dimension. There have been many such cases; one of the more notable occurred in France, near Valenciennes, in September 1954.

Marius Dewilde was 34 years old, married, the father of a family. He was a metal worker in the Blanc-Misseron steel mills on the Belgian frontier. He had the reputation of being a reliable man at his job, a good worker, certainly not a visionary.

At the time he lived with his family in a little house isolated in the midst of woods and fields, about a mile from the village of Quarouble. In front of the house there is a small fenced garden. Alongside this garden runs the National Coal Mines railway track from St. Amand-les-Eaux to Blanc-Misseron, and grade crossing No. 79 is next to the house.

This is M. Dewilde's story taken down by an investigator:

'My wife and son had just gone to bed, and I was reading by the fire. The clock over the kitchen stove said 10.30 p.m., when my attention was attracted by my dog Kiki's barking. The animal was howling loud enough to wake the dead. Thinking there must be a prowler in my yard, I took my flashlight and went out.

'When I got to the garden, I noticed on the railroad track, less than six yards from my door, on the left, a sort of dark mass. Some farmer left his cart there, I thought at first, I'll have to tell the station men first thing in the morning to take it away, or we'll have an accident. Farmers do sometimes use the railbed as a road during the harvest, because in that marshy country the motor roads are in poor condition.

'Just then', M. Dewilde went on, 'my dog came up to me, crawling on her belly, and all at once I heard hurried footsteps on my right. There is a path there that we call the 'smugglers path', for they sometimes use it at night. Kiki turned towards the sound and started barking again. I turned on my flashlight and directed its beam toward the path.

'What I saw had nothing to do with smugglers. Two creatures such as I had never seen before were not more than three or four yards away from me, right behind the fence, the only thing that separated us, walking along one behind the other toward the dark mass that I had noticed on the track.

'The one in front turned around toward me. The beam of my light caught a reflection from glass or metal where his face should have been; I had the distinct impression that his head was enclosed in a diver's helmet. In fact, both creatures were dressed in one-piece outfits like the suits that divers wear. They were very short, probably less than three-and-a-half feet tall, but very wide in the shoulders, and the helmets protecting their 'heads' looked enormous. I could see their legs, small in proportion to their height, it seemed to me, but on the other hand I couldn't see any arms. I don't know whether they had any.

'When the first seconds of stunned amazement had passed, I rushed towards the garden gate, intending to go around the back of the fence and cut them off from the path, to try to get hold of at least one of them.

'I was no more than six feet from the two forms when I was blinded by an extremely powerful light, like a magnesium flare, coming through a sort of square opening in the dark mass on the tracks. I closed my eyes and tried to yell, but I couldn't, it was just as if I had been paralysed, I tried to move, but my legs wouldn't obey me.

'Feeling that I was going crazy, I heard, as if in a dream, a yard away from me, the sounds of steps on the cement block that stands in front of the gate to my garden. The two creatures were going toward the railroad.

'Finally the beam of light went out. I then recovered the use of my muscles and ran toward the track. But the dark mass that had been standing there was rising from the ground and hovering lightly, like a helicopter. I had been able, however, to see a kind of door closing. A thick dark steam was coming out the bottom with a low whistling sound. The craft went up vertically for about a hundred feet and then, still gaining altitude, turned toward the east in the direction of Anzin. When it was some distance away it took on a reddish luminosity. A minute later it had disappeared completely'.

When he had partly regained his senses, M. Dewilde went and woke his wife, then called a neighbour. Then he ran to the nearest police station, in the little village of Onnaing, about a mile away. He was so agitated and his speech seemed so confused that the police took him for a lunatic and dismissed him. Then he ran to the police commissioner's office and succeeded in winning Commissioner Gouchet's attention.

Realising at once that something extraordinary had happened, the commissioner dismissed without hesitation any notion of a joke or a hoax. M. Dewilde's fear was too unmistakable.

Commissioner Gouchet's report set off a triple enquiry, conducted jointly by the regular police, air police, and the Department of Territorial Security, also somewhat similar to the American F.B.I. The following morning these three organisations took down M. Dewilde's deposition once more. They came to the same conclusion as Commissioner Gouchet and discarded any idea that the story was a deliberate invention.

Then they returned to the spot on Monday, 13 September, to try to explain the phenomenon itself. The first hypothesis they considered was that of a clandestine helicopter carrying contraband. But this explanation had to be abandoned; telegraph wires would have prevented any landing in that spot.

Then the three enquirers looked for any traces and for further witnesses. They first examined, foot by foot, the land adjoining the railway tracks, where the two supposed creatures might have left footprints. They found nothing that might have confirmed M. Dewilde's story. But they said in their report that the ground was hard and the absence of footprints did not disprove the story.

They found their search along the railroad track more productive. In five places on three of the wooden ties were identical depressions, each about an inch-and-a-half square. These marks were fresh and sharply cut, revealing that the wood of the ties had been subjected at those five points to heavy pressure, as if they had supported a very heavy weight. Furthermore, the marks were symmetrically placed. Three were in a row in the middle, all on the same tie, and about 18 inches apart. The other two were on either side of the line formed by these three, and were about 28 inches away from it.

Questioned by reporters about these imprints, one of the air policemen working on the investigation replied: 'A machine that landed on feet instead of on wheels like our own craft would leave prints exactly like them'.

To explain these prints it was first suggested that they might have been left by workmen screwing on track bolts. But no work of that sort had been done for a long time, and the prints were freshly made. Neither then could it account for the prints' geometrical arrangement.

The railroad engineers, when consulted by the investigators, said they had calculated that the pressure revealed by the prints corresponded to a weight of thirty tons. Later some 25ft of track was removed for more stringent analysis. Also, closer examination at the site of the alleged landing revealed that some of the ballast between the lines had been subjected to extremely high temperature. The stones were brittle and calcined.

We may now ask, who among us is going to suggest that a humble man could go to these extraordinary lengths to perpetrate such a hoax, or even that this was a case of misinterpreted natural phenomena? If there are those who think the flying saucer explanation is untenable, then, equally, I for one find the above alternatives positively amusing!

There are many similar cases where ground effects were left by saucers, therefore we can conclude that such craft are real, solid contraptions, whatever their *modus operandi* in space may be.

Noise and Rotation

Predominately UFOs are silent, but in dealing with the various descriptions of noise which they are said to emit on very rare occasions, the investigator is instantly aware of the two main levels of noise. That is, they are either extremely noisy or they emit a low intensity noise, usually heard at close quarters, and within this range can be fitted all kinds of descriptions, which on closer examination finally emerge as meaning much the same thing.

Before we examine some of these cases, however, it may be as well to form an idea of various levels of noise. For instance, most powerful motor-cycle engines are noisy at any time, but such an engine noise may be hardly noticed a few yards away in a busy street. On the other hand, the same engine started up two blocks away in the dead of night may be very noticeable indeed. Therefore, when we describe the level of a noise, the background noise is a necessary adjunct. Even some advanced UFO students with an engineering background have hastened to explain a genuine UFO landing as a mistaken helicopter, simply because it left ground effects, such as flattened grass, and because the witness claimed to have heard a *humming* noise.

In order to clarify the issue somewhat it will help to have an appreciation of the units in which sound is measured, that is, decibels. Noise is the word we use to describe pressure fluctuations in the atmosphere at frequencies in the audible range, which are usually somewhere between 30 cycles per second to around 15,000 cycles per second. But the ear has an incredibly wide range of sensitivity, the loudest noise it can tolerate being about a thousand million million (10^{15}) times the quietest sound it can detect. But the energy in the pressure fluctuations is used as a measure of that noise. Therefore in order to keep the numbers reasonable, it is necessary to use a logarithmic scale. Thus the level of noise is defined by the ratio of the energy in its pressure fluctuations to that in an arbitrary figure of .0002 dynes/sq cm, which is approximately the level of the threshold of hearing. The logarithm of this ratio (to the base 10) gives the noise level in 'bels', hence 10 times this number produces the level in 'decibels' (db). Therefore, doubling the noise would mean an increase of 3 db at this level.

Naturally the sensation of loudness varies from individual to individual and this is not linearly related to the logarithmic scale. For instance, in order to produce the *sensation* of doubling the loudness of an aircraft producing 100 db, an increase of 8 db is needed.

Noise level in db	
160	Permanent damage to the ears.
150	Acute pain to the ears.
140	Threshold of feeling, conversation impossible.
130	Aero jet engine 50ft away.
120	Aero piston engine 50ft away.
110	Jet airliner at 500ft altitude.
100	Pneumatic drill 10ft away.
90	Inside tube train.
80	Motor horn.
70	Heavy traffic.
60	Inside bus.
50	Quiet car passing.
40	Private office.
30	Ticking clock.
20	Quiet garden.
10	Wristwatch at close quarters.
0	Threshold of hearing.

Fig 50. Some common noise levels on the decibel scale.

Fig 50 gives an idea of how some common noises are related on the decibel scale, but more applicable to the case in point is the fact that an airborne vehicle of some 10,000lb weight, out of ground effect and using a 14ft diameter rotor, would create an *aerodynamic* noise of 90 db at 500ft distance, comparable to sitting inside a tube train in fact, to say nothing of the additional noise of the engine. Therefore, when a witness in a country area says he saw a disc a hundred yards away hovering over grass which was flattened beneath it, and the machine made a *humming* noise, it would be more prudent not to hasten to explain it in terms of present man-made machines. Neither would it be wise to assume some freak local effects which deadened the noise, for although it might just be possible, it would be begging the question to offer this explanation over a number of cases. Bearing these facts in mind, here are a few typical UFO cases.

We saw how in the instance of Mrs John Riead at Tamoroa, Illinois, she described the UFO as making a 'spluttering noise—like someone pulling into the driveway'. The object was accompanied by five or six loud booms and three flashes. Then the lights in her home went out.

Van Cortlandt Park, Broadway. A truck driver for a rubbish disposal company, Franck C. ——, was talking to a bus driver. Saw a UFO over the park. It was described as being a 'classical' saucer shape. 'Its flat disc-like base was *spinning*, emitting a *purring* or *thrumming* sound'.

Rene Gilhan, who received the facial burns, said: 'It went straight up and headed west, the light became more intense, and a *whirring* noise like a high-speed electric motor gaining revolutions'.

When Olden Moore was interviewed he said: 'I didn't hear any sound until it started to settle slowly to the ground. Then I noticed a *whirring* sound, something like an electric motor, only a little deeper'.

Yet again Joe Martinez, of Santa Fé, whose wristwatch stopped, said: 'When it got right over the car, moving rather slowly, we heard a distinct *humming* sound'.

A large disc-shaped mass which gave off a light so brilliant that it woke a witness and his wife in Eastern France; they got the impression it was revolving on its axis. It turned into a horizontal position, until it looked cigar shaped, *changing colour as it did so*. In the complete silence of the sleeping village the couple could hear a distinct 'sort of buzzing noise'.

An Indiana farmer who thought he heard a car driving up to his house looked out of the window and was amazed to see a white luminous object, larger than the apparent size of the moon, pass over his barn at about 500ft altitude. It emitted a 'powerful, whirring, *humming* sound', unlike an aeroplane or jet motor.

M. Dewilde, who saw the strange contraption on the railway track near Quarouble, thought he saw a thick dark steam coming out of the bottom with 'a low *whistling* sound'.

Remember Herr Linke, who claimed to have seen a saucer near Hasselbach, said: 'Then I noticed that the whole object was rising slowly from the earth. The cylinder on which it had rested had now disappeared inside the centre and reappeared again through the top. The rate of ascent now became much greater, and at the same time my daughter and I heard a *whistling* sound, rather like the noise made by a falling bomb, but not nearly so loud'.

'A big metallic object, circular and rather flat,' terrified a witness and set all the dogs in the neighbourhood howling. As it took off from a wheatfield 'a slight *whistling* sound was heard'.

A group of high school girls at the Weatherly High School in Lansford, Eastern Pennsylvania, saw an oval-shaped thing which swooped down upon them in 1957. They claimed that it had about four red lights on its rim, which was blurred by its rapid rotation. Said the girls: 'It made a *sizzling* noise as it went by'.

This witness is 19-year-old M. Roland: 'I was riding my motorcycle between La Bégude, three miles south of Lyon, and Feyzin. I was about 200 yards from the fort at Feyzin when suddenly a white light from the sky swept the road in front of me. I stopped and watched the light, which had become motionless. I then discovered it was coming from the top of a dark mass, about 10 yards above the ground and 50 yards away from me; the mass was motionless and seemed to be elliptical in shape. I looked at it for a moment, and then I heard a low sound, like a *sputtering fuse*. Sparks came out underneath the object, which rose at tremendous speed.'

An Indiana housewife called the police to investigate a UFO hovering over her house. Two officers drove to the spot, saw a white and red light about 500ft up. One officer spoke of 'a series of beeps like the *squeaking of a dry bearing*' and '*a thumping sound*'. Other people in the vicinity also heard the noise. The craft then 'banked like an aircraft' and sped away.

Two hunters in Pittigliano, Tuscany, Signori Bacherini and Formiconi, heard a *series of detonations*. On looking up they saw a 'round white craft' moving slowly across the sky with lateral oscillations. It then accelerated and disappeared.

In New Mexico, Mrs Lilian Stickney reported 'an unusual whirling spinning object' with horn-like projections. It travelled low and made a noise 'equal to several large trucks passing over a hard-surfaced road.'

New Zealand, April 1960. An Invercargill woman, who does not wish her name to be published, told a *Wellington Evening Post* reporter that at about 5.10 a.m. on 5 April she saw a strange object travelling low over the estuary towards Bluff. She had been awakened by a deafening roar and looked out to see the object which appeared circular in shape, with a number of flashing lights. Other reports of this tremendous roaring noise at about the same time were made, but no other sightings.

The woman described herself as having been sceptical of flying saucers, but is now convinced that 'there is really something in it'.

The Newcastle on Tyne *Evening Chronicle* of 13 October, 1964 carried the following account:

'Unidentified flying objects were sighted over Gateshead last night (Monday, 12 October).

'Several people say that they saw glowing pink discs in the sky, and their descriptions tally.

'Mr Arthur Toogood, a 38-year-old electrician, of Whitehall Road, said today: 'I was reading a book when I heard this sort of humming sound—quite loud it was—and when I looked out of the window I saw these things like full moons, one bigger than the other.'

So we can now sum up this aspect as follows. Most of the saucers do make some kind of noise at close quarters. And going by the many descriptions it is logical to conclude that such noise is mechanical in origin. When the power is increased for rapid take off, the noise probably gets higher in pitch and fades. This could mean that saucers do make noise all the time, but that it reaches extremely high frequency, much higher than the audible range. Therefore, the existence of both 'noisy' and 'silent' UFO reports does not necessarily constitute an inconsistency, far from it, it is what we might expect to occur. It could explain, for example, why animals are frequently terrified by the passing of a saucer, for it is well known that they are susceptible to much higher auditory frequencies than are humans. It is this kind of confirmation among sightings which serves to illustrate the accuracy of witness reports rather than contradict them.

Descriptions of humming 'like a flock of birds', 'like a swarm of bees only much louder' are more common than any other noise, with the exception perhaps of the 'swishing' we have already dealt with. Most of these latter noises suggest very strongly a mechanical movement of some kind, 'like an electric motor' or perhaps a 'rotating generator' may be nearer to the truth. Here also we find corroborative evidence among the sighting reports. For many of them give descriptions of high-flying discs which, although noiseless, appear to revolve rapidly, thereby agreeing with near proximity observers, who claim to have seen a rotary motion accompanied by noise. Among some of the sightings already quoted there has been mention of rotating parts on the discs; here are a few more:

Switzerland, September 1954. Mme Périat, a resident of Porreutruy, looked at the sky early one morning and saw a bright red circular object. It was moving rapidly across the sky at a speed comparable to a jet. It looked as if it was rotating. Other witnesses saw it and when Mme Périat's husband called the military authorities, they confirmed that an unidentified object had just flown across Switzerland; they would reveal nothing more.

A mechanic riding a motor-cycle coming from Beaune, France, saw a luminous aerial vehicle. It passed over him emitting a greenish-white colour, and as it got closer he observed 'a sort of revolving motion'.

One of the most authenticated cases involving a grounded saucer took place on 27 September, 1954 at a little village called Prémanou, Eastern France. All the witnesses were young children, Raymond Romand, 12, his sisters Janine and Ghislaine (9 and 8) and their little brother Claude, four years old.

They were playing in the barn that night. Outside it was quite dark and raining, when suddenly their dog started to bark. Raymond went outside to investigate and almost ran into something shaped like a 'lump of sugar standing on end'; he said it seemed to be split at the bottom. Eerily it reflected the light from the barn. Not unduly concerned, the lad first threw some pebbles at it, which bounced off with a noise 'like as if they had struck tin'. Then Raymond took his toy pistol and shot a rubber-tipped arrow at the object, but with no result. Plucking up his courage, the boy had advanced so as to touch it, when he was instantly knocked to the ground by 'an ice-cold invisible force'. Scrambling to his feet, the terrified boy retreated into the barn.

Hearing Raymond's yell, Janine looked out in time to see the thing going off with a queer kind of waddling gait.

The huddling children had then peeped through the door and, seeing nothing, fled to the farmhouse, where young Claude excitedly pointed to a large red luminous ball of light which was hanging in the air in the meadow below the farm. The four children had gone to bed that night terrified, not saying a word to their parents. Noticing their whispered remarks the next day, their school teacher had managed to coax Raymond into telling her the story. The children were quite convinced they had seen a ghost. Questioned separately, their stories tallied exactly.

The following day police asked the children to show them the position of the red ball. 'Over there in the meadow', they said, leading the way. At the indicated spot, the investigators found strange marks, despite the heavy rain.

'Over a circular area about 12ft in diameter, the grass was *flattened* counter-clockwise, fixed in the motionless pattern of a whirlwind. Meadow flowers within the circle looked as if they had been put through a press'. Also within the circle, which was clearly defined, were four holes arranged in a square, shaped as though four triangular objects, four inches across, had sunk into the ground. They made a forty-five degree angle towards the centre.

At one side a flagpole had a 6in strip of bark torn from it, five feet above the ground, a little to one side of it there were another two marks like the triangular ones, but they were elongated as if something had dragged and bounced a little before coming to rest.

Everyone who interviewed the children were satisfied beyond doubt that this was no childish prank and, in any case, living as they did in an isolated mountain farm some 3,000ft high, the children had little chance to hear of flying saucers.

Remember it was Mrs Myra Jones, who saw the disc over the roof of the car being driven by her husband along the Leicestershire-Derbyshire border, who said: 'There were dark spots around the rim of the base and the whole thing seemed to be tilted and to be revolving'.

Then again, M Kawecki, who saw the UFO rise out of the waters of the Baltic, said: 'I was certain that there was some rotating movement involved. I could not make out whether the spikes were moving or the dark streaks gyrated under them. But I had no doubt that one or the other rotated'.

The reader will notice the similarity of the 'spikes' with the description of the strange contraption which hopped merrily over the road in Vins, south-eastern France, *i.e.* 'This lower cone was made up of a sheaf of multi-coloured luminous rods, which were in rapid movement'.

Note also in the Romand case the grass was flattened in a *counter-clockwise* direction, 'fixed in the motionless pattern of a whirlwind'. Well, from children to pilots, let us see what the list has to show.

2 February, 1955. An Aeropost airlines plane bound for Merida from

Maiquetia, piloted by Capt. Dario Celis, co-pilot B. J. Cortes. They said: 'At about 11.15 a.m. a round green glowing ' apparatus ' approached the plane; it was rotating in a *counter-clockwise* direction. Around its centre was a reddish ring emitting flashes of brilliant light. Above and below the ring were markings not unlike portholes'. As Capt. Celis banked his aircraft towards the UFO, it instantly whirled downwards, levelled off and sped away at a terrific speed. Capt. Celis tried in vain to report the incident by radio, but found it inoperative.

We turn now to Kawango, a small village 50 miles west of Kampala, near the Katonga River in Uganda, where a local resident, Mr L. A. Pellissier, claimed to have seen an object 20ft or more in diameter. He said he first saw the UFO when it was hovering motionless at about 1,500ft. 'It was of a light green colour and appeared to be made of glass' (note); he added, 'it made a light humming noise'.

Mr Pellissier, who is an engineer, described how after a minute or two the object descended and passed close to him. He thought it was going to land; instead it increased speed and disappeared in the direction of Lake Victoria. A number of Africans also saw the object.

Said Mr Pellissier: 'The object seemed to rotate, and the noise was a kind of light humming rather than a hissing, and was certainly one of some form of rotation'.

Now while as yet we cannot be certain that there is any significance in the direction of rotation, it does help to establish the correlation among sighting reports. Notice in the description given by Capt. Celis, the UFO sounds as if it approached his plane on the same elevation, for he says 'it instantly whirled downwards'. From which we assume that the disc was rotating in a counter-clockwise direction when viewed from above, which tallies with the markings on the grass in the Romand case, *i.e.* 'the grass was flattened counter-clockwise'. So one would deduce from this that viewed from the underside it would be equally true to say that the discs in these cases rotated in a *clockwise* direction. Now let us take a look at the following case. It happened in Scotland at the little village of Balmaclellan, in Kirkcudbrightshire. It was 7.30 p.m. one night in November 1955 when 32-year-old Maurice Brazier, a mechanic with the Forestry Commission, saw something literally out of this world. He said the villagers scoffed and joked at first because they thought his story fantastic, as did the police and his intimate friends, until they finally realised he was deadly serious. Brazier, a teetotaller with an excellent record of wartime service with the R.A.F., was driving home from Newton-Stewart along the deserted hill road to New Galloway, when he saw a light on the hillside. Taking the light to be the headlamps of another car, he expected to meet it at the corner ahead. But instead of a car, a large illuminated object, which he first took to be a helicopter accompanying a mountain unit, appeared instead. As the UFO approached, it assumed a very large elliptical shape fully 60ft in length with bluish lights along the side. By then it was at no more than 40ft altitude.

Badly shaken, Maurice Brazier stopped his van and jumped out. Then, when only 20 yards from him, the thing banked away to his left, revealing its underside from which he could see that the shape was that of a 'huge double saucer with lights along the outer rim which appeared to *revolve clockwise*'. The inner drum was a dull metallic colour that reflected lights. 'If ever I had a fright, it was that night,' Brazier said.

His wife and family testified to his alarmed state and white face. Meanwhile others in the district were having similar strange experiences. While the Reverend William Peebles, of Balmaclellan, said he knew of a watch that was being maintained for such things. He said: 'Mr Brazier is the last man in the world to be imagining things like this'.

Later, police confirmed there were no helicopters flying in the area that night.

So far, then, we are justified in saying UFO sightings continue to be corroborative and it seems some saucers or parts of some saucers appear to rotate, and, furthermore, some rotate counter-clockwise. As I have said we cannot be sure that this is true of the majority of cases, it would be most interesting to ascertain if in fact this was so. It may also be true that even saucers which have been seen at close range bearing no visible evidence of rotation, may encase revolving parts within them, for witnesses of this type of UFO have claimed to have heard noise suggesting this. Be this as it may, we know that in the majority of cases parts of the enigmatic craft rotate.

Now at the beginning of chapter 8 I said rotation would be one of the more obvious ways to stabilise a device which was 'weightless' or suspended by a gravitational field, and we might loosely refer to the craft as resembling a balloon. But remember we cannot balance such a device by giving it a low centre of gravity as we can the balloon, for every particle of the disc's structure is weightless and there can be no such thing as a low centre of gravity. So the problem is a bit more difficult.

We shall see in the next chapter how there may be alternative ways to stabilise a saucer, but let us first consider the spinning top balancing system.

Whole volumes are, of course, available to the student who wishes to study this phenomenon but, because of its special relationship, a word or so about it here will not be amiss.

When pondering the subject of rotary motion and the forces set up when the plane of rotation of a rotating mass is disturbed, I have constantly recognised the difficulty of trying to visualise such spatial meanderings. It is all very well to tell the student that when a couple is applied to a revolving gyroscope, for instance, it will change its plane of rotation, Fig 51(a), but this is still difficult to understand without further elaboration. Neither is it generally known, for instance, why a spinning top balances, or precesses, but in actual fact the two effects are similar, Fig 51(b). Neither do I recognise that it is essential to introduce mathematics to illustrate the phenomenon, although obviously mathematical treatment must be adopted for further analysis. Consider one particle at the top of the wheel in Fig 51(a), which at any given instant is moving in the direction of X. Now it follows that when the couple f is applied, the particle is asked to move in two directions at that instant, and clearly it cannot do this, but it can and does take up the resultant path R. A similar argument can be applied to all the other particles which make up the revolving mass of the wheel and the result is that the gyro rotates in the direction shown. The case of the balanced spinning top is similar, as a glance at Fig 51(b) will show, we have merely substituted weight for the applied couple f in Fig 51(a), and there results a stabilising couple about the support point as shown.

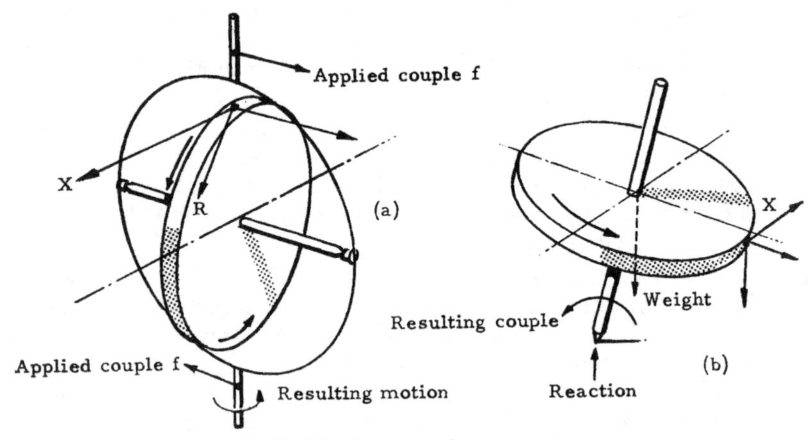

Fig 51. The dynamics of a spinning top.

The same can be said of a hovering disc which was rotating in the air. If it were subjected to a side wind, that is, if the wind reached it as at low field intensity, then it would be subjected to aerodynamic forces, which unless counteracted would cause it to precess or wobble. There are, of course, plenty of UFO sightings which might substantiate this.

Now let us examine a little more closely what may take place in such a condition. In Fig 52(a), a bi-conic shaped disc is rotating about its central axis. We shall consider that the whole thing is rotating, but the following holds true even if only the rim or any other part were rotating. Let us suppose the disc is supported by and inclined to an air stream from which it derives lift, where it will be seen the centre of pressure occurs nearer to the leading edge as with any inclined aero form. But the centre of gravity of the disc occurs at the geometric centre; therefore the disc will rotate, or stall. Should the centre of gravity coincide with the centre of gravitic lift, as with the G field craft, then this would cancel out and the whole thing tends to rotate about the point X as shown in Fig 52(b).

Now in turn this immediately has the effect we were previously discussing; it sets up a couple at right angles and the disc banks, port or starboard, depending on the direction of rotation. Such an effect the author developed as a means of controlling and stabilising an aerodynamic disc, for it obviates the necessity of control surfaces. All that is required in the rolling plane are two equal revolving masses. If one is speeded up only a little, while the other is slowed down in direct proportion, on an inertia principle, then the resulting couple will roll the aircraft left or right as desired. For hovering, one of the rotating masses must be stopped, for a contra-rotating top is difficult to balance, due to the above-mentioned resultant forces cancelling each other out, unless some quickly responsive automatic system was installed which could bring about the instantaneous adjustments required. Whereas at high speed when the G field is extensive, aerodynamic effects on the machine would be inoperative.

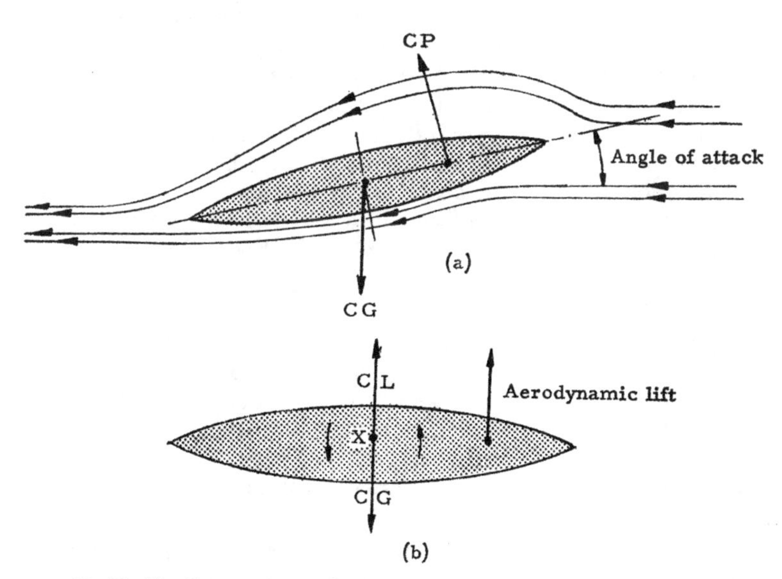

Fig 52. The Gyroscopic couples set up on an aerodynamically disturbed spinning disc.

But could rotation be employed to stabilise a G field machine, to tilt and orient it in any desired plane? For clearly there would be the need to do so. Indeed, it could. It is a well-known effect in astronautical technology, in fact orbiting satellites can be orientated by what is known as the inertia principle. It is simply a means of rotating a flywheel of comparatively small mass by an electric motor, the torque of which turns the housing, or the vehicle which contains it, at much slower revolutions in the opposite direction. In space one simply starts the motor up and stops it gradually when the required orientation is achieved. By having the device fixed in a universal gimbal, it is clear that the effect can be brought about in any desired plane. The reason why reaction jets are used in space capsules to do the same job is simply a consideration of the bulk and weight penalty incurred by the above alternative. Therefore it may be said, if a saucer is weightless, be it in space or in the atmosphere, and if the device employed a rotating mass system for stabilisation, then it is logical that such a rotary system might be employed for control also.

Before we leave the gyroscopics of a rotating disc subjected to aerodynamic forces, there is one other aspect we should consider. Briefly it is what is known as the 'Magnus' effect after its discoverer, the principle of which is shown in Fig 53(a).

If a rapidly rotating thick disc or cylinder is subjected to a moving air stream, an aerodynamic force is generated in one direction. This is due to the difference in pressure created by the relative velocity of the airstream on either side of the revolving mass as illustrated. There is a very well-known kite which employs the same principle to derive lift from a rotating cylinder. From this it will be apparent that if a conical or thick disc is rotated rapidly while moving at fairly high forward speeds, then in addition to the gyroscopic forces already mentioned, it would be subjected to the Magnus effect, which might cause the machine to proceed in a series of rolling turns or *spirals* as in Fig 53(b). But the Magnus effect is to be associated with fairly high airspeeds and therefore not likely to apply in the case of a fast moving G field disc.

But from the purely gyroscopic point of view it is not difficult to imagine what would happen if a disc, part of which was rotating at high speed, hit an obstacle, and we are dramatically reminded of the bucking gyrations and threshing antics of the UFO which struck the palm tree on the banks of the Peropava river in Brazil. There are several cases like this where discs have been seen behaving similarly after striking an obstacle.

No doubt it will have occurred to the reader that the rotating parts of a UFO may perform more than one function, for instance, a large high-voltage generator. Certainly there would seem to be enough evidence to support this. The author has never seen, nor heard, a flying

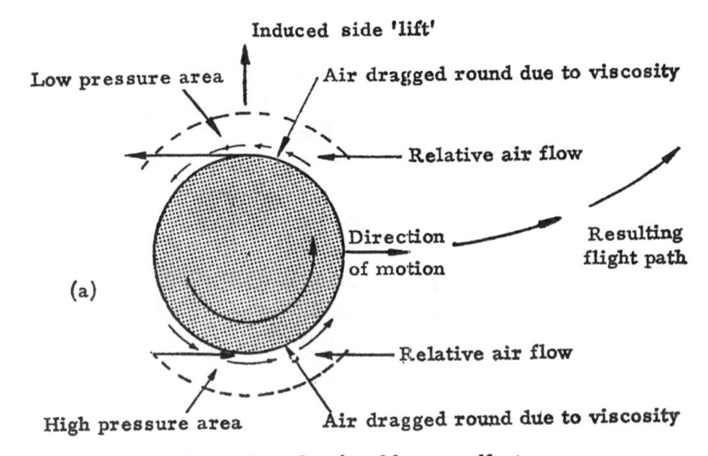

Plan view showing Magnus effect

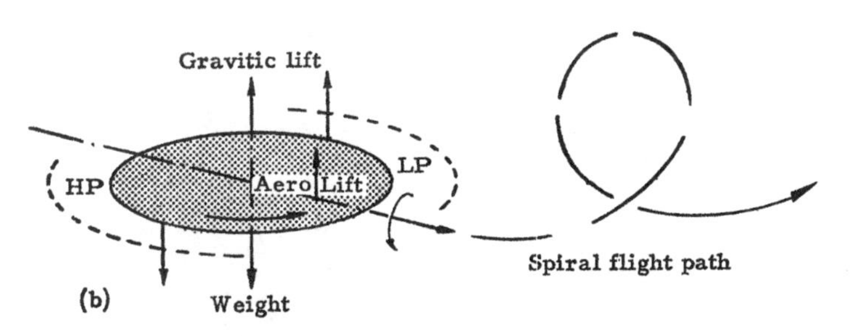

Combined Magnus and Gyroscopic couple effect.

Fig 53. The combined aerodynamic effects of a rotating disc in a moving fluid.

saucer, but I have frequently been reminded of witnesses' testimonies when listening to a simple Wimshurst electro-static generator running and, for that matter, watching it in the dark. A common description comes to mind, 'It sounded like a swarm of bees', for instance, while I do not have to remind the student of the smell of ozone which accompanies high static discharge. Needless to say, both the effects are common among sightings. Anyone who has seen these electric machines operating could not fail to recognise the similarity to UFO reports. In a dark room, the revolving disc is brightly lit with shimmering bands of blue, while around the rim can be seen small points of red flashing lights as the ionised air breaks down.

Now this brings me to an interesting point about revolving masses and UFOs in particular. That is, just how do witnesses tell that a brightly-polished disc *is* revolving? To this end I have attempted some simple experiments.

First, a chrome disc with no surface markings was suspended and witnesses were invited to ascertain from a distance whether or not it was rotating. It was obvious, if the plane of the disc was kept still, there was no way at all of telling if it was revolving or not. On the other hand, if the plane of rotation was wobbled a little, then this presented a distorted effect and the observers received the impression that the disc *was* rotating. But more obvious still, if the disc had surface markings, such as slots, projections, etc., then these blurred and again the observer would say 'it appeared to be revolving', Plate 25. In other words, reports of discs flying fairly high which appeared to rotate absolutely corroborate reports where witnesses have claimed to have seen UFOs low down and described surface markings, slots, projections, etc.

15

Analysis Five

Stability and Orientation of the Space Craft

IF noise and rotation can be associated with control, then the following suggested method of orientating a G field suspended vehicle might be additionally interesting.

So far we have talked of creating this field as a point source, but little has been said about the mechanics of producing it. This I have reserved for the summing up of this work, but the basis of the unit layout may be described here. Fundamentally this is a simple principle and can be likened to an ordinary optical system in which light or heat radiating from a filament F_1 is reflected by a curved mirror and brought to a focus at a greater radius F_2 as in Fig 54(a). If the filament is withdrawn into the dish to position F_3, then the rays will be focused to the greater distance at F_4.

In a similar manner rays are produced within the main field inducer of the ship, 'reflected', and beamed to a concentrated focus, which in turn modulates the C rays of space and creates an intense point source or local gravitational field, into which the ship itself tends to move, just as though it were 'digging a hole' into space, and in effect that is exactly what it is doing. Any aeronautical engineer, physicist, dynamicist or the like who, reading this, dismisses the idea as an attempt to 'lift oneself up by the shoe strings' is betraying his complete ignorance of certain facts. To those I would say, with no sense of bigotry, that time will reveal the truth of these words. Neither do I wish to set myself up as a prophet, but I do regret the unenlightened prejudice which exists in the minds of some of those who take it upon themselves to lead mankind.

The need to change the focal length of the point source will be dealt with later; here we are primarily concerned with a possible technique of controlling the vehicle in space. Again this is fundamentally simple and recourse to diagram Fig 54(b) will facilitate further discussion. To return to the filament analogy once again, it is another equally well-known effect in optics that if the source of light at F_1 is displaced to one side at F_5, then on reflection the light will be displaced in the opposite direction to a focal point at F_6.

Similarly, should the field inducer of a craft be mounted in a gyroscopically stabilised gimbal, then if set in a 'normal' position relative to the earth's surface, the field inducer will tend to remain thus irrespective of any tilt of the craft itself. But we said earlier that if the vehicle tilts,

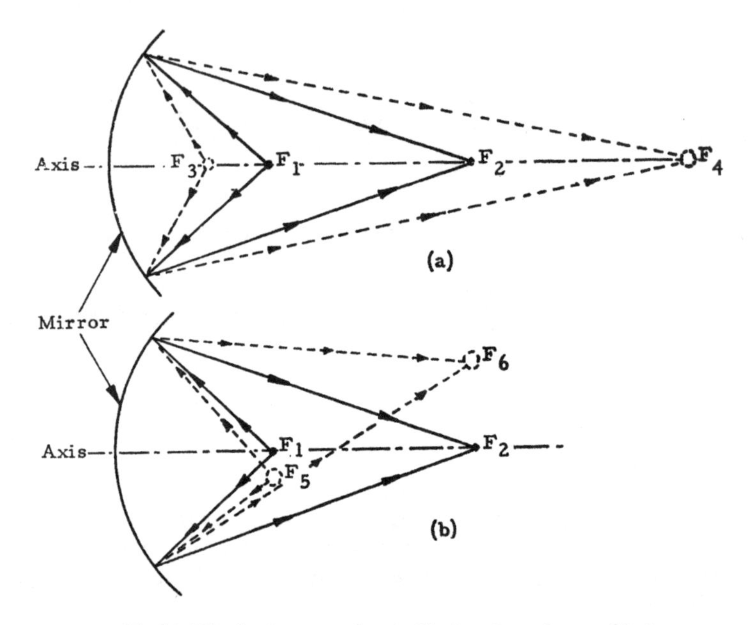

Fig 54. The fundaments of optically focusing a beam of light to a variable focal length.

then it will move in the direction of the tilt, due to the displaced point source and the side thrust vector, also in forward flight, constant altitude condition, thrust must be increased to compensate the loss of lift of the vertical vector. Bearing these facts in mind we can go one stage further.

First the vehicle tilts by the introduction of other factors, but the field inducer does not. This produces exactly the same result as in the light reflector analogy, *i.e.* the G field point source is effectively displaced to one side. A glance at Fig 55 will reveal that this is mechanically convenient, for while the field will still act about the centre of mass of the vehicle, nevertheless some parts of it will come closer to the point source than others and due to the inverse square law effect will experience an automatic corrective trim as indicated by the respective lengths of X_1 and X_2. Again note that for this idea to work efficiently it is best that the vehicle be circular in plan form. In some cases, designers might not bother to house the field inducer, with the result that when seen on edge from the ground a canted disc might reveal a peculiar vertical structure projecting from the centre.

In some instances where additional factors are introduced, we could expect to find the combined effects of tilt, side thrust vector and corrective trimming to give rise to a gentle sideways rocking motion. UFO researchers will be well acquainted with this; the following quotes are typical:

'The last disc dropped much lower than the earlier ones, to the level of the new bridge, where it remained still for an instant, *swaying slightly.*'

' 'That's a queer colour for a haystack,' I said to Yves. ' Look at it '. I was puzzled, and all of a sudden I noticed that the haystack was moving a little, with a slight swing back and forth, *like an oscillation.* 'Say, look —that's no haystack,' I exclaimed to my companion. Then we both rushed through the fields towards the mysterious object. To get there we had to cross an uncultivated field, then a field of beets. We had just reached the beetfield when the object took off on a slant, travelled

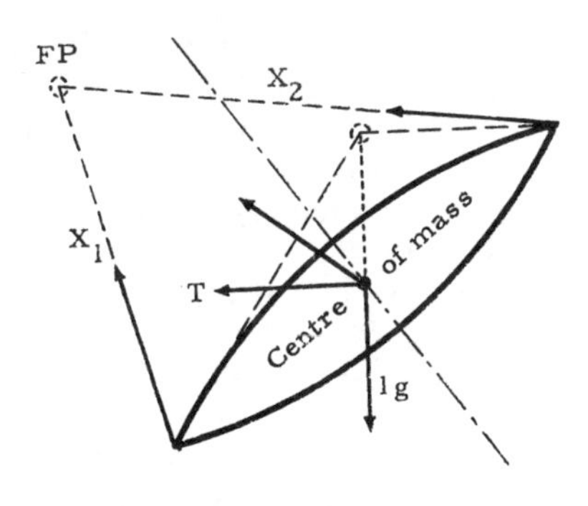

Fig 55. Asymmetrical forces about the centre of mass, due to displaced focal point.

diagonally upward for about fifty feet, and then began to go straight up. We watched it for perhaps three minutes altogether, and then it disappeared in the clouds'.

'He examined it for a long time through binoculars, and could make out its circular form quite clearly, also that it was flat, luminous red, swayed *when it turned,* and then changed to an incandescent white colour'.

'Looking up, they noticed a round white craft, moving slowly in the sky with a *lateral oscillation* as it advanced'.

'They saw a circular object appear suddenly in the north. It was flat, grey, and appeared to be metallic; it slowed, stopped and remained motionless for about thirty seconds, during which time *it swayed back and forth slightly*'.

Now this oscillation of saucers is not uncommon, but neither is it prevalent among sighting reports. But it is interesting, for designers of our own VTO aircraft have to meet this hovering stability problem and, as we saw in the first chapter, the effect has to be compensated either by the pilot or an automatic stabiliser. We saw how with such aircraft, stability is carried out to a remarkable degree. We might then ask, how is it that a vehicle which might conceivably be hundreds of years ahead of any device produced by our technology, still suffers from poor stability. Or is this of no consequence to the crews of visiting space ships? Certainly in a G field ship there would be no physical ill effects in such circumstances, though one would think there would be unpleasant optical ones!

Fuel Requirements and the Prime Mover

If the acceptance of extra-terrestrial space ships is an impossible task for many a sceptic, then it is equally true to say the acceptance of humanoid space crews is an impossible task for many a UFO researcher. While to have only a fundamental appreciation of astro-physics makes it an even harder task for those who do accept such visitors. On the other hand I would say with all humility that it is comparatively easy for the layman, unencumbered by scientific know-how, to accept the idea of visiting space people very much like ourselves. But it is also true that once the trained investigator has arrived at such a decision, having done so much 'back-pedalling', there can be no further turning back for the seekers who have trodden the lonely roads of original research. They may not be able to 'explain' the startling truth, and I for one find it good to be in the company of those who wish not to blind themselves to this truth, however scientifically disquieting it may be!

Now I have continually, and I hope not over-tiringly, correlated the technical aspects of some contact, or near contact claims, with the theory being outlined in this book. I have done this, not as the sceptic, or even as some UFO researchers may claim, out of wishful thinking, because, as I have already said, I cannot explain such visitors in terms of accepted data, but simply because I feel there is too much of a pattern among such sightings to be denied or ignored. I must therefore ask of the reader indulgence in making the following obvious, though nevertheless important observation.

If we are correct in anticipating the truthfulness and technical accuracy of some claimants in that our visitors are humanoid or very much like ourselves, then it is logical also to expect that their simple requirements would be much as our own, *i.e.* they would need water to drink and even perhaps water as a source of hydrogen for the prime mover in their ships. We may speculate on the latter in greater detail later, but first let us again turn to the sightings file for confirmatory evidence.

To begin with let us take note of the fact, that nearby, and to the south-east of the Aston Clinton road where Mr Wildman had his encounter with the hovering disc, is a large reservoir. In fact, subsequent investigation of this case revealed that the disc was in all probability moving away from that area.

When assessing the value of the following I would ask the reader to notice the several different aspects already mentioned in previous cases. This report was published in the *Steep Rock Echo*, July 1950, Sawbill Bay, Ontario:

'In the dusk of 2 July, 1950 I and my wife had drawn up our boat on the sandy beach of a tiny cove in Sawbill Bay, where we had gone fishing. Cliffs rise on all three sides of the cove. Small trees and bushes concealed us and our boat from the sight of anyone overhead in a plane, had there been one round that evening. We had snacks and a thermos flask of tea and, as the dusk was drawing on, we talked of going home. Suddenly the air seemed to vibrate as if from shock waves from a blasting operation at the local iron mines. I recollected, however, that the mines were too far away for that. I had an intuition to climb ten feet up a rock, where there was a cleft that gave on to the bay.

'I was amazed at what I saw. As I peered through the cleft, taking care to make no noise, I could see out on the bay a large shining object resting on the water. It was in the curve of the shoreline, about a quarter of a mile away, across the top end of some narrows. I got down from the cleft and sped back to my wife. She was startled as I came running up. ' Why, what on earth is the matter? ' she asked. ' Come and see if you see what I see, ' I said, grasping her by the arm. ' And make no noise or show yourself '. I drew her by the hand to the cleft. We both peered through it.

'The shining thing was still resting on the water. It looked like two saucers, one upside down on top of the other. Round the edge were holes like black ports, spaced about 4ft apart. We could not see the underside, because the bottom of the thing was resting either on the water, or close to it. On top were what looked like open hatches, and moving round over its surface were ten little figures. They looked queer, very queer. Rotating slowly from a central position, and about 8ft up in the air, was a hoop-shaped object. As it rotated, to a point directly opposite to where my wife and I were peering through the rock cleft, it stopped, and the little figures also stopped moving. Everything now seemed concentrated on the little opening through which we were peering. We were about to duck down, as we thought those midget figures might see us and take alarm, when, on the opposite side of the cove, a deer appeared, came to the edge of the water and stood motionless.

'We again peered through the cleft of the rock. The little figures and the previously rotating circle were aligned on the deer. But now the circle moved to the left. We ducked down, counted twenty and took another peep. The thing was gyrating and the figures moving; but the deer didn't seem to trouble them. We ducked down, supposing that a ray had been projected toward the rock from the thing on the water. Maybe the rock was a barrier and kept it off us.

'It looked as if the whole machine was worked from a central point below the circling ray. The operator was a midget figure on a small raised stand. He wore what seemed to be a red skull cap, or perhaps it was red paint. The caps worn by the others were blue. I should say the figures were from 3ft 6in to 4ft tall, and all were the same size. We could not see their faces. In fact, the faces seemed just blank surfaces! It was odd that the figures moved like automata, rather than living beings.

'Over their chests was a gleaming metallic substance, but the legs and arms were covered by something darker. These figures did not turn round. They just altered the direction of their feet. They walked on the angle, or camber, of the surface of the disc, and the leg on the higher side seemed shorter; so that the compensation—real or apparent—provided against any limp. As I looked, one of the midgets picked up the end, or nozzle, of a vivid green hose. He lifted it, while facing one way, and started to walk the other way. And now the air hummed in a high-pitched note, or vibration. Maybe water was being drawn in, or something was ejected. I do not know if something was being extracted from the water of Sawbill Bay.

'Next time we peered through the rock-cleft, we found that all the figures had vanished, and the machine was about 8ft up in the air. I noticed that the water of the lake, near where the thing had rested, was tinged with colour combined of red-blue-gold. The disc I reckoned was about 15ft thick at the centre and some 12ft at the edges. *It tilted at an angle of near* 45 *degrees. . . . Now, there came a rush of wind* . . . a flash of red-blue-gold, and it was gone, heading northwards, and so fast that my eye could not follow it. It was now quite dark. We decided to call it a day, and got into our boat and went out into the bay where the saucer had rested on the water. I had aligned two trees to estimate its size, which, I think, was 48ft. I went back there again, on another day, and as we came through the narrows I heard a rush of wind, and again something flashed above and beyond the trees. What it was I could not see. My wife was scared. She said she would never go there again.

'A day or two later, I spoke to a friend at the mine, and told him what I had seen. He suggested we both go to the cove on a fishing trip. We had cameras, but after we stationed ourselves at the rock-cleft for three evenings running, nothing happened; and on the last evening we moved quickly along the shore. We patrolled the bay for three weeks, when, one evening, as we were in our outboard motor-boat, and a strong wind was cutting across Sawbill Bay, chopping the water, *we saw the disc*! It was in the same spot. I surmise that, as the wind was up from them, they could not hear our motor chugging. I swung her round into the wind, and my friend got the cameras out. But it is difficult to hold a motor-boat into the wind on choppy water, while trying to take a photo. Indeed, the wind was so darned cold, that my fingers went numb, and I could not manage both the helm and the camera. The boat see-sawed up and down so much that my friend could not focus the camera.

' 'And now', said he, 'I've seen what you saw, and see!'

'But before we got close up to the saucer, I saw the little figures vanish into the hatches. They had seen us! The rotating mechanism vanished, and the hose reeled in like a flash of green lightning—so fast did they work! *There came a regular blast of air* and the saucer whizzed off like greased lightning. But my eye was quick enough to see that a little figure, close to the water's edge, was only half-way to the hatch. He must have operated the end of the green hose, or suction pipe. . . . Our own engine stalled and then ran hot: so we got home late, and our wives were terrified. We had to promise never to go saucer-spotting again!'

Here is another: On the evening of 29 June, 1962 a Mrs P. Roberts, aged 50, of London, saw a large round object glowing a deep violet, hovering over the Kilburn area. It darted backwards and forwards, then hovered again apparently over the nearby Brent Reservoir. Other witnesses watched the UFO.

And again; September 1955. Bush Pine, N.Y. Frank and Eileen Bordes were fishing on a ruffled reservoir near Bush Pine one dark and breezy night in September, when the rhythm of the lapping waves was interrupted by a loud splash and a gurgling sound. Eileen Bordes peered into the darkness towards the noise and suddenly saw an iridescent pink, mushroom-shaped object rise about two feet above the water, then descend and drop beneath the surface. Badly scared, she described it to her husband who was disentangling fishing lines in the bottom of the boat, begging him to row back to shore, whence she and Frank Bordes peered into the blackness lit only by the twinkling stars above. About 100 yards distant they could make out a brightening light. Then two long parallel lights became visible below the first light they had spotted. It seemed as though they came from a partially submerged elongated object some 15ft long and around which there was a good deal of turbulence.

Intrigued by the thing, Frank persuaded a very apprehensive Eileen to climb back into the boat in order to make a closer investigation. They rowed along the shore with a distinct feeling of being watched, and each time they made as if to approach the thing, it would speed towards them. When they retreated it would stop, and gradually retreat, too.

During the whole of this time the submarine turbulence continued. Then, finally, the object made off down the lake at speed until its lights faded in the distance.

Later, commenting on the phenomenon, Frank Bordes said an odd thing about the object was the way it changed course, without turning round. He dismissed the suggestion that it may have been another boat fitted with lights. He said it was faster than any boat he had ever seen.

Still more; September 1962. W. Nyack, N.Y. J. McVicker, an ex-navy electronics officer, reported two 'silver dollars' which he said appeared to rotate. Later that evening five boys spotted a disc which hovered over the Orandell Reservoir. They said the UFO touched down on the water, then took off silently and at very high speed. Later still, two other youths reported a very bright light which moved back and forth over the reservoir, following which a loud explosion was heard.

Sufficient now to say, if space ships are visiting this planet, with occupants very much like ourselves, then sooner or later they will land for water, and we could expect sightings such as the above to occur.

The reader will realise that a theoretical appraisal of the possible technique employed by the UFOs to generate the propulsive field would be beyond the scope of this book; even so, we can make a fundamental approach in simple stages of deduction.

To begin with, we might consider that the G field technique could be employed by the space craft to generate something in excess of 'escape velocity' from the planet and, having achieved this, conservation of fuel would be maintained by letting the vehicle coast on through space. Then, on nearing the atmosphere of its planetary destination, the drive would be used in reverse to brake the velocity of the machine, similar to current rocket procedure. Alternatively, it is more logical to consider that the smaller craft join one of the cigar-type carriers en route for any planetary voyage.

Many researchers may feel that both these huge craft and the discs employ a super developed atomic engine of some kind, and argue that there has been left evidence of radiation by grounded UFOs. Yet others may point out that this is not always the case and anyhow it is usually radiation of very short duration.

Some are convinced that the saucers employ the rotary motion partly for generator purposes, while nuclear fusion is used for their energy source, either by the direct conversion into electrical energy or mechanically by a thermo-regenerative system. That is, heat from the 'fusion' process used to drive a turbine, the propelling fluid of which is passed through a cooler and thence back through the pile again to collect more heat.

From fundamental considerations, such an installation can be extremely bulky, and although this might be improved upon, it is doubtful that a small compact unit which would fit into a small craft would be feasible.

On the other hand, the thermionic generator which could convert heat from nuclear fusion into electrical power is also unlikely, for physicists know enough about this state of the art to show that such a process is likely to be so inefficient as to render it impracticable. But it is important to add that these conclusions are based on power requirements assessed on *known* technological standards in the aero-astronautical domain, the

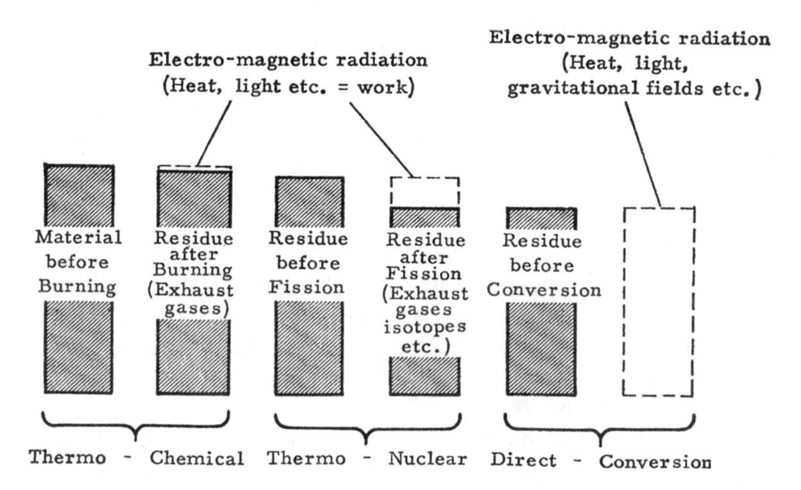

Fig 56. The three main mass-energy conversion levels shown diagrammatically as it would be impossible to show the true relationship in this scale.

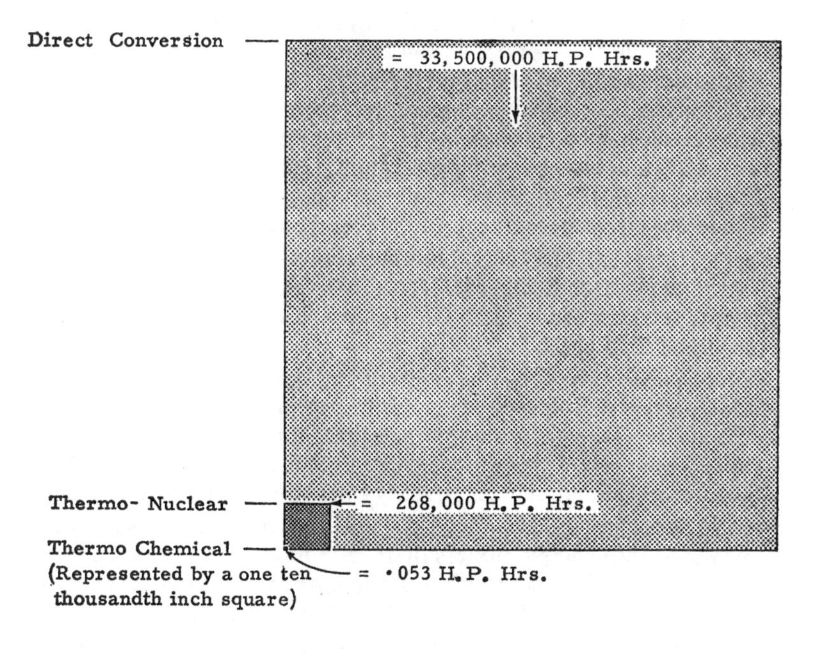

Fig 57. Theoretical energy relationship at the three main levels of exchange, for an expenditure of 1 gram.

jet aircraft and the rocket. Employed to develop a 'field' of force, however, the fusion motor as a prime mover may not be ruled completely out of court.

That there is some kind of a motor unit used by the discs, I think, has been established among the many reports, and there is no point in tiring the reader more by quoting them here. Nevertheless, I am bound to say I for one am convinced the energy used by UFOs is something way ahead of anything known to our technology, atomic or otherwise. We shall discuss what kind of energy this may be later; to do so now would be tantamount to putting the cart before the horse.

However, in order to prepare ourselves for what is to follow, it may be as well to consider some hard facts about matter and energy in terms of power requirements for space flight. First let us reconsider more fully some of the points raised in chapter 6.

There are three main levels known to man at which energy can be extracted from matter. Two of them are currently fairly well understood, the third is not. To deal with these in order, we have first the *thermal* process in which radiation is emitted by the chemical change known as burning, over the parts of the electro-magnetic spectrum which give us light and heat.

The second level is the *thermo-nuclear* process in which yet more heat is liberated by nuclear reaction, the level at which even the waste products left over from the thermal chemical process is still capable of yielding an enormous amount of comparable energy. In other words, the waste products given up by that screaming jet fighter as it streaks overhead, and the latent energy in the exhaust residue of a Titan rocket blasting spacewards are still capable of yielding enormous quantities of energy, nuclear energy, which if unleashed would fly many more fighters and send the Titan and many more like it, to the moon and back many times over.

The third level is the direct conversion of mass into energy, a technological wonder of the future. This is the domain in which Dr Burkhard Heim spends much of his life in contemplation, the final level of exchange on the physical plane Fig 56.

Earlier we compared the efficiency of these last two levels of energy exchange by pointing out that the fusion of only one gram of hydrogen into helium will yield 200,000 kilowatt hours in the form of heat and radiation, and how this liberated energy was the result of the 'mass defect' being the minute difference in mass when the helium nucleus is formed.

On the other hand, when one gram of ordinary mass is *completely* converted into energy, no less than 25 million kilowatts become available for one hour! The relationship of these three levels of exchange is shown graphically in Fig 57. Now exactly what does this mean in terms of space flight; can we relate it to the saucers so as to further establish their technical feasibility? I think we can, but first, in order to examine this part of the evidence, we must assume that our scientists have mastered the direct conversion of mass into energy and the G field spaceship propulsion principle. Indeed, we might presuppose that an understanding of one will automatically shed light on the other.

Consider the case for a hypothetical lunar spaceship of 30 tons weight. For simplicity's sake we assume that the vehicle will be accelerated to the moon at constant acceleration for half the journey, then decelerated at the same value for the remaining half, though of course this need not necessarily be so for this type of spaceship. For instance, depending on the type of mission, the crew may decide to accelerate at a constant 2g for the major part of the trip, and then decelerate at say 10g for the remainder, in any event, neither they nor the vehicle would experience the least disrupting stresses. But for this particular exercise however we shall neglect the accelerations of the earth-moon system and suppose the craft to be operating in a constant 1g acceleration/deceleration field over the entire journey and set this out as follows:

Mean distance of the moon=238,857 miles.
Half distance of the moon =119,428 miles.
Acceleration of ship =1g (32.2ft per sec. per sec.)
Weight of ship =30 tons (67,200 lb.).

Now according to the laws of motion we have:

$$V=\sqrt{2fS}$$

where V=Final velocity in feet per second.
f=Acceleration, *i.e.* 32.2.
S=Distance in feet.
W=Weight.
t=Time.

and substituting we have:

$$V=\sqrt{2\times 32.2\times 119{,}428\times 5{,}280}=201{,}800\text{ft per sec.}$$

But V also $=ft$

$$\therefore t=\frac{V}{f}=\frac{201{,}800}{32.2}=6{,}250\text{ sec.}=1.735\text{ hr.}$$

Now 1 h.p. hour is equal to 550 foot pounds of work × 3,600 sec., or 1.980×10^6ft lb.

But the work done=WS=67,200 × 119,428 × 5,280ft lb.

$$\therefore \text{H.P. hr. expended}=\frac{67{,}200\times 119{,}428\times 5{,}280}{1.980\times 10^6}=21.4\times 10^6$$

And 1 kW is equivalent to 1.34 h.p. Therefore we can write

$$\frac{21.4\times 10^6}{1.34}=16\times 10^6\text{ kW/hr.}$$

We know that one gram of matter *totally* converted into energy could yield 25 million kW for one hour. Which means the mass consumed in the liberation of 16×10^6 kW/hr. would be $\frac{16\times 10^6}{25\times 10^6}=0.64$ grams!

But this calculation was for only half way to the moon, and because it will take just as much energy to decelerate the machine at 1g as it does to accelerate it at 1g, we can assume the mass consumed to be exactly twice, *i.e.* 1.28 grams. Which in a word means, if totally converted there is sufficient energy in an ordinary mass of 2.56 grams to send a thirty-ton space ship to the moon *and back* in a little under seven hours!

The acceleration—mass conversion graph in Fig 58 shows the straight line relationship. From it the following startling facts emerge.

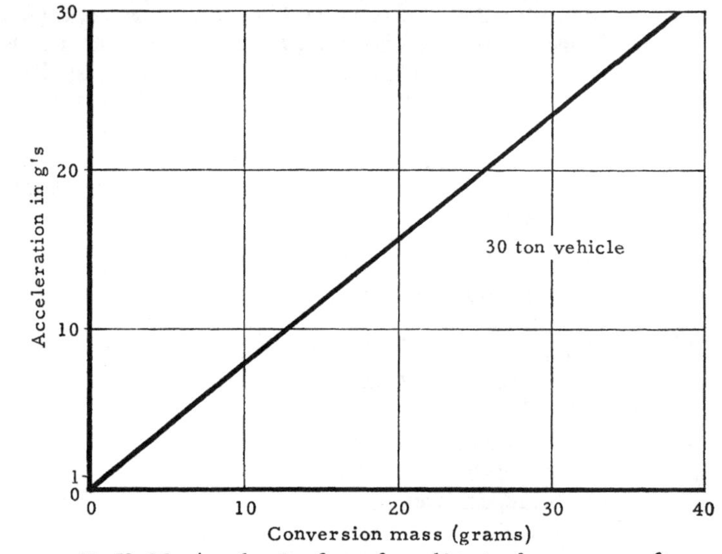

Fig 58. Mass/acceleration factor for a thirty-ton lunar space craft.

It would take just over one hour to send the same vehicle to the moon at a constant acceleration and half-way deceleration of 10g for the total conversion of approximately 12.75 grams of matter. .78 hour for the same ship accelerating at 20g at a consumption of 25.6 grams, while if we like to be rash and expend the exhorbitant amount of 76.8 grams, we could send the ship to the moon and *back* at a comfortable 30g in a little over one-and-a-quarter hours, or to be precise, 1.272 hours. In other words, during the trip, it is quite possible that the crew might drink considerably more fluid than the 2-3 oz. conversion mass required by the ship to accomplish the entire journey.

Perhaps even more fantastic by modern rocket technological standards, the same 30-ton ship could be accelerated to the planet Mars at its nearest (35,000,000 miles) at a constant acceleration/deceleration of 50g for the dematerialisation of approximately 20.5 pounds of mass, and would take a little under six-and-a-half hours to do it! This means, that at the half-way point, the ship's velocity would be roughly 3,270 miles per second, but even this is only 1.76% of the velocity of light.

Plate 25. The only way in which an observer could tell if a polished disc was revolving would be if it wobbled or had distinctive surface markings. This being consistent with witnesses' reports which claim viewing a saucer at close quarters.

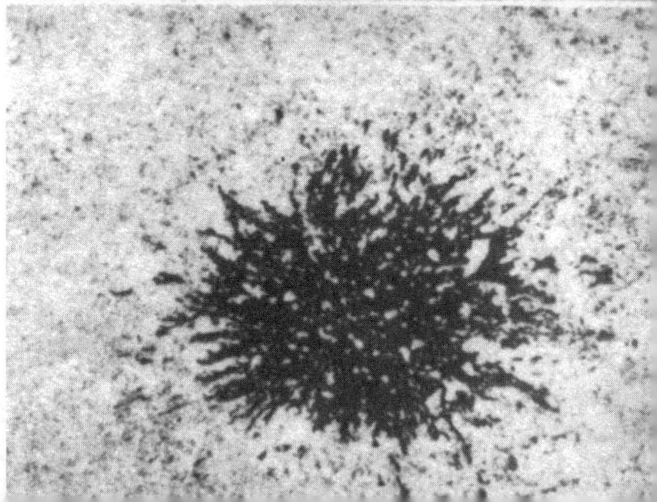

Plate 26. Effect of downward air stream on film of flour. This pattern is often produced by hovering helicopters, but the same effect has also been produced by saucers. This simulation therefore is the opposite effect to the one portrayed in iron filings in Plate 19.

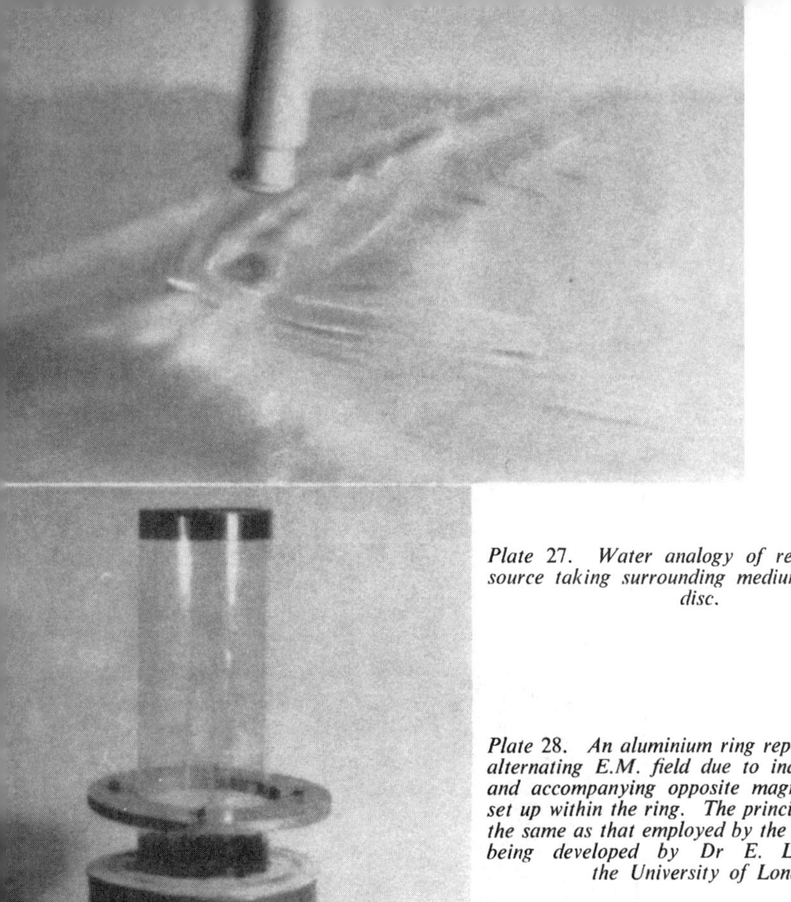

Plate 27. Water analogy of repulsion point source taking surrounding medium along with disc.

Plate 28. An aluminium ring repelled from an alternating E.M. field due to induced current and accompanying opposite magnetic polarity set up within the ring. The principle is exactly the same as that employed by the linear motors being developed by Dr E. Laithwaite at the University of London.

Plate 29. Norwegian photograph by Mrs W. Felton Barrett. Photographic experts find this object difficult to explain as a lens flare. Note increased brilliance below the ring. This could indicate the location of the point source.

As in the previous examples, no account has been taken of the gravitational effects of the Earth-Mars system in order to simplify the problem. Also they are based on a 100% efficiency in the conversion process which may, of course, never be quite realisable even by the ships visiting the earth. Indeed, the UFO records are full of descriptions containing evidence which suggests that this is so. For we can only conclude that UFOs are not brilliantly lit for our benefit and that it is more logical to suppose this luminance to be a by-product of the lift-propulsive system, comparable to the exhaust gases of more primitive machines. If this is so, then we could expect changes of luminosity to accompany varying field intensities. And just as craters may be formed by a rapid take off, so we might also expect spectacular lighting displays, and witnesses' comments like the following:

'An object shaped like a lampshade. . . *a flash of light*, bright as the sun . . . it went straight up into the sky'.

The above is typical; here is a more recent case from Scotland's *Dundee Courier* of July 1965:

'Reports of a flying saucer landing yesterday highlighted a rush of sightings of unidentified objects over Latin America.

'A youth saw a four-legged object touch down for two minutes in broad daylight on Saturday on a beach just across the River Plate in Uruguay, according to the reports.

'He was quoted in Buenos Aires as saying the object shot skywards in a *blinding flash* after a tug hooted a siren'.

Be these optical effects representative of radiation losses or not, we cannot but help being overawed by the fascinating prospects in astronautics which may be awaiting mankind. For instance, astronomical navigational aids would take on different dimensions, for at such short transit times, the ship might be optically beamed on to its planetary destination.

Some physicists may argue that even if such a breakthrough occurred, it would be impossible to realise such energy exchanges purely on physical grounds. For as we saw in chapter 3, there is a limit set to the temperatures at which mechanical prime movers will function. This is true, but we are not suggesting that energy of this magnitude would be suitable for bigger and better rockets, far from it, but rather the direct conversion would be in the form of radiation, gravitational radiation. In this way present physical limitations may be of little account.

Incidentally, it is of interest to note that if 33,500,000 horse-power hours are theoretically available in a mass of 1 gram, then are the claims of some adepts, to be able to levitate, quite so preposterous as some sceptics would have you believe? The amount of mass lost by the body to produce a 1g levitating field could be quite fantastically small. Something like .001 grams for one minute levitation in fact.

From the foregoing it is quite feasible that visiting space craft to this planet would require water, and UFO report files include many sightings, near to, or over reservoirs, or stretches of water. We cannot be certain there is a connection; even so, the pattern continues.

There have, of course, been various clues left by departing UFOs, such as scorched grass, oil deposits, and so on. On other occasions witnesses have seen smoke or steam gushing from the discs, as for instance in the case of Marius Dewilde, who saw the grounded saucer on the permanent way near Quarouble, France. It was he who said, '*A thick dark steam* was coming out the bottom with a low whistling sound'. I would remind the reader that the ballast between the tracks was fused as though by intense heat.

The following bizarre incident occurred on 8 November in the 1957 saucer wave in the United States, at the tiny hamlet of Holly, West Virginia. Hank Mollohan and eight other members of his family watched a peculiar craft behaving oddly just above a nearby ridge. 'It looked about 40ft long, more or less the shape of a hot dog bun, with five or six portholes on each side. There was *fire and blue smoke* coming out of these portholes. It would swing like it was trying to get in under the timber, then it would back up, and a red blaze of fire would fly—like they were having trouble with it. Then it just dived out of there all at once—we thought it had fallen or landed. After this we could see movement—it looked like there were people around it. We definitely saw a man, right under the place where we had seen it come down. Whether he was from this or if he was a hunter, we don't know'.

Now I have no wish to overburden the reader with too many incidents, but even so, this is partly an attempt to publicise the technical corroboration among such sightings. Therefore I would ask patience as we close this section with these final clues. This time the phenomena are sparks—sparks and UFO pyrotechnics in general; but with it I shall add my conclusionary evidence for this chapter. So once again let us turn to southern France; this time to Marignane airport, where M Gachignard, a customs officer, saw a UFO land and take off again. This is a well-known case, which has been checked, examined and checked again; there is no need to repeat the background story here. Sufficient to say that all who cross-examined M Gachignard were satisfied as to his scrupulous honesty and reliability. I shall let him take up the story from here.

'At about midnight (Sunday/Monday, 26/27 October) there was a touch of mistral, but pretty soon the sky clouded over again, as if there was rain about. Towards two o'clock in the morning, I was in the hangar. I had been on duty since ten o'clock. I was wide awake, having slept during the day. I'd just bought a snack, some bread and cream cheese. I went outside to eat it on a bench. These benches are on a concrete terrace below the hangar. The terrace is separated from the runway where the planes park by some cement troughs in which flowers are planted. After finishing my snack I'd intended to go to the traffic control office to find out if the mail plane from Algiers was going to arrive at 2.20, as I'd been told. I had been misinformed. This plane does not run.

'The whole airfield was in view, but though it was dark, I knew every corner of it, and anyhow it is never pitch dark in these parts. Here in the south you can almost distinguish outlines. The runway to the hangar behind me was faintly lit up by the letters in a red neon tube, thirty metres long and three high, of the word 'Marseille' (the airport).

'It was just after 2.3 a.m.—the Nice-Paris mail, scheduled to leave at that time, had just taken off—when suddenly I saw on my left a small light which seemed to be flying toward me, following the runway. It was not very bright, but quite visible and distinct in the darkness. It appeared to be travelling at the speed of a jet coming in to land, something like 250 kilometres an hour. At first I thought it was a shooting star and I'd got the distance and speed wrong. The background was lost in the darkness and I couldn't see where exactly the sky began.

'But somewhere about a kilometre away, to the left of the runway, there's a building called the 'Double Cask' on account of its shape, and I saw the light, apparently still approaching, pass over it at just about 10 metres. Its course was absolutely straight and it was gradually coming down without any sign of wobbling. Pretty soon it passed in front of me and then I realised that it wasn't a shooting star but something which was actually flying. It all happened very quickly; I had no time to think.

'The light had hardly passed when it touched down, and without slowing down suddenly stopped dead about 100 metres to my right. Instantaneous pull-up from 250 kilometres an hour! Just as the object settled on the criss-cross runway I heard a dull muffled sound, like someone slapping down something flat on the ground. That was the first thing I heard. There hadn't been a sound as the thing came in.

'I then realised that it wasn't a plane, because it hadn't reduced speed or taxied. Only fifteen to twenty seconds had passed since its first appearance, and there it was under my nose. It wasn't a plane, but it wasn't a light either, because I had heard a noise. It was something solid.

'I got up quickly and walked towards it, prompted by curiosity no doubt, but also because it's part of my job.

'It took me about thirty seconds to get half-way and during that time I discovered that the light was part of a more substantial affair.

'It stood out faintly against the lighter background of the yellow Météo building. This building was between me and the landing ground, which is always lit up; otherwise the spot where the object was would have been lit up also.

'The object was dark, darker than its surroundings. What sort of thing was it? I hadn't the slightest idea and even though I've been questioned about it over and over again, I still haven't. It might have been metal and then again it might have been cardboard. Checking with the known distances and the dimensions of the building behind it, all we could say was that it was one metre high and three metres long. It was shaped like a rugby football with very pointed ends which emerged from the darkness, thanks to the faint neon light. They tapered to a very marked degree. The underside of the object was in complete darkness, so that I could not see if it had any wheels. I couldn't see it at all, so I just don't know. The upper side was also in darkness, and I couldn't make out anything there either. The only thing I'm certain of is that the light I saw from the very start came from four windows, all perfectly square, with sides about 20 to 30 centimetres long. They were placed in line, but the line wasn't straight, but curved, following the upper curve of the cigar.

'The four windows were exactly in the centre of the machine, so that the last on the right and the first on the left were at the same distance from the tapering ends. But they were grouped in pairs; there was the same distance between the windows of each pair, while the distance between the two inner windows was greater. The two outer windows seemed to be at a slight angle.

'Through these windows I could see a strange light flickering. It was neither steady nor bright, but ghostly and soft, almost milky at times. It reminded me of lights appearing and disappearing behind windows, which make things look blue and green against a light background. Anyway, the light wasn't strong enough to show up the dark parts of the object and it never varied. It didn't change when the object moved. But it never stopped 'throbbing' like the movement of waves.

'I noticed all this when I was walking up to the object.

'But suddenly, when I was not more than 50 metres away, I saw a *shower of sparks*, or rather a stream of tiny white *glowing particles*, shoot from under the rear end on my left. They did not give me enough light to enable me to make out what the object was like. The fiery jet was directed towards the ground.

'This was all over in a flash, yet while it was going on the cigar took off so suddenly and violently that I lost my nerve and instinctively took five or six steps back. I wondered what was going to happen, and if the machine would burst into flames or run over me. I honestly believe I was in danger. I couldn't see 'them' because the shadow of the building almost hid the machine from me, but 'they' could see me easily enough with the moonlight behind me.

'The shower of sparks and the take off were accompanied by a slight noise, like a squib on 14 July. There was no airstream, no blast, no preliminary downward tilt. It's true I was 50 metres away. But in hardly two or three seconds the object had disappeared in the opposite direction to that from which it had come. Its terrific speed on take off was as noticeable as its low landing speed. It did not seem to accelerate at all. It was stationary one moment and travelling like lightning at the next. I've no idea what its speed was. Its ascent wasn't steep, as on arrival. The machine slipped through the 30 to 40 metre space between the Operations and Runway-Control building, which is in line with the criss-cross strip where it had landed.

'As soon as it had taken off, I would have lost it but for the stream of fiery particles gushing out from the rear, as the windows and lights were invisible from where I was. I could see that when it passed between the two buildings it was still very low, lower than their roof tops, which are about 30 metres above the ground. Next moment the light disappeared above the pond at the side of the airport, across the road'.

The customs officer stood there in a state of utter bewilderment, trying to make up his mind if he had been dreaming. Frantically he looked around for anyone who might have seen the thing, but he was alone on the runway.

Back at the hangar everyone was asleep, for at that hour the traffic was at a standstill. Finally, he contacted the police, who declared that 'M Gachignard looked like death'.

October 1954, France again; this time near Villers. Two people, Mme F—— and her daughter, both experienced an unpleasant twitching of the eyes as they were preparing for bed. Unable to sleep, the daughter rose and went to the window. There she saw a luminous mass resting near a hedge. The girl called her mother and both saw that the illuminated mass was, in fact, white on top and dazzling red below. The object threw out a *cascade of sparks* from time to time.

I would now like to offer a small contribution to the record of sightings by giving a brief description of one seen by my son Gary. As UFO sightings go it has nothing special to offer, save for the fact that it was later officially substantiated. It happened in the summer of 1955 at St. Albans, Hertfordshire.

At approximately four o'clock one morning I was dramatically awakened by Gary standing, as I thought bathed in moonlight, at the foot of my bed, urging me to rise and see something odd. He was very excited and ran to the balcony window. Joining him I could see nothing, but he insisted we hurry into his room for a better vantage point. We did so, and with a 'there it goes' he pointed towards a nearby tree. I was just in time to catch a glimpse of a bright halo of light disappearing behind a tree, then there was a bright flash followed by darkness.

He then told me his story. 'I suddenly found myself awake, I don't know why, looking at the moon shining through the curtains at the window. Then I found myself thinking, the moon shouldn't be moving like that and this woke me up properly. It was so bright I could see it shining through the curtains and it slowly rose higher and higher, rocking a little from side to side.' Going to the window Gary then saw the UFO more clearly. 'It was glowing white and fuzzy-looking, and shaped something like a banana. Then it stopped rocking, tipped over and started to slip downwards. Then *silvery dust* seemed to sparkle and drift away from it. Then I called you'.

The following morning I called the editor of a leading national paper in case others had seen the UFO. He advised my son 'to take more water with it'. Gary was thirteen at the time!

Now it would be natural for the reader to believe that Gary had been influenced to some large extent by my interest in the subject. But in fact the boy was more interested in natural science, almost to the exclusion of other topics such as space travel and saucers.

Reconstructing the event, I was able to make measurements from the position of the bed and through the window. A compass bearing and a map placed the line of sight directly through the middle of the Handley Page Airfield at Radlett three-quarters-of-a-mile away. Gary's rough estimate of distance placed the object immediately over the runway. From this I calculated the altitude and estimated the angle of descent, and again referring to the map found a place of intersection on the ground. This turned out to be a nearby heath land. We visited the spot and found a large area of burned bracken which looked fairly recent. I was not able to check on the real cause of this.

I had almost forgotten this event when, three months later, two colleagues of mine, both engineers working at Radlett aerodrome, visited me and somewhat furtively told me of a 'leak' going round their firm. It appeared that on the same morning that Gary stood at the foot of my bed, the night staff at Radlett aerodrome had seen a large glowing thing take off vertically from the runway! The whole thing had been hushed up and I was asked to be discreet. The reader will imagine the looks on my friends' faces when I told them the sequel. But there we have it: sparks on take off, 'silvery dust which drifted away'.

I must thank the reader for his indulgence while I told this little story; after all, it is the one I know best. But we must not lose sight of the original intention, and that is—what are these sparks? Are they evidence of the prime mover? Is it some kind of ionic discharge? I think probably not. We cannot be certain, but it could be something we might expect from all that has been said.

Let us imagine a saucer taking off. We have seen how in some instances it will take aloft loose materials—yes, no doubt, you the reader are way ahead of me. Some of this material will be thrown about, perhaps subject to enormous vibratory oscillation. Heated by electro-magnetic induction and brought to an incandescent state where the kinetic energy of some of it will be sufficient to overcome the binding force of the G field and finally be ejected as a *cascade of sparks.* It is what the theory demands and without being asked or prompted, the witnesses, 'these ordinary folk' have supplied the facts, the corroborative facts in a thousand sightings!

PART THREE

Same Country
Different Track

16

Analysis of a Theory to Fit the Facts Six

ANY examination of such a highly speculative topic as flying saucers demands a strict retention of the facts, no matter how complementary or otherwise these may be to current pet theories. Indeed throughout my own investigation into the phenomena, I have constantly been aware of the alternation between encouragement, born of corroboration, and the often disquieting and inevitable disillusionment when a few reported facts fail to tie up. Naturally, one can always assume complacent possibilities when this happens—for a time—but eventually we have to go right back and begin all over again.

Now I have tried to keep to these facts, and so far the ideas expressed in this book and elsewhere are, to a very encouraging extent, largely supported by such facts. Many of the conclusions may be wrong, the whole theoretical conception may be wrong, and I could be left with nothing but the task of reconciling so many fantastic coincidences. But I believe implicitly in the contra-gravitational idea of spaceship propulsion; to me this is an inevitability, a matter of time before such a technique becomes a practical reality for mankind. If flying saucer reports continue to support this conception, then I shall continue to accept this as the logical explanation, until the case is proven otherwise. Such an attitude is truly scientific. I believe it is honestly reasonable.

Apparent technical inconsistencies among UFO reports are indeed few, but here and there they continue to appear. I say apparent, because on closer examination they are eventually revealed as corroborative. Therefore, in accordance with the dual nature of this book, I propose to re-examine some phenomena in terms of a theory to fit the facts in order to show that even then there is corroboration among the sightings. In so doing, I think the reader will agree, we may be drawing a little closer to a more accurate mechanistic interpretation of the flying saucer. This far we may go; the rest we shall have to leave to the back-room physicists. As a good friend musingly put it, I may be merely analysing the steam from a kettle, without having the slightest idea of the latent heat which produces it. I have no illusions that this may be true, but if in doing so I can encourage others to look for the hidden cause, then my efforts will not be found wanting.

Some Inconsistencies

To begin with, the well-informed reader will know that many sightings of saucer landings are not accompanied by crater phenomena, and I think I have adequately covered this in assuming a gradual lift off in some cases and not in others, and/or provision being made for some of the craft to be raised from the ground by an undercarriage.

We have seen evidence in which such appendages have left imprints or indentations in the ground to mark the spot. But there have been left other ground effects, some of which I have already quoted.

For instance in the Romand case discussed in the last chapter, we saw how the grass was *flattened* counter-clockwise and 'fixed in the motionless pattern of a whirlwind'. Meadow flowers within the 12ft diameter circle 'looked as though they had been through a press'. Now at first hearing this may not appear inconsistent with the G field concept, for such a craft could have landed there and we might visualise part of the rotating disc swirling around and flattening the grass much the same as a garden rotary scythe, but for the fact that the machine in this case obviously sported something like an undercarriage, viz., 'Also within the circle, which was clearly defined, were four holes arranged in a square, four inches across, sunk into the ground. They made a forty-five degree angle towards the centre'. Note the torn flagpole, also the other two triangular marks, elongated, 'as if something had dragged and bounced a little before coming to rest!'

What caused the whirlwind pattern in the grass if the vehicle had been raised above it? Could it be an aerodynamic effect such as made by a helicopter? Perhaps, but we have seen that to produce such an effect aerodynamically requires a good deal of energy which would certainly cause considerable noise. But in this case, as in most others like it, there was nothing like such a noise. Once again I would draw the reader's attention to the background of the Romand case. I would also suggest it is well worth re-reading.

Crushed flowers and grass! There are many more. Here is another case which also includes a description of a saucer occupant, and again I have selected it because of its technically corroborative nature. This dramatic sighting occurred in Chabeuil, about eight miles east of Valence, southern France, in September of that epic year of 1954.

M and Mme Leboeuf, of Valence, were visiting their grandfather in Chabeuil. Later, in the afternoon, they took their dog for a walk and Mme Leboeuf wandered about one hundred yards away to gather mushrooms in the nearby wood. After a little while the dog began to bark and howl miserably. Turning, Mme Leboeuf saw the little animal standing at the edge of an adjacent wheat field confronting something she took at first to be a scarecrow. She went closer to investigate and saw that the 'scarecrow' was in fact wearing some kind of 'small diving suit' made of translucent plastic material, three feet tall or a little taller, with a helmet that was also translucent. Suddenly the startled woman realised that something inside the diving suit was watching her through the blurred transparency of the helmet. She had the impression of two eyes, larger than human eyes. She had no sooner realised this when the diving suit moved towards her 'with a kind of quick waddling gait'.

Mme Leboeuf had until then felt little more than mild surprise or curiosity. But the rapid approach of the 'diving suit' terrified her. Uttering a cry of horror she fled to the comfort of a nearby thicket, from where, looking back, she could no longer see the thing, but the dog continued to howl pitifully and all the dogs in the district joined in.

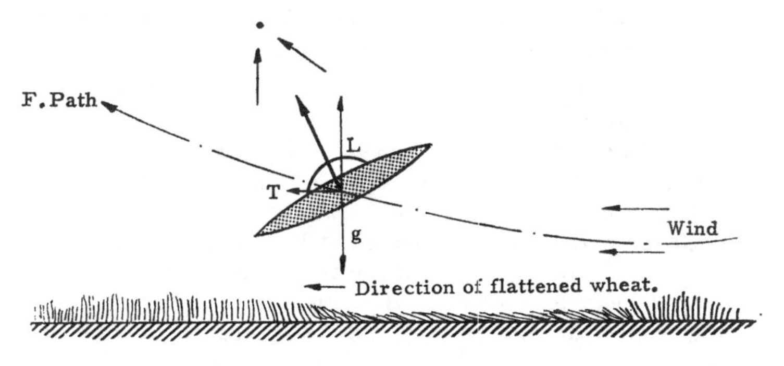

Fig 59. Wheat flattened due to forward G field component and wind effect.

Then, from behind some nearby trees, rose a large metallic object, circular and rather flat, emitting a slight whistling noise. It moved across the wheat field, then suddenly veered off and up at a tremendous speed. Hearing the commotion and his wife's cry of terror, M Leboeuf, accompanied by people from a nearby cemetery, came running. They, too, had heard the strange whistling sound.

Gradually the dogs ceased their howling and practically the whole village was soon on the spot. They found, among the trees from which the 'aircraft' had risen, an 11ft diameter circle, where shrubs and bushes had been crushed. At the edge of the circular depression stood several acacia trees. From one of them hung a 3½in thick branch, smashed downward by a great pressure from above. Hanging 8½ft above the circular patch, yet another branch was stripped of its leaves, while in the *path of the object* as it took off over the field 'the first few yards of *wheat were flattened out* in radiating lines'.

Mme Leboeuf was found in a state of nervous shock. For the following two days she suffered a high fever and the little dog was still trembling with fright three days later. Do what they will, the sceptics of flying saucer stories like this cannot offer plausible alternative explanations, for clearly in a case like this, fraud cannot be entertained seriously. For even a boyish prank which might shock a human, would certainly not reduce a little dog to a state of physical tension for a period of days. No, a prank would not do this, but an extremely high-frequency force field might!

Now we might conclude that such ground effects as the above are strictly in accordance with the supporting point source, or G field conception so far outlined in the previous pages, but is this necessarily so? To elaborate, Fig 59 shows such an effect in progress, where the increased field strength of the moving disc reaches the ground below and although not having the strength to create a tapering crater, might conceivably form such a tapering pattern in wheat or grass. The damage to bushes and trees we might also conclude to be caused by the weight of the descending vehicle itself. So far so good, there are other cases like this, but let us take a look at the following, herein is a subtle difference.

It happened in May 1964 near the small town of Hubbard, about twenty miles north of Salem, Oregon, U.S.A.

It was the usual task of ten-year-old Mike Bizon to turn out a cow into the field at 7 a.m. That particular morning she seemed very nervous, turning and backing as he tried to lead the animal out. He had almost reached the barn door when he saw a bright silver object in the middle of the wheat field. He thought it was several yards across and four feet high, and constructed of some very bright metal. There appeared to be a cone-shaped object on the front and the apparatus stood on four equally shiny legs.

The boy said the thing was making a beeping noise which continued as it began to rise slowly to the height of a telephone pole, 'then it whooshed straight up'. Mike's mother later said that he had 'run into the house, pale as a ghost, and so scared his lips were quivering'. The boy wanted his mother to go to the field, but she preferred to call the deputy sheriff.

Ray Mortensen, a carpenter, was the first to investigate the scene. He found the wheat at the spot 'flattened out like the petals of a flower, *as if a terrific wind blast* had flattened it out from the centre'. The pattern was even and regular in all directions, not the sort of disturbance one would expect if an animal trampled the wheat. Where young Mike had said the legs of the craft rested, were three dinner-plate-size patches. The fourth, barely discernible, occurred on a hard spot of ground caused by a tractor. Like many others the case was investigated by officers from an Air Force Base.

Note, there was no noise, which would obviously accompany an aerodynamic effect on wheat like this. The machine was raised above the ground by a supporting undercarriage which made impressions in the ground. *Therefore it could hardly have been the weight of the object which flattened the wheat down.* Now here is the subtle difference between the last two cases. Whereas we could attribute the radiating lines in the wheat at Chabeuil to the forward G field vector effect of the moving vehicle, we cannot similarly accommodate the radial distribution of the wheat at Hubbard. For it would appear the vehicle in the latter case descended and took off vertically and left an impression 'as if a terrific wind blast had flattened it out from the centre', indeed, much the same as a helicopter, but without the noise! Plate 26 portrays the effect more graphically, the pattern here being formed by subjecting a film of flour to a downwardly directed air jet.

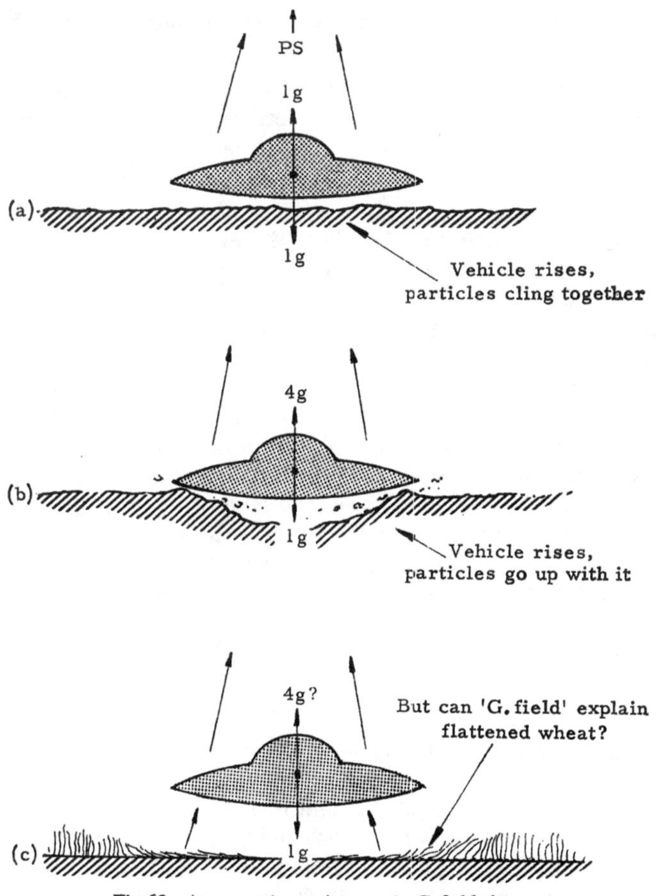

Fig 60. Apparent inconsistency in G field theory.

Now let us consider the above sequence sketches. Fig 60(a) represents a G field vehicle ascending slowly with a 1g vertical component. We have already seen that in this condition, even if the craft was not elevated on legs, the ground immediately beneath would probably hang together due to inherent cohesive forces. And it has been shown that in the condition Fig 60(b) with a 4g plus vertical acceleration, as in a rapid take off, ground in the immediate vicinity might be subjected to very high stresses which would overcome normal cohesion and the soil would literally 'fall' upwards along with the machine. In the last chapter we saw how a disc hovering over water caused the surface immediately beneath to 'boil and be sucked up'. While, on the other hand, condition Fig 60(c), such as that which occurred at Hubbard, would appear mechanically inconsistent with the theory. For how could a powerful vertical accelerating force radially flatten the surface texture beneath and *outwards*?

Indeed, before we are through with this chapter we shall come across several instances where there has been left evidence of high downward pressure having been exerted by a departing UFO.

Neither is our task made any easier by the knowledge that there might be several approaches to achieve 'anti-gravity'. For example, a lighter-than-air balloon climbs mile after mile into the atmosphere, so does an aeroplane, and a rocket; the approach may be mechanically different, the end product in terms of miles above the earth is the same. Therefore, in trying to analyse the technological aspects of our subject we would be wise to first try to correlate different phenomena from one point of view, until the pattern appears to break down. Then, and only then, can we wander off into alternative lanes. This I have occasionally allowed myself to do, by first admitting that there is an inconsistency, which albeit may be far outweighed by the rest of the amassed evidence, then endeavouring to relate this to a second pattern. Then according to philosophical ruling, if this second hypothesis is found to be more embracing, then it must take precedence. Indeed, as we develop the theory in this chapter with the aid of further electro-magnetic analogies, the latter may appear to be the case, but in the end we may find that this is but a parallel route.

A Magnetic Analogy

One day when pondering on the possibility of producing an induced alternating electro-magnetic field in space, similar to the aluminium repulsion ring in Plate 28, I found myself sidetracked into imagining an interesting alternative way of developing a free d.c. type magnetic field. By this I mean a highly concentrated bar magnet type field devoid of the customary encumbering copper coil. It was merely a tempting little mechanical exercise, but it was partly responsible for the conclusions arrived at in the present chapter. But first a word to students. In describing the following magnetic device, I do not intend to imply that this is how saucers may work. But in keeping with procedure elsewhere in this book, I offer the following idea merely as a descriptive stepping stone.

The first step in the exercise was to conjure up a way in which the above-mentioned magnetic field could be generated in space, free of a physical device, *i.e.* generating coil, etc. In order to describe this we should seek the aid of the sketches in Fig 61. Sketch (a) shows an

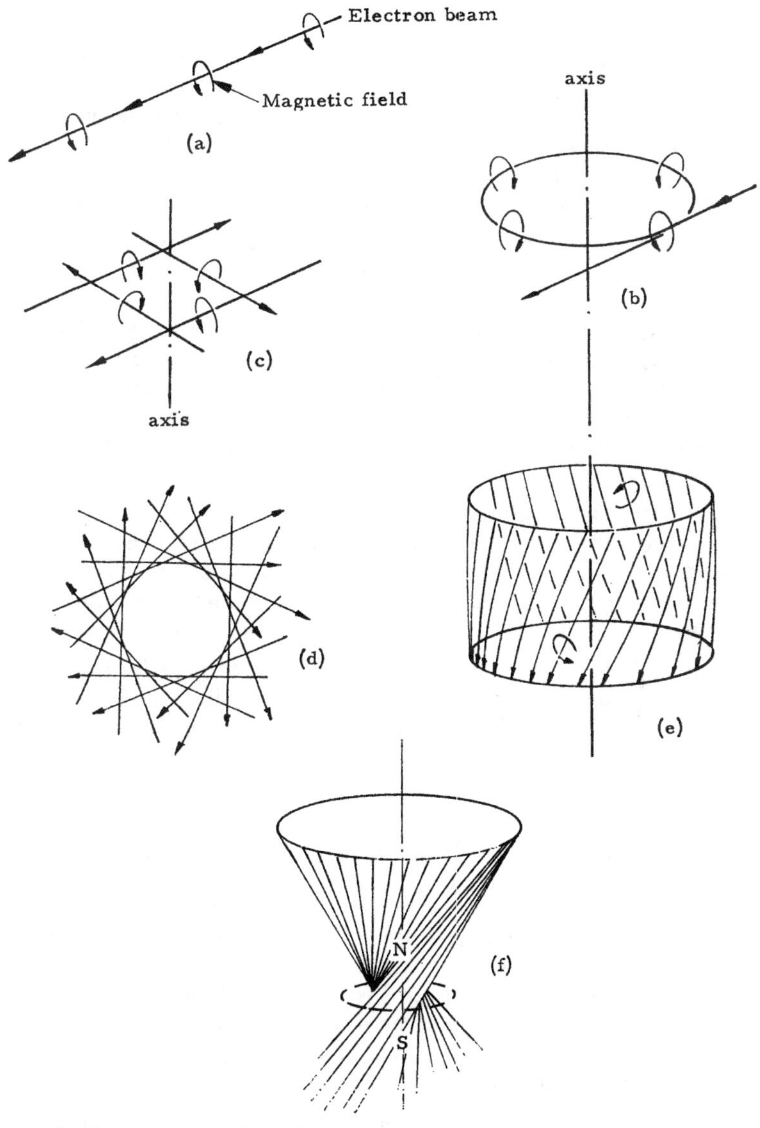

Fig 61. Magnetic analogy of a 'mass-less' magnetic field in which a solenoid is formed by a helical focused beam of electrons.

electron flow in a wire with accompanying magnetic field. Fig 61(b) shows the resulting direction of the magnetic flux when the wire is looped, that is, the field is lined up parallel to the axis of the loop. In other words, a simple coil magnet of one turn is formed.

Now it follows from this that many separate straight wires carrying a current arranged as in Fig 61(c) should produce the same result; four only are shown, for clarity. Fig 61(d) shows the resulting coil effect in plan view. The same might be said of Fig 61(a), (b), (c) and (d), if the current-carrying wires were replaced by pure high-energy electron beams, save for the fact that (d) would never be realised in actuality because of the mutual repulsion of the electrons. But if the beams were given an axial or helical component as in Fig 61(e), this objection would be obviated.

It is a short step from there to appreciate that an inward or conical pinch would produce an even greater concentration of flux at the 'focal point' formed by the centrifugal component of the electron stream partially off-setting the mutual repulsion as in Fig 61(f), the axis of the resulting magnetic field being perpendicular to the plane of the rotating component as in any common solenoid or coil. Therefore, with the exception of the negligible mass of the electron flow, we might truly call the result an isolated magnetic field with bar magnet configuration.

The next stage in the exercise is to imagine the electron beams generated and controllably emitted from a source tangentially, thence intercepted and brought to a focus by a dish-shaped reflector; see two views, Fig 62(a) and (b). It follows that if a strong magnetic field of opposite polarity were to be generated about the axis of the dish, repulsion should take place between the dish and the isolated magnetic field.

Now if this field had a generating coil then it and the dish would be repelled from one another, but it has no coil so how do you repel a mass-less magnetic field? The answer is, of course, you do not; what would take place is an acceleration of the electrons forming the field, and we are back to the principle of the particle accelerator or electronic rocket.

But here now is the whole point of the analogy. Suppose we imagine the high-energy electrons to be accelerated towards the upper limits of the velocity of light, then according to relativity there would be an accompanying increase in mass. And as the mass ratio of the dish unit and the electron beam approached unity, these two would separate at a shared velocity according to Newton's third law of motion. At the velocity of light the ejection mass would reach infinity and the condition would be tantamount to the dish trying to repel the whole universe, and requiring an equal amount of energy to do it!

But we must not lose sight of the original intention. Suppose for one moment the device was not using electrons to create a repulsive magnetic

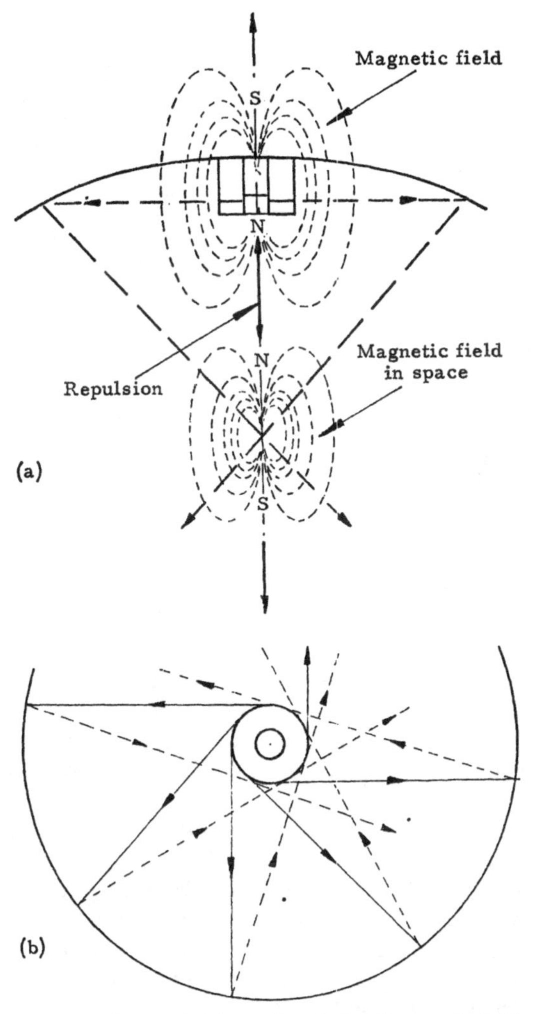

Fig 62. Analogous repulsion of an isolated magnetic field.

field, but in fact was ejecting a stream of concentrated rays which produced an effect opposite to the G field, *i.e. repulsive* space, as may be predicted by the *Unity of Creation Theory*? Truly it might be said of such a vehicle, that it was trying to 'repel' the whole universe. It would be in effect as if the dish magnet was trying to repel a coil of the same field strength, to which was attached the entire universe. As the universe would not budge, all the resulting energy would be stored in the dish system and converted into motion. The efficiency of such a system can be appreciated if we consider the mass-less coil to be continually replaced the

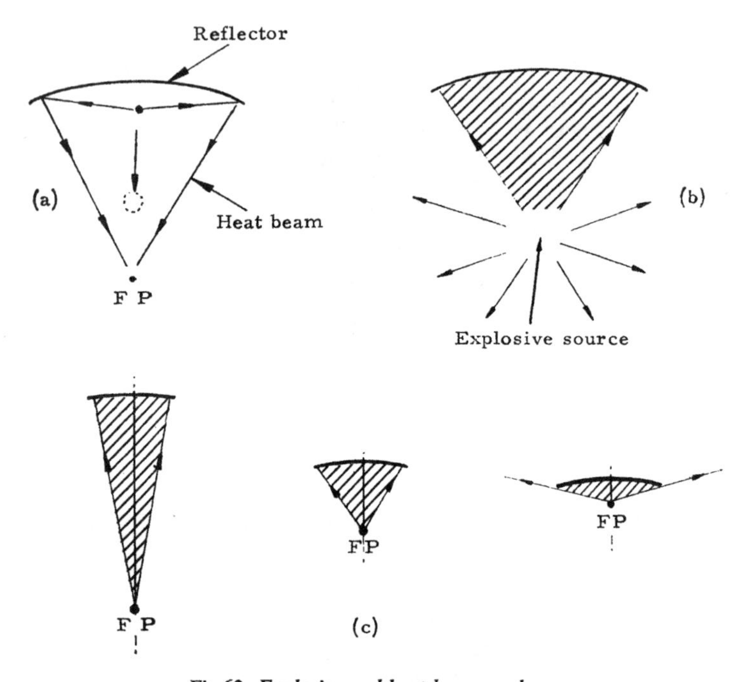

Fig 63. Explosive and heat beam analogy.

same distance from the dish, but still retaining the anchorage or solidity of the universe. From such basic deductions a theory to fit the facts was born.

It is my earnest wish not to offend more technically minded readers by offering such mechanical analogies, but I am constantly reminded how easy it is to be misunderstood when describing some abstruse explorations into dynamics. One instance, I feel, would not be out of place.

In trying to convey this same idea at lectures, I have sometimes resorted to the example in Fig 63. This is almost self-explanatory, as follows: Heat from a central source is radiated, reflected and brought to a steady focus by the parabolic reflector. If a highly combustible charge were to be dropped from this dish, on passing through the intense heat at the focal point, it would explode radially, as in Fig 63(a), while the shaded areas of the explosive force, Fig 63(b), would propel the vehicle upward, the process being repeated indefinitely. It will be apparent that by adjusting the heat emission source up or down relative to the reflector, the focus, and therefore the explosion, can be moved nearer or further away. Therefore a corresponding area of the explosive force can be intercepted as required to control the thrust, as in Fig 63(c). In this case also, the device is nothing but a rocket, and an inefficient one at that, but it

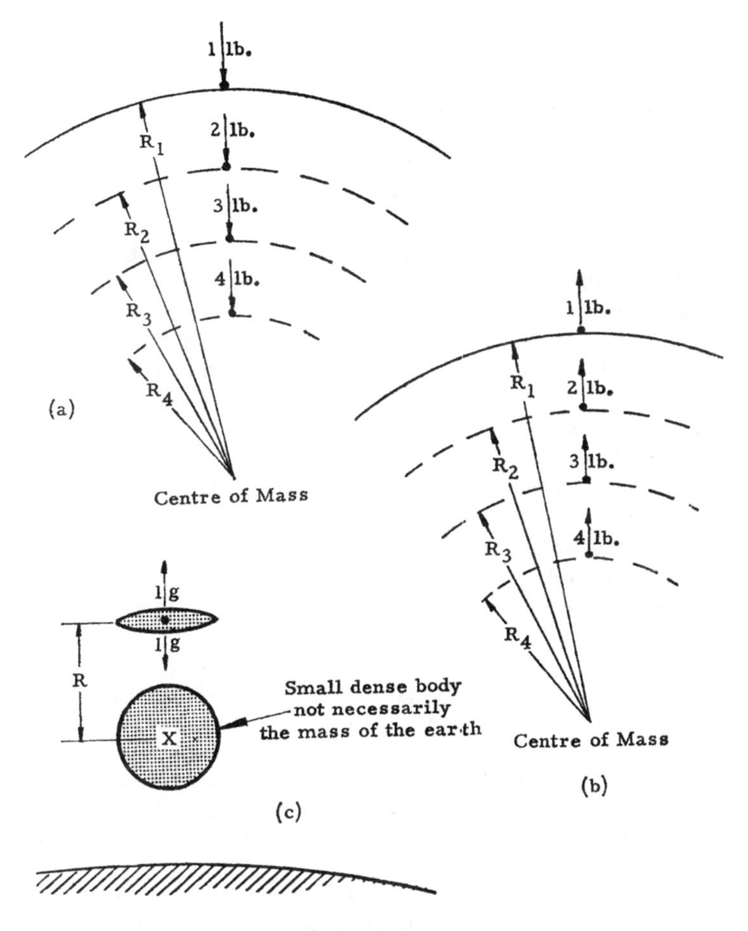

Fig 64. The R field antithesis of the G field theory point source as developed in Chapter 8.

does serve to illustrate the subtle difference; that is, in the case of the rocket, the explosion or propulsive force takes place within the system. Whereas the whole point of the analogy is to portray an identical propulsive force which originates and is controlled from within the system, but nevertheless takes place in an isolated space *outside* it. In describing this at lectures I have on one or two occasions mentioned a spontaneous inflammable substance such as magnesium. The reader will imagine my consternation a day or so later to see quoted in the press and elsewhere, that Mr Cramp suggested flying saucers are propelled by magnesium, despite my efforts to stress otherwise!

The R Field Theory

No, I do not suggest that saucers are propelled by magnesium, but I do feel the above analogy suggests an interesting possibility, and I would now like to express it in terms of gravitational *repulsion.*

In developing the idea of the gravitational point source in chapter 8, we considered a body the mass of the earth, diminishing in size and mass to produce an acceleration of 1g on a closely situated body, and the sketch showing this has been repeated here for convenience. Fig 64(a) Similarly we could surmise a completely reversed condition to exist, as in Fig 64(b), in which a gravitationally repulsive earth exerted a repulsion of 1lb force on a body at the surface, diminishing values of R for a constant mass, producing increasing force or repulsion on the body. And, as with the case for the gravitational field, in order to maintain a constant 1lb force or repulsion at decreasing radii, so also must the mass decrease proportionately. From this we arrive at the anti case for the conditions imposed in chapter 8, *i.e.* in which a small dense repellant body is situated between a space craft and the earth as in Fig 64(c). The next step is to visualise the shrinking or changing of the small mass to a condition of anti-gravitationally modulated space as indicated by the *Unity of Creation Theory,* and there we would find, not gravitational phenomena, but repulsive space! Does this not remind us of the work of Burkhard Heim?

Generated from within the craft the repulsive point source would of course be the very antithesis of the G field theory and would in fact also repel the earth, but would have no more noticeable effect on its huge bulk than does the downward blast of a Titan rocket. But, and this is important, if a craft employing this principle were near to the surface, then we would expect local ground effects, blast effects. Now let us see how far the gravitic repulsion theory fits the facts. First let us consider the vertical take off condition.

You will remember it was young Mike Bizon, of Hubbard, who said the thing in the wheat field first began *to rise slowly* 'then whooshed straight up'. Later, carpenter Ray Mortensen found the wheat 'flattened out like the petals of a flower, as if a terrific wind blast had flattened it out *from the centre*'. Fig 65 shows that in fact this is just what we might predict with an anti-gravitational propulsion device. There are many cases; here are two.

Here is a quotation from Oklahoma newspapers:

29 July, 1952. 'A man, white as a sheet, walked trembling into a police station at Enid, Oklahoma, and told the desk-sergeant, Vern Bennell, that it looked as if a saucer had swooped down on him with the intention of abducting him! The man was Sid Eubank, employee of a sales department of a photo studio at Wichita, Kansas, aged 50. Eubank told the sergeant: 'I was almost swept from my feet on the highway last

night when a huge flying saucer swooped down at terrific speed and stood directly over me, on U.S. highway No. 81, between Bison and Waukomis, south of here, Enid, Okla. The object appeared suddenly out of the night and the tremendous *pressure* it exerted threw my automobile off the road. It was a huge round ball and stood right over me. Then it completely reversed direction, vanishing in a few seconds in the west' '.

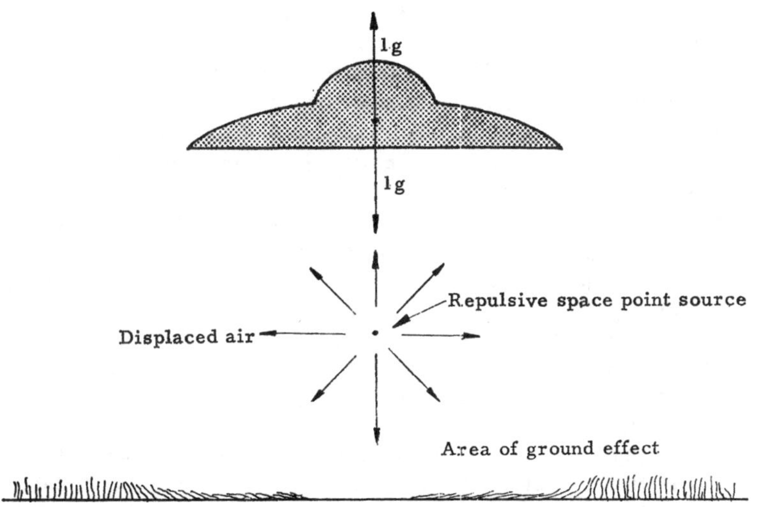

Fig 65. Disc hovering at 1g *vertical acceleration due to repulsion field.*

Remember the case of Frank Williams at Whithorral, Texas, U.S.A., quoted in chapter 7? It was he who said that the landed UFO gave off a pulsating light which caused his car headlights to fluctuate. Finally the machine had risen swiftly 'with a noise like thunder'. Earlier, at 10.50 p.m. and close to the same spot, there had been another sighting.

Officer A. J. Fowler received a phone call from a 'terrified' farm hand, Pedro Saucedo. He and a friend, Joe Salaz, were driving on route 116 about four miles west of Levelland when they saw a flash of light in a field. 'We didn't think much about it', Saucedo said, 'but then it rose up out of the field and started towards us, picking up speed. When it got nearer, the lights of my truck went out and the motor died. I jumped out and hit the deck as the thing passed directly over the truck with a great sound and a rush of wind. It sounded like thunder and my truck rocked from the blast. I felt a lot of heat'.

Dunn, North Carolina, U.S.A., 7 November, 1957. Lester Lee, an elder in the Primitive Baptist Church, said: 'I heard a noise like a dynamite blast and looked up. There was a flash of light as bright as the sun, and I saw an object shaped like a lampshade hovering over my potato patch. It went straight up into the sky'.

Now this is just what we might expect from a *rapid* repulsion field take off, ground effects and noise, aerodynamic noise caused by the sudden intensification of the repellant point source. Should this be of sufficient order, then it would create an explosive displacement in the air, producing a noise 'like thunder' or 'dynamite blast'.

At the beginning of chapter 12 we discussed some centrifuge tests in which grass and soil samples were subjected to negative g effects, and it was shown that we could expect craters to be formed beneath a departing saucer if the ground effect was in the neighbourhood of 3g to 4g. This was the more difficult experiment to conduct, for the samples had to be anchored into a test cell in such a way as to emulate, as near as possible, normal ground cohesion, when the g forces were inverted.

But centrifuge tests on samples which are subjected to *increased* or positive g are comparatively simple. Plate 30 shows some results. First the sample is photographed at normal terrestrial conditions, the accelerometer reading zero. Plate 30(a). Then at a downward pressure of plus 3g the grass is beginning to be flattened as in Plate 30(b), while at plus 6g and 10g, Plate 30(c) and (d), most of the grass and foliage has disappeared beneath the marker line. It must be pointed out that this test, as with the G field counterpart, represents a vertical component of the field, that is, that portion of the downward force immediately beneath the point source. Grass situated further away, although subject to a decreasing force, will be flattened radially outwards, perhaps even more so due to the greater leverage of the horizontal repulsion component.

Earlier we saw that it was possible for a saucer to take off slowly without leaving ground effects, and this we could justifiably explain by the gravitational point source theory. But now we have to ask the question, one theory or two, which is it to be? Well, according to Philosophical Economic ruling we are more likely to be nearer the truth if we seek one.

Broadly, then, on the one hand we have UFO phenomena which leaves craters and another which exhibits pressure, but they have something in common, they both leave ground effects on *rapid* take off. We have seen that sometimes there has been evidence left of some kind of an undercarriage and we have examined cases where witnesses claim to have seen such appendages in operation. By using this as a starting point, let us see if the two effects can be reconciled by one cause.

17

Analysis Seven

R Field Craters

ONE of the most unique cases on record when a witness saw a kind of undercarriage in operation was that of Herr Linke, who, with his step-daughter Gabriele, saw the grounded saucer near the village of Hasselbach. We dealt with this case in chapter 9, but a study of the VTO sequence sketches in Fig 66 may help us further.

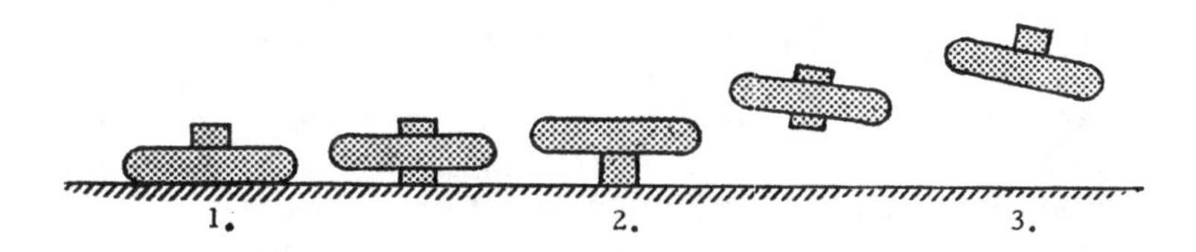

Fig 66. VTO sequences of Herr Linke saucer. If part of the disc did rotate then it might well cause the ground effect left in the Romand case, i.e. the grass was flattened counter-clockwise.

First, you will remember, the disc of the saucer rested on the ground with the cylindrical portion projecting through the centre, as in phase 1. Next, the disc was lifted from the ground to the top of the cylinder, which then acted as the central pedestal mentioned in the other cases, phase 2. This in turn was now retracted up into the disc, finally to appear projecting above the upper surface, phase 3, so that the machine was now suspended and ready for forward flight. On the ground there was a slight depression no doubt caused by the weight of the machine as it rose on the pedestal.

Now from this, it is obvious that for some reason it was necessary for this part of the machine to be raised up. Herr Linke thought that it appeared to be rotating, certainly there was a humming sound, but *was* that the *only* reason? Or was the disc first raised *mechanically*, so that once it was a short distance above the ground, another kind of lift took over? According to the theory outlined in these pages the latter was the case. But if the craft was being lifted away by the gravitational point source conception, then as we have seen a slow take off would not require a ground clearance. Can we now offer a further explanation for this manoeuvre, an explanation which will reconcile the phenomena of craters *and* downward pressure? I think perhaps we can, whether or not it is the correct explanation only time will tell, but remember in this instance we are setting out to find a theory to fit *all* the facts.

When discussing the gravitational repulsion point source in terms of downward pressure as in Fig 65, we saw how this would quite easily account for radiating pressure lines in wheat, etc. By this time no doubt some readers are way ahead of me already, but let us continue. In our mind's eye, let us imagine a saucer 'balanced' on a fixed focal length repulsion point source, at slightly less than 1g, for otherwise, as with the G field ship it will rise through aerodynamic effects. But then reduce the power still more, and the vehicle will descend, with it the focal point. Lower still and the ground beneath suffers pressure as the focal point approaches. Grass is laid outwards flat, and loose debris hurled away from the central blast effect. But I have tried to show that matter is modulated space and the point source is a further modulation of space; it will not be interrupted or quenched into non-existence like a cigarette end stamped into the earth. It will go on downwards like a focused X-ray beam biting deep into the soil, and a phenomenon which split seconds ago announced its presence to the homely worms as something which threatened to crush their tiny burrows from above, is now transformed into an abrasive grinder, which tears them apart, and a crater is formed, Fig 67(a). Thus that part of the economy rule is satisfied, *i.e.* one theory to explain *both* effects.

Here is a more recent case which lends support to the theory:

From the June 1965 *Colchester Gazette* comes the following:

'Until Army bomb experts visit Rockingham's Farm, Layer Marney, the mystery of the crater in the barley field will remain unsolved.

'The sudden appearance of the crater, about 5ft by 3ft 6in, was reported to the police by the farmer, Mr J. W. Black. He found it in a field of young barley, with the *crops around the crater flattened.*

'A puzzling feature of the crater is three 3in holes, arranged in a triangular shape near one side of it, that extend 3ft into the ground. (We shall see shortly that there could be good reason for those displaced marks.—Author.)

'The police are unable to offer any clue as to what had caused the hole, but among the theories advanced so far is that it was caused by a meteorite or some object dropping from an aircraft'.

Yes, there would be a crater, unless a thoughtful designer of this strange intruder arranged for the point source to be arrested before it lowered that far, and one necessity for an elevating undercarriage is established as in Fig 67(b). Should the craft be hovering over water and the point source descend below the surface, the reverse of the Saude case would occur and a 'crater' would be formed in the water, the perimeter of

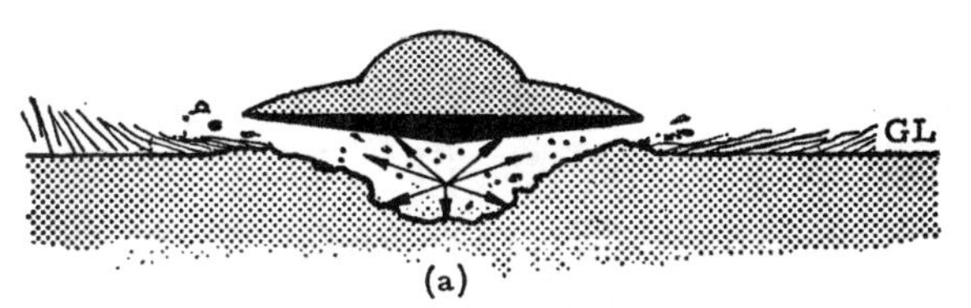

(a)

Formation of a crater.
Repulsion FP below ground level.

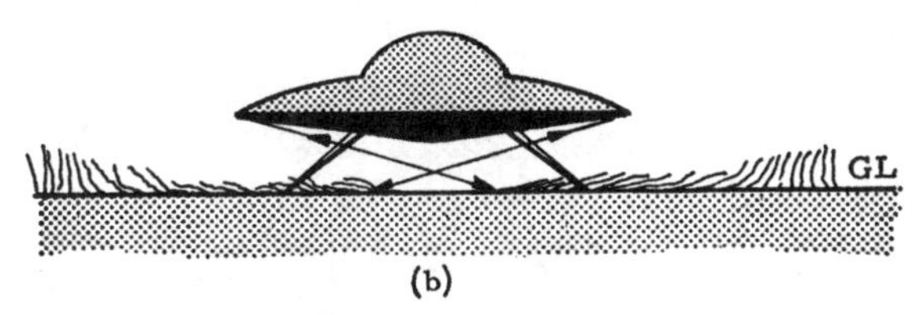

(b)

Downward pressure.
Repulsion F P above ground level.

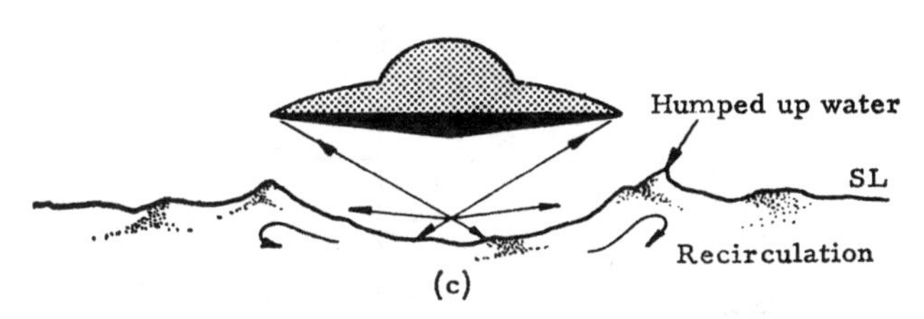

(c)

Fig 67. Surface effects due to a repulsion point source.

which would be 'humped up and pressed back', the resulting recirculation causing the water to froth and boil, according to other UFO witnesses' statements. Fig 67(c)

Space modulated waves which would negatively unbalance or repel matter would almost certainly do so according to the inverse square law. Also, as in the case of the G field, a repulsion (R field) would act on the centre of mass of the vehicle, there being a differential set up within the structure according to the focal length used. Therefore there will be similar effects on other nearby bodies. Fig 68(a) may help to emphasise why it would be desirable to keep the repulsion point source some distance above ground level, why it might be a good idea to keep some distance from a hovering saucer producing ground effects, and why from a design point of view, it would be necessary to make the length of the focus adjustable.

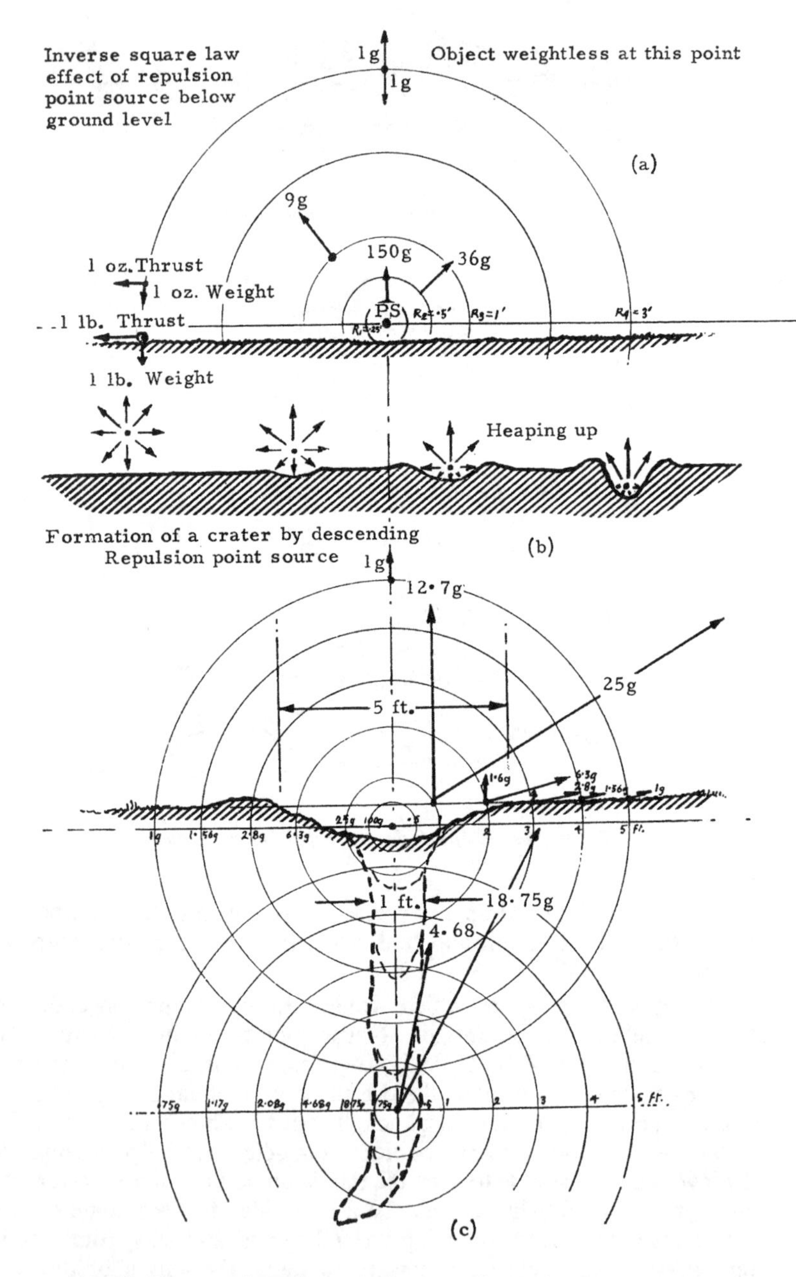

Fig 68. The natural development of a crater due to descending point source.

In the sketch, an extreme case is shown where the point source is down to ground level, the 1g radius being situated at 3ft, that is, if the vertical component of this radius passed through the centre of the mass of an object, saucer or otherwise, then it would be rendered weightless. On one side of the circles the radius from the F.P. is shown in feet and parts of a foot down to .25ft, and the corresponding repulsion g values, derived from the inverse square law for the appropriate radius, is shown on the other side of the circles. From this we can see that loose particles on the surface of the ground will receive a sideways thrust from the F.P. according to their proximity and mass. For instance, a stone situated at the 3ft radius and which normally weighs an ounce will experience a 1-oz radial force. Similarly, a stone of 1lb weight placed at the same radius, will experience a 1lb lateral force. But as friction is a function of weight, both of these bodies would only roll gently away with less force than if they were exposed to the air blast from a hovercraft.

But on the other hand, suppose the 1oz and 1lb stones had been situated at the 3in radius from the point source in the diagram? There we find forces of a totally different magnitude at play, for the stones will now experience a force of no less than 150 times their normal weight, that is 150oz, or over 9lb, and 150lb, respectively. Apply the inverse square law at even smaller radii and we find 'pressure' effects quite sufficient to cause a crater if the F.P. were just penetrating the ground. While on the other hand, if it were formed a foot or so above the surface then no noticeable ground effects would be manifest, certainly far less than a hovercraft in the same attitude.

The analysis of the Gravitational and R field craters could well represent a separate work in itself, but due to the evidential value of such phenomena, it is important that we should make at least a cursory examination, and here again recourse to sketches will afford a good deal of space saving.

The sequence sketches in Fig 68(b) illustrate the physical effect which a descending repulsion point source would have on soil. First the surface immediately beneath will be subjected to downward pressure, and a slight horizontal 'push', both components increasing as the point source gets nearer. So that the general effect will be, to both press the soil downwards in the centre and scoop and heap it up a little at the sides in the process. Therefore, even if the point source remained fixed a short distance from the surface, a shallow crater would be formed, but if it continued the descent, the 'digging' process will also continue and the crater will get deeper.

Fig 68(c) shows a general analysis of the forces at work. The concentric circles at 1ft intervals represent different repulsion levels indicated by the inverse square law, starting with 100g at 6in radius from the point source, up to 1g at the 5ft radius. If we now bisect the focus with the intersections of the circles at ground level, we can plot the magnitude and direction of the accelerations of the particles at these stations. From which the corresponding vertical acceleration components can be read off as: Soil above the centre of focus will experience a vertical acceleration of 100g. Station 1ft, will receive about 25g; Station 2ft, 6.3g; Station 3ft, 2.8g; Station 4ft, 1.56g; while Station 5 will experience a vertical acceleration of only 1g. Therefore, allowing for the horizontal shearing movement and bearing in mind that centrifuge tests indicate that soil would become detached at just over 3g vertical acceleration, then we can plot a rough approximation of the crater formation as shown in the diagram.

From this it will be apparent that if the intensity of the point source is reduced while still descending, then the crater 'digging' process will continue much the same as the above. But due to the g levels intersecting the sloping sides of the crater, the digging and piling up will have an additive effect and the walls will become steeper, finally developing into an elongated funnel, the depth of which will depend on the point at which the field was collapsed.

In fact the result would be almost identical to that caused by a jet of air descending on to sand. First a slight depression, then an ever deepening shaft, the diameter of which is largely governed by the compactness of the medium.

Although it is impossible to treat the vectors fully at such small scale, such a shaft is developed in Fig 68(c) together with two g levels and resulting acceleration values.

Now we can say that should the R field craft (descending at some point above ground level) move to one side a little, then the 'funnel' will be similarly curved or displaced. And further, should the descending vehicle sway, then rise and descend in the other direction, we can imagine its path being traced by the point source somewhere beneath the soil, and we would not be surprised to discover curvilineal and even forked shafts left in the ground. Should the vehicle be taking off and for some reason the point source were situated deep in the ground beneath, then the field must be generated fairly slowly and not too intensive. Then the same general pattern will follow and the funnel 'digging' process will be reversed, from the bottom upwards, finally terminating with the customary crater and/or skinned grass turves which fall back and inwards when the machine has departed.

I will not have to remind UFO researchers that such craters are common and because of their shape the local authorities and some well-meaning UFOlogists are quick to explain them away as old bomb shafts, subsidences, etc. This despite the fact that so many appeared mysteriously in England and elsewhere over a space of a few months! The following are typical:

On 7 October, 1963, the Hastings *Evening Argus* carried the following report: 'Farmer Alfred Gadd peered down the mysterious 20ft deep hole in his barley field at Fittleworth and scratched his head. 'Never seen anything like this before', he said. 'But I suppose someone will start saying it was made by a flying saucer from outer space'.

'Bomb disposal experts—who also investigated the Wiltshire turnip field riddle—were called in, but were non-committal. No full-scale digging operations were started, although the soldiers left nothing to chance, squelching across the muddy stubble to probe around the hole with a detector. They found nothing. Still, they are coming back to have another look, because a 1,000lb bomb could have buried itself up to 40ft deep in soft earth.

'Mr Gadd, aged 65, of Fitzleroi Farm, thinks the 2ft wide hole could have been caused by an unexploded German bomb during World War II. 'I heard them fall but I didn't hear them go off', he said. 'We have been ploughing this field for 30 years and the hole appeared after the combine harvester had gone over the spot to cut the barley".

'Bomb or no bomb, it won't stop him working in the field. The only clue that there is a hole is a bale of straw sticking up in the middle of some of the most isolated countryside in West Sussex. And that's been put there to stop some unsuspecting rambler from falling down the hole'.

The Nottingham *Guardian Journal* on 26 November, 1963 carried the following report: 'A Bomb Disposal squad from Sussex were yesterday drilling in a field at Home Farm, Belton, near Grantham, after a tractor-man had reported a hole appearing in the ground. Home Farm is owned by Lord Brownlow. Three of the squad have been sleeping in a tent only a few yards from where they have been drilling.

'Capt. J. E. Rogers, Officer Commanding No. 2 Troop at Horsham, said last night: 'So far nothing has been found to suggest that there is actually an unexploded bomb in the field. We have been drilling in the vicinity and inserting metal locators in an effort to trace any metal which could indicate the presence of a bomb'.

And here are two nearer-to-home occasions published in *Space Link*, the journal of the Isle of Wight UFO Investigation Society.

'Mystery' Hole on Island Farm.

IWUFOIS carried out their first group investigation of a mysterious hole which appeared suddenly on the land of Mr Ray Peach, Puckwell Farm, Niton, on the night of 16 January, 1964.

Mr Peach said that he had to stop ploughing his 13-acre field, Ridges, when the hole appeared. It was approximately 18in in diameter. It widened out under the surface to 3ft across and went almost straight down for about 12ft, narrowing slightly, then turning and disappearing round a bend. Fig 69(a) and Plate 31. The sides were slightly irregular and a large stone or flint jutted out of the wall about 6-7ft down.

A Bomb Disposal squad from H.M.S. *Vernon*, Portsmouth, made a preliminary investigation without committing themselves to the cause

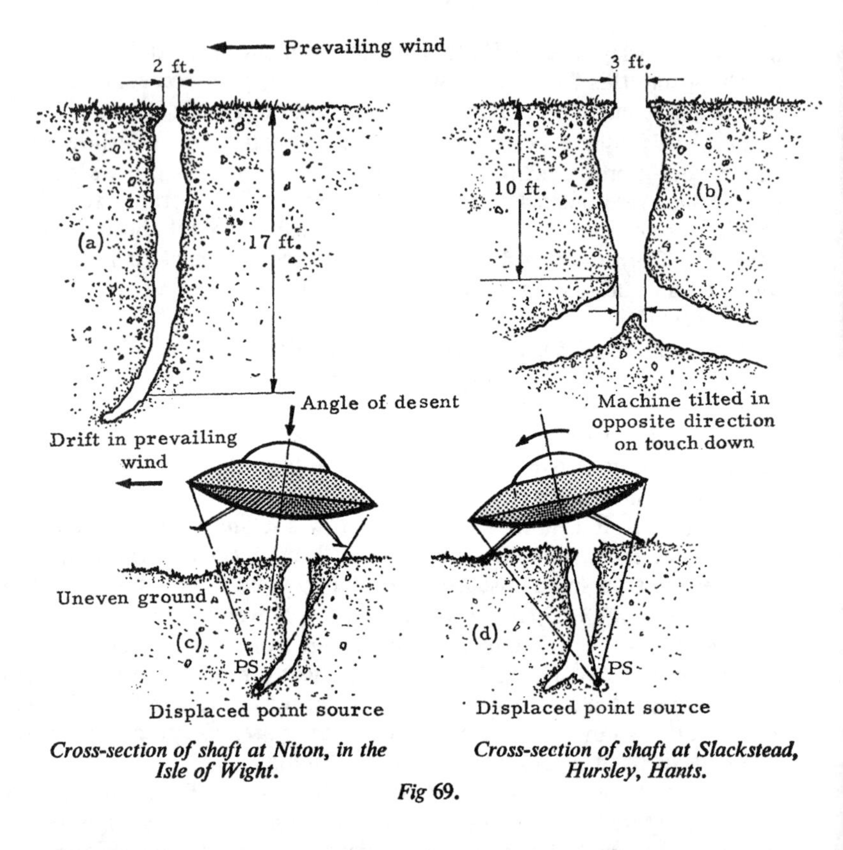

Cross-section of shaft at Niton, in the Isle of Wight.

Cross-section of shaft at Slackstead, Hursley, Hants.

Fig 69.

of the hole. A further more ambitious investigation by an Army Bomb Disposal unit was then made. The group started digging, only to give up with no firm conclusion after producing a hole 10ft square by some 20ft deep. Conclusion: The Bomb Disposal squad decided that it *could* have been caused by a natural movement of the ground and again *possibly* caused by an unexploded bomb. A few days before the discovery of the hole, two boys claimed to have seen a circular, silent machine taking off at Niton.

Another hole investigation by the Southampton Aerial Phenomena Investigation Group: Two men almost fell down yet another mysterious hole which suddenly appeared at Slackstead, on Hursley Estate, Nr. Winchester, on 12 February, 1964. A Bomb Disposal unit said an 'old bomb' had caused the hole. The shaft, which was branched at the bottom, was 10 to 15ft deep and 2 to 3ft in diameter. Fig 69(b) and Plate 31.

And yet another, this time in the United States: According to the Philadelphia *Sunday Bulletin* of 6 September, 1964 and the Philadelphia

(a) (b)

Meadow grass subjected to increased g in centrifuge test cell. As with the negative g tests, a marked response occurs at 3g.

(c) (d)

At an increased g load of 6-10g, the grass and foliage is considerably pressed down as UFO witnesses have claimed. Photographed while cell was in motion at lineal speeds of up to 40ft per sec.

Plate 30.

Photograph looking down shaft found in December 1963 at Slackstead, Hinsley Estate, Winchester, England.

Plate 31.

Photograph looking down shaft found in February 1964 at Niton, I.o.Wight.

Plate 32.

UFO photographed at Beaver County, Pa., U.S.A. in August 1965 by James Lucci. This might well be visual evidence of the R field.

Evening Bulletin of 7 September, 1964 police authorities and public at Glassboro' (New Jersey) were greatly mystified by weird marks found in an oak forest four miles north-east of that town.

Glassboro' Police Chief Everett Watson described the marks as a circle of charred earth some 20ft in diameter, with a hole 2½ft deep at the centre, the hole being surrounded by a small series of mounds of burnt material and what appeared to be metal scrapings. There were also three marks, arranged to make an equilateral triangle with sides 27ft long. The marks, which looked as though they had been made by the legs of a huge tripod, were 2in in diameter and 6in deep. A few broken tree limbs were found in the vicinity.

The police were called to the spot by Mr Ward Campbell, of 30 Delsea Drive, Glassboro', whose two sons had found the marks. The boys had been fishing in a lake nearby, and they had met a man who had told them that shortly after dusk on the previous Friday (4 September) he had seen first a glow in the sky and then a glowing object which landed in the forest about half-a-mile from the lake.

The affair caused tremendous excitement locally, and hundreds of people went out on the Sunday afternoon to view the mystery marks. The newspaper reports conclude with the statement that the Glassboro' Police, completely baffled, had taken 'samples of earth and other materials' and had sent them to McGuire Air Force Base for examination.

From what has been said, it would seem that the bottom of the shaft represents either the beginning or the end of the point source formation, or both. And it is not difficult to imagine the saucer drifting down vertically, then due to the now almost negligible field, the wind momentarily taking hold and causing the machine to drift a little, and with it the displacement of the much-narrowed and almost-petered-out funnel. (Note how in the Niton case the curve of the funnel coincided with the direction of the prevailing wind, Fig 69(c).) At the instant of impact, however, the landing pads may contact unlevel ground; consequently when the point source is formed for the lift off, it has been displaced to one side, and we have another possible reason for the forked shaft at Slackstead, Fig 69(d). Notice also in the Rockingham Farm crater at Layer Marney, considered previously, there were ground effects which might bear evidence of displacement, quote: 'A puzzling feature of the crater is *three 3in holes* arranged in a triangular shape *near one side of it*'.

Now we have considered reasons why a G field craft, having caused a crater on lift off, might take some of the debris aloft, can we relate this same phenomena to the R field concept? Indeed so; in fact it may already be obvious. For we have seen that in a case where the P.S. forms a crater, the ground particles immediately above it are exposed to a

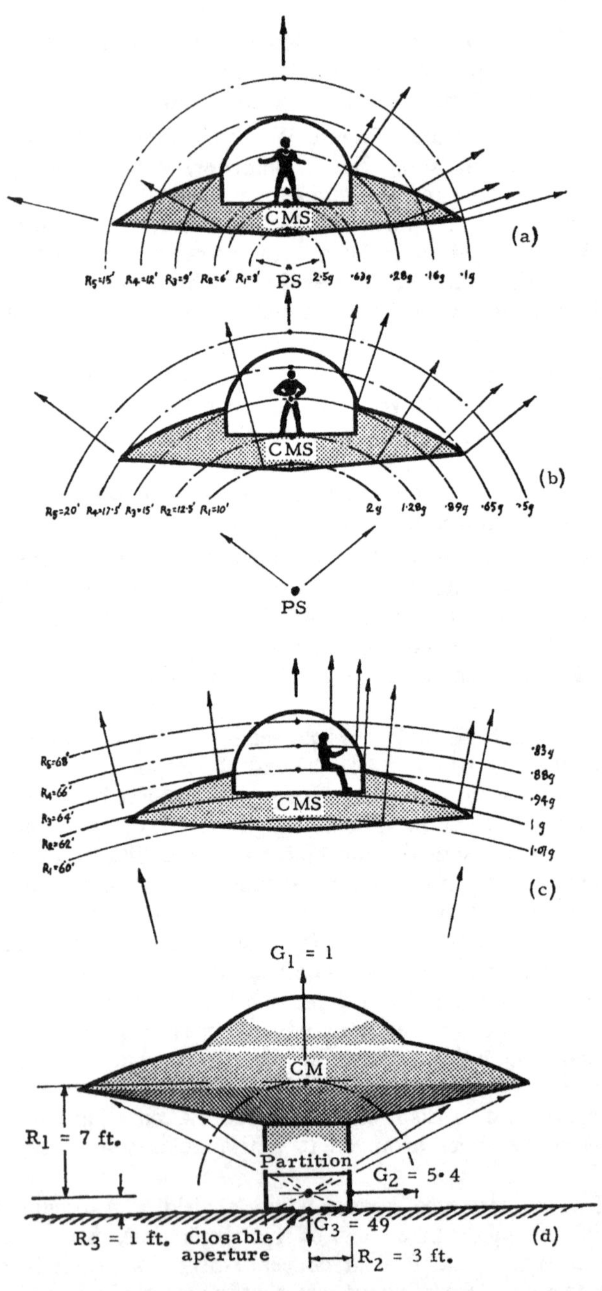

Fig 70. G differentials in R field ship due to different focal lengths, and forces on retractable cylinder at lift off.

stronger part of the field than the machine itself, and therefore are accelerated at a higher velocity than the craft, hence they rapidly overtake it and appear to cling to the under-surface as in the case of the G field. So far so good, but are there any other factors to the R field which support witnesses' claims? Indeed, there are many, we can only examine a few. Often I have deliberately tried to break many of these cases down to find any weakness. Sometimes I think I have succeeded, only to find that I have been wrong, as indeed the whole of these conclusions may well be erroneous; the reader must be the best judge.

The Undercarriage

In discussing the inverse square law relationship to the R field, I pointed out that the vehicle and everything in it would be subjected to varying values of acceleration. Fig 70(a)(b)(c) illustrates diagrammatically changes in these levels, accompanying changes of field intensity and/or focal length. We shall see later that there is a comparatively simple means of rectifying this state of affairs, but for the moment let us examine the crater and ground effect problem from the engineering point of view.

If we were designing a gravitationally propelled machine, what could we do to eliminate the ground effect? Well, first we would make sure the F.P. never reached below ground level. How? By fixing a limit on the focal length and giving the craft a leg-type undercarriage. Good, but allowing this to be possible, suppose the strength of the field had to be fairly high for a rapid lift off, due to the inverse square law, the aerodynamic and ground effects could still be spectacular. Yes, but remember, the point source can be created below ground level, without any nullifying tendency on the levitation of the machine, for, as with gravity, there can be no shielding of the R field point source, it will act on and penetrate all matter. Therefore, on the same premise, you the reader would in all probability suggest that a protective shroud be built around the point source, so that nothing could come within a certain distance of it. Better, it could be an exhausted cylindrical chamber, containing low pressure. You might feel pleased with your idea and go on to suggest that such an appendage could play a dual rôle, the lower evacuated part of it could be partitioned off and the remainder used as an undercarriage, a pedestal type undercarriage as in Fig 70(d). Conveniently placed as it is immediately beneath the central cabin, what better way for gaining access to your aerial vehicle! Though note, the occupants of the UFO in Herr Linke's case did not. Fig 71.

Even so, we are in a position to do a few sums on this part of the new design; see how it would work out in practice. But first it must be emphasised the values of g are *repellant* or *negative* in this issue, and will be designated Gr in this present context.

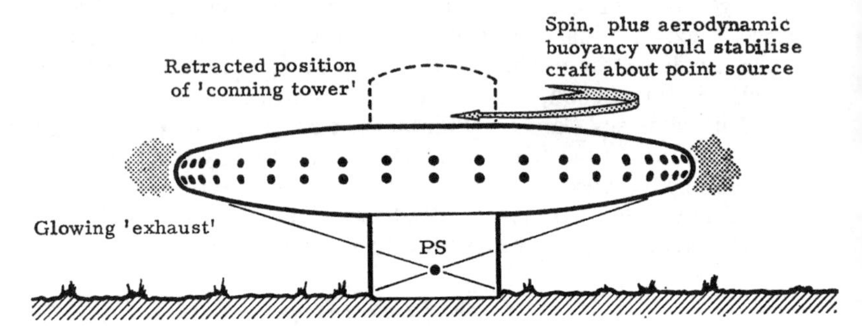

Fig 71. Herr Linke case discussed in terms of the R field theory. Occupants of the craft did not enter the lower part of the central 'conning tower' in this case.

The 30ft diameter disc in Fig 70(d) is supported some 6ft off the ground by a cylindrical pedestal. The 1g radius (1Gr) which passes through the centre of mass of the ship we shall estimate at being 7ft, while the distance from the R field point source is only 1ft from the base of the cylinder and the ground, the radius of the cylinder, say 3ft. We shall designate these R_1, R_2 and R_3 respectively. Assuming 1Gr at R_1=7ft, from the inverse square law we find the remaining two values for Gr:

$$\text{viz. } Gr2=\frac{R_1^2\,Gr}{R_2^2}=\frac{7\times7\times1}{3\times3}=5.4\ Gr$$

$$\text{and } Gr3=\frac{7\times7\times1}{1\times1}=49\ Gr.$$

This in effect says that the walls of the cylinder in this circumstance would experience a force of approximately 5.4 times its normal earth weight. Therefore, if constructed of light alloy this force should not be formidable, while only a cursory examination is sufficient to reveal that the bursting or hoop stresses set up would present no problem at all from the engineering point of view. And in any case this would be offset somewhat by the low internal air pressure. On the other hand, the downward acceleration Gr3 on the bottom of the cylinder is much greater, equivalent to some 49 earth gravities in fact, which again means that if the end structure of the pedestal normally weighed only 50lb, then it would experience a downward force of something like 2,500lb, which must be subtracted from the lift force. So that if the total weight, and therefore the lift, of the vehicle was about 30 tons, then the effective lift of the R field would be in the neighbourhood of 29 tons. Now this is a pretty formidable lift penalty to pay in order to render more safe the lift off of the ship, so the next obvious step would be to move the cylinder out of the way. Once it was well clear of the ground, what easier than to make it retractable?

As the retracting cylinder moved towards the point source, an iris type vent must be provided in the disc floor to lessen the downward thrust and prevent the metal from volatilising due to the enormous forces there. At this stage the power of the field would be closely controlled, also air would enter the cylinder, which in turn would disintegrate unless the process was a slowly governed one.

I intimated earlier in chapter 6 that, given a starting hypothesis to work on, any engineer who had never made a study of UFOs would end up designing a typical flying saucer; I would like to think that I have established the point. Of the sceptic then I would ask again, how are we to explain these strangely prophetic 'hallucinations', 'misinterpretations' and 'fairy tales' of UFO witnesses? For they continue to make good sense. With all good grace and humility I would hasten to add that I welcome my task rather than yours. But let us return to the R field theory; there is much more to come.

18

Analysis Eight

Crew G Differential

IN chapter 11 we considered a G field arrangement which would furnish the crew of the space ship with a comfortable 1g differential in all phases of flight. The repulsion field concept takes this problem more in its stride and even the briefest examination will reveal that the 1g ship/crew differential could be provided without recourse to a secondary field inducer, though it must be understood, we are dealing at present with the centre of mass only. The differential spread throughout the pilot's body will still apply, but is neglected here for the time being. In the accompanying diagram Fig 72 an extreme case has been chosen where an R field saucer is hovering only 4ft above ground level. There are mechanical difficulties likely to be encountered in this condition, which although solvable would spoil the story if discussed at this stage; therefore, it is best to neglect this factor and assume the difficulty overcome.

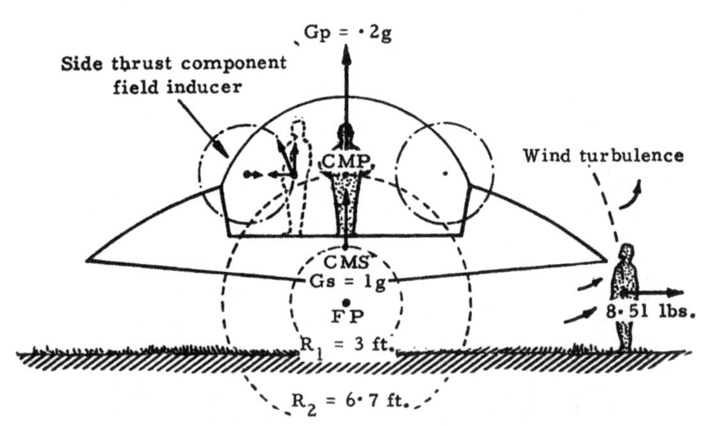

Forces acting on hovering R. Field type saucer

Fig 72. In order to maintain a 1g differential on the crew when the craft was being taken aloft by a 2g acceleration, with a FL of 15ft (R1), then (R2) would have to be no less than 21ft. Consequently there would be a drastic increase in the depth and therefore overall bulk of the ship.

The centre of mass of the vehicle is 3ft from the F.P. (R_1) and the centre of mass of the pilot is 6.7ft from the F.P. (R_2). Assuming a lift factor of 1g through the centre of mass of the craft, then the lift component acting on the pilot will be as before:

$$G_p = \frac{R_1^2 \, G_s}{R_2^2}$$

and substituting we get $G_p = \frac{3 \times 3 \times 1}{6.7 \times 6.7} = .2g$.

Therefore, the pilot has lost 20% of his weight, which, if he weighs 170lb, amounts to approximately 34lb. This is by no means intolerable, nor inconvenient; in fact, a trained person would barely notice such a reduction. In this condition grass immediately beneath the F.P. would be pressed down a little and, as we have already seen, loose stones might move away. Out further still towards the perimeter of the disc there would be little disturbance, perhaps a slight wind could be felt. Standing by the craft, an average-sized man would experience a lateral force, of some 8½lb, tending to push him away from the ship, but this would be barely noticeable, and in any case he would in all probability attribute it to the wind. Later on towards the end of our story we shall see there is a means by which even this force will be eradicated.

As with the G field point source, there would be an inherent side thrust component acting on the crew, but in the opposite direction, *i.e.* while an inward force acts towards the centre of the ship in the case of the G field, an *outward* component is experienced by the crew in the case of the R field. Again this can be nullified by the waist position side thrust field inducers, but similarly these would be of the opposite sign, that is R field instead of G field. Naturally we are only considering these conditions basically here. No doubt properly engineered, some of the inherent variations in field strengths would be quite easily overcome and the crew would feel no sense of restraint whatsoever.

As with G field propulsion, a high increase of the power factor for the same value of R_1 in Fig 72 would mean increased or even prohibitive structural g differential, to say nothing of the discomfort of the pilot. Consider the case for an 8g lift off.

$$G_p = \frac{3 \times 3 \times 8}{6.7 \times 6.7} = 1.6g$$

Therefore, while the machine reacts to an acceleration of 8g, the pilot reacts to an acceleration of only 1.6g, that is, in this case he experiences the uncomfortable force of 6.8g.

Now this can be overcome comparatively easily and, as I have said, without the aid of the secondary field inducer, by simply increasing the focal length of the field. The accompanying table, diagram and graph, Fig 73 and Fig 74, show the different focal lengths required to

Gs on c.m. ship.	Gs on c.m. crew	Diff. Gs.	F. Length (R_1)	F. Length (R_2)
1	.106	.894g	1.77ft	5.44ft
2	1	1	9.17	12.87
3	2	1	16.55	20.25
4	3	1	23.98	27.68
5	4	1	31.32	35.02
6	5	1	38.75	42.45
7	6	1	46.20	49.90
8	7	1	53.60	57.30
9	8	1	61.00	64.70
10	9	1	68.40	72.10

Note, for the pilot to maintain 'normal' weight in the ship at any of these G values the physical distance between his c.m. and craft c.m. remains a constant 3.7ft. But this is a purely arbitrary figure and would depend on the physical dimensions of the craft.

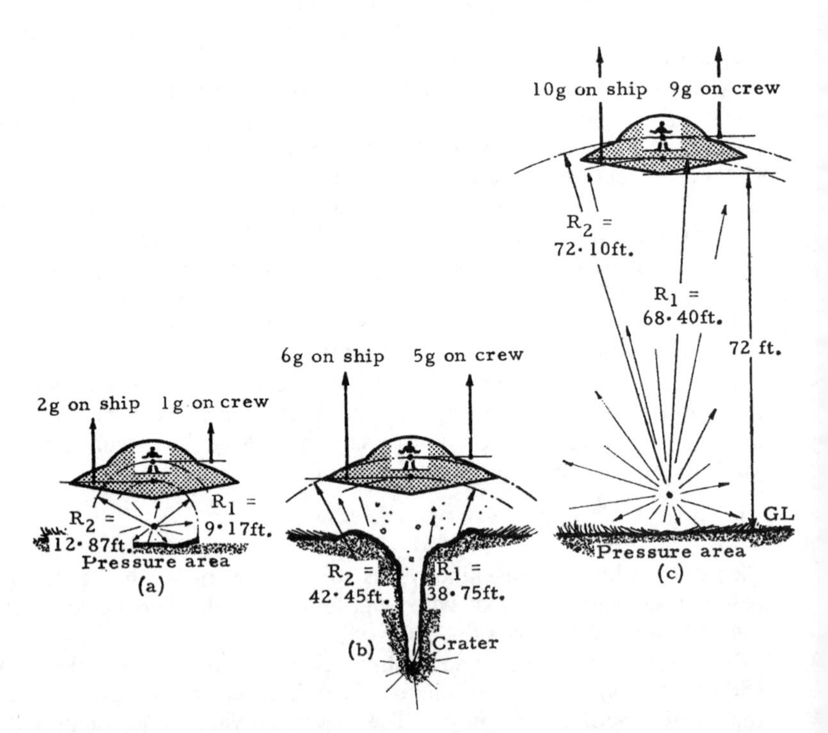

Fig 73. Factors for the variable focal length, to maintain 1g differential on lift off.

produce a constant differential of 1g between the ship and the crew (terrestrial condition) assuming an initial F.L. (R_1) of 3ft.

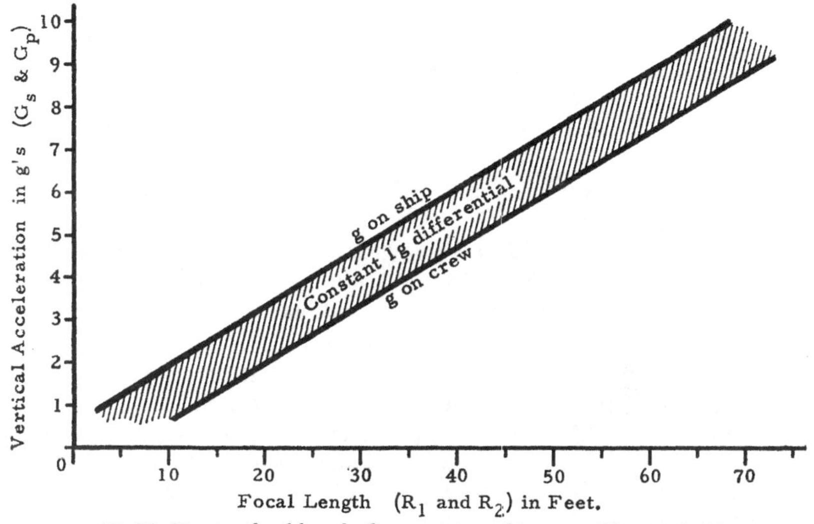

Fig 74. Varying focal lengths for a constant ship-crew differential of 1g.

I would draw the reader's special attention though to the exception, in this instance, where there is only a differential of .894g, that point at which the pilot loses 10.6% of his weight, *i.e.* when the craft is hovering just before *lift off* or *landing*, for we shall consider this factor from another aspect later on.

Sketches Fig 73(a)(b)(c) illustrate more graphically how the R field theory is complementary to the above requirements. Note how conveniently short the focal length need be at even a 2g lift off, to maintain the 1g differential, Fig 73(a). The alternative for a quicker take off from that position would be increased power for the same F.L. resulting in a high structure and crew differential, yet no crater. Or high power and a longer F.L. as in Fig. 73(b), which will offer a quicker getaway but, as we have seen, will also produce a funnel-type crater, particles from which will overtake the ship with the risk of damage. On the other hand, should the pilot first ascend as in Fig 73(a) it will take about 2.12 seconds to reach an altitude of 72ft, at which point he can then increase power and adjust the F.L. to 68.40ft, producing a vertical acceleration of 10g on the ship, a comfortable 1g differential on the crew (normal earth weight) a loud bang due to air displacement and a flattened pressure area on the ground below, Fig 73(c).

Earlier in this chapter I quoted several applicable cases which support this theory, but here is another for extra measure. For including so many

I ask forgiveness of the UFO researcher, but to the scientific snob and scoffing UFO sceptic at large I would say: 'If you still doubt, then I shall be quite relentless in presenting such technically corroborative evidence, I shall drive you on and on until your case is made to look more utterly foolish than you have tried to make mine. For each time you snigger at many good folks' honest testimony, each time you laugh at their failure to make you believe them, so the more wanting will your plea, 'Well I always did feel there might be something in it', be!'

It was about 10 p.m. on 11 October, 1954 at a place called Montbazens (Aveyron), France. Six men were working in a garage workshop run by M Carrière, when he asked his son Bernard, a boy of 17, to bring him a tool. In order to do so the boy had to pass a window and in doing so he noticed a bright light which appeared to come from an adjacent field. Thinking it was a fire, he called to the others.

All the men ran out, then stopped in amazement. For there, parked beside a neighbour's house, was a disc-shaped object of about 4 yards diameter, emitting a powerful red light. They hesitated to approach the thing, but one M Gardelle, went closer for a better look. He had only gone a few yards when the disc rose noiselessly from the ground. Then, when it had risen *some yards from the surface*, it had disappeared with a terrific burst of acceleration. M Gardelle staggered back clutching his hands to his face. When the others reached him, they found him 'choking and gasping for breath, *stunned as if by a violent concussion*'. As if in fact the breath had been *knocked* out of him? (Author.)

Remember the case of Signor Zuccaàl at Cidinella we saw in chapter 9? It was he who said: 'I felt myself struck and lifted up by a sharp gust of wind'. And again in the Sawbill Bay case where the witnesses spotted the crew of a saucer actively engaged in some mysterious exercise above the lake: 'Now there came a rush of wind, a flash of red, blue, gold and it was gone.'

If on occasion saucers might be dangerous to humans, what kind of hazards might they involve for those who crew them? No doubt there are many, but a more obvious one comes to mind. What would happen, for instance, if an R field focus went wrong, and the P.S. was generated at high power within the ship? Well, every molecule within its structure would experience an outward pull. And perhaps significantly, if the power were to be suddenly generated beyond the point of molecular cohesion, the whole thing would volatilise, molecule, nay, atom by atom! This might prove to be quite a novel way of obliterating an expendable space craft. Certainly it would prove more practical and far more effective than an installed conventional explosive to do the same job. It so happens that there are an adequate number of cases on record known to researchers which might suggest this. A 1960 issue of *The Times of India*, published in Bombay, carried this one.

A report that a flying saucer manned by 'four little spacemen' had landed in Mozambique, East Africa. *The Times* quoted Lisbon papers of 7 April and referred to the Portuguese news agency, Lusitania, as the source of the information. The despatch from this agency stated that inhabitants of Beira, on the Mozambique coast, had seen an orange saucer-shaped object in the sky emitting a sharp whistle. It landed a few seconds later and soon afterwards was destroyed by a loud explosion, the inhabitants said, adding they they had seen four small creatures of human shape running away from the machine.

In chapter 12 we analysed in terms of the G field theory the claims of some witnesses who said they had stood near to, or even beneath, a hovering saucer. And we found this to be in agreement with identical predictions allowable by the theory. So now we shall look at these same witnesses' claims from the Repulsion theory point of view.

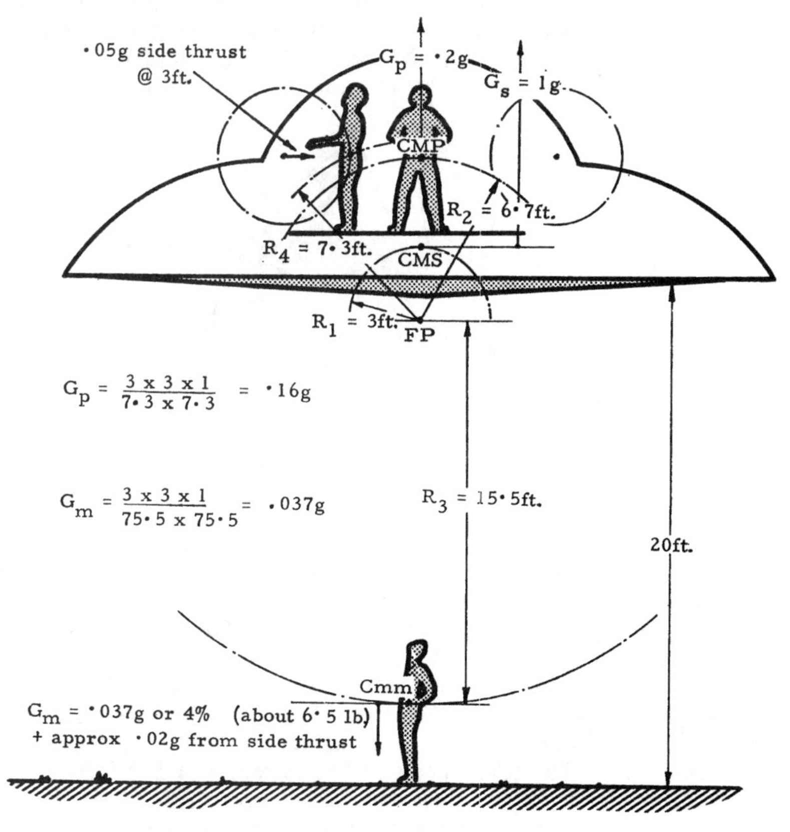

Fig 75. A craft employing a short focus R field might hover above a witness producing no other effect than a slight increase in weight.

Already we have seen how a man might well stand alongside this type of saucer when it was hovering a few feet from the ground, but what would be the effect on him if the disc were to be suspended, say, 14ft above his head as in Fig 75? Well, in the first place we know that any physical effects the field did have on him would be in the form of an *increase* in weight, the opposite in fact to his G field counterpart in similar circumstances, considered earlier in chapter 12. By assuming R_1 and R_3 to equal 3ft and 15.5ft respectively, and applying the inverse square law again, we get downward g on the man Gm to be:

$$G_m = \frac{R_1^2 \, G_1}{R_3^2} = \frac{3 \times 3 \times 1}{15.5 \times 15.5} = .037g$$

or about 4% of the man's total weight, which again represents about 6.5lb if he weighs 170lb. To this we must add the vertical downward component of the side thrust inducers on the man, and we find in this instance the average side thrust would have to equal something like .05g at 3ft radius. This would produce a vertical downward force at the centre of mass of the witness of approximately .02g. Therefore, the total increase in weight of the man would be no more than about 9½lb.

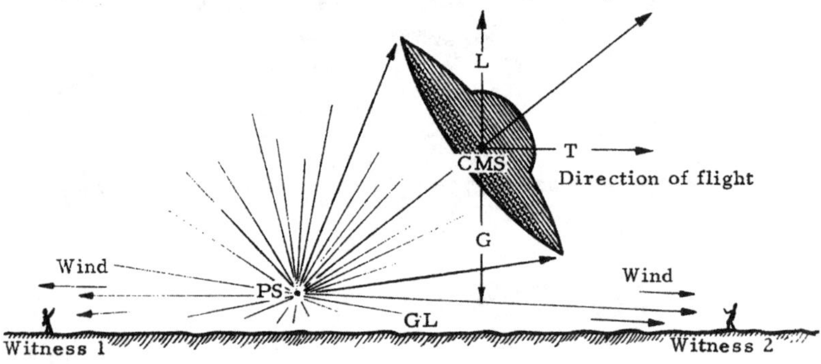

Fig 76. The R field operated ship must be tilted for forward flight for exactly the same reason as the G field counterpart.

The Confirmative Tilt and the Effect on Moving Cars

Before we leave the local ground effects of the R field saucer, we must go on to the next mode of operation, that is, forward propulsion. The G field theory predicts that the forward or leading edge of the disc must be tilted downward, helicopter fashion (and for the same reason), *i.e.* to produce a forward thrust component. And so with the R field ship, there is no fundamental vectorial difference, Fig 76. The ship receives a thrust away from a lower displaced point source, instead of an 'attraction' to a higher one. The components of lift, terrestrial g and forward thrust, are identical. It follows then that we can now predict some additional factors. I shall deal with the more obvious as though the reader were piloting the vehicle himself, bearing in mind the conclusions arrived at thus far about this type of craft. Remember also we are dealing with facts arrived at by calculation.

You are hovering your craft at a few feet above ground level using the short focal length beam in order to maintain the minimum weight reduction for you and your crew. Even so you are now some 34lb. lighter than you would be standing outside the ship on the ground. You wish to turn the ship through several degrees and a slight turn of the axis control gently absorbs a fraction of the energy in the spinning mass of the main rotor. Watching your view carefully you ease back the rate of turn and press forward on the stick. The nose of your little craft dips slightly and soundlessly as you move slowly forward. Opening the throttle more, you are conscious of returning weight as the vehicle climbs. The earth begins rushing past and you feel slight effects of acceleration. Watching the crew accelerometer, you reach out and adjust the beam focal length, you are now well clear of ground effects and you ride with the controls set at normal, so the focal length will adjust itself automatically, compensating any further increase in throttle setting. Your weight now returns to 'normal', the slight feeling of acceleration disappears and, save for the porthole view and the tell-tale instrument dials, you have no awareness of movement. A momentary flick of the throttle records a responsive 8g on the clock and a brief streaking panorama manifests below, yet you feel absolutely nothing.

At this stage, we leave this imaginary flight for a while and return once more to the scene of your take off.

Had there been two witnesses, positioned one ahead of and one behind your flight path, they would tell a similar story, in so far as they may both have felt a rush of wind as we could expect from Fig 76.

'A tilt and a rush of wind'; how often have saucer witnesses stated this?

At Sawbill Bay. 'It tilted at an angle of near 45 degrees. Now there came a rush of wind—it moved so fast my eye could not follow it.'

Miner Black said: 'The disc hung in the air for a few seconds and went off at an angle of 45 degrees with a hiss.'

It was truck driver Pedro Saucedo at Levelland whom earlier in this chapter I quoted as saying: 'I jumped out and hit the deck as the thing passed directly over the truck with a great sound and a rush of wind.'

Among the UFO files there are many witnesses' reports almost identical to this, a rush of wind away from a retreating saucer, and a rush of wind as it approaches. If the blast rocked Saucedo's truck, it would certainly flatten wheat or corn in the direction of the UFO's travel. And again our case is not made any easier by the knowledge that this latter ground effect might be predicted by both the G field and the R field theory. Earlier in Fig 59 we analysed this effect in terms of the G field. Here in Fig 77 we see how the phenomena could be equally applicable to the R field theory, that is when the craft are tilted and moving. Remember our theoretical deductions suggest that the obvious way to obtain forward thrust is to tilt the lift vector of the machine, and no account of corroboration among sightings would be complete without a mention of the following classical observations of UFO phenomena:

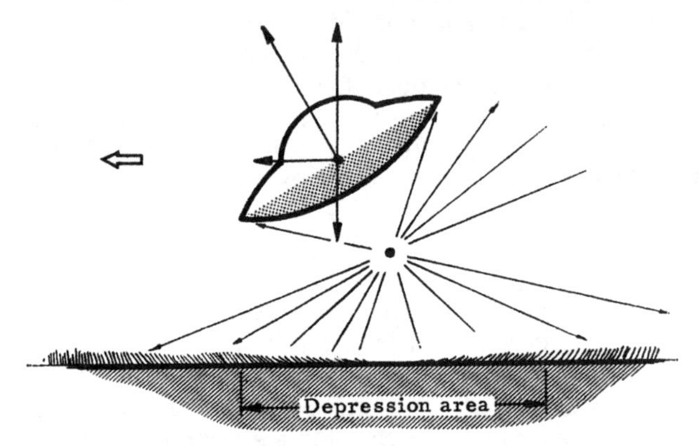

Fig 77. Chabeuil case reviewed in terms of R field theory.

On 30 June, 1954 near Oslo, Norway, three planes carrying scientists and technicians on a scientific expedition were flying through the moon's shadow during the solar eclipse. No less than 50 of these personnel saw and filmed two enormous silvery discs swooping down from the cloud layer some 15 to 20 miles away (the distance was assessed by the fact that the craft were lit by sunlight). Many of these trained observers detected the metallic appearance and spin of the UFOs through binoculars, and quote: 'As they sped along the horizon keeping an exact distance from each other, one slightly above the other, both with *forward edge tilted down*!' The general showing of this colour film alone would convert many a sceptic.

Cars and hovering saucers; we have examined a few cases from the electro-magnetic standpoint, but now let us look at two of these cases in terms of the R field theory. In chapter 8 we saw how Mrs Myra Jones saw a saucer flying immediately above her husband's car. 'There were dark spots around the rim of the base and the whole thing seemed to be tilted *slightly* and to be revolving.' Note there was no mention of any braking tendency on the car in this case. On the other hand, Mr Ronald Wildman said his car *lost power* and I pointed out that there may have been other reasons for this in addition to electro-magnetic ones. Now the question arises: Is there an inconsistency in these two cases, or can they be explained by either the G field or R field theories, or both?

First then let us re-examine both cases from the G field point of view. In the case of Mrs Jones, the operative word to note is *slightly*, for a slight tilt means a *small* forward thrust component of the hovering vector, which the UFO would require, travelling as it was at the relatively slow speed of a car. A small point perhaps to the layman, but a mighty big one to the trained engineer.

Fig 78(a) is a reconstruction of Mrs Jones' sighting, in which we assume there may have been slight errors in judgment of distances; even so, from the following we can arrive at a fair assessment of the case. I have made several assumptions, such as, the saucer was hovering at 1g

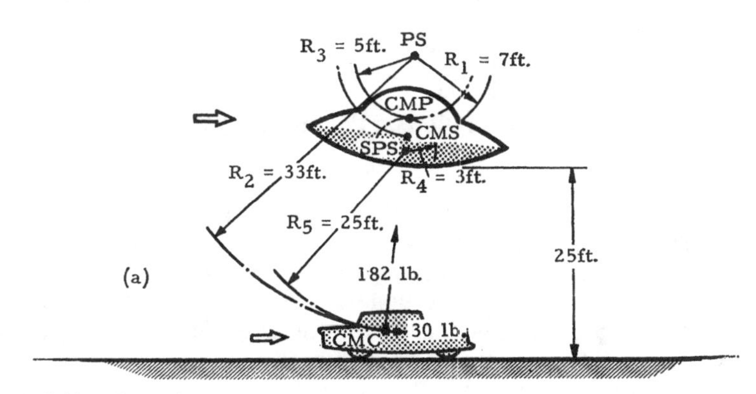

G field analysis of Mrs Jones' sighting on Leicestershire-Derbyshire border.

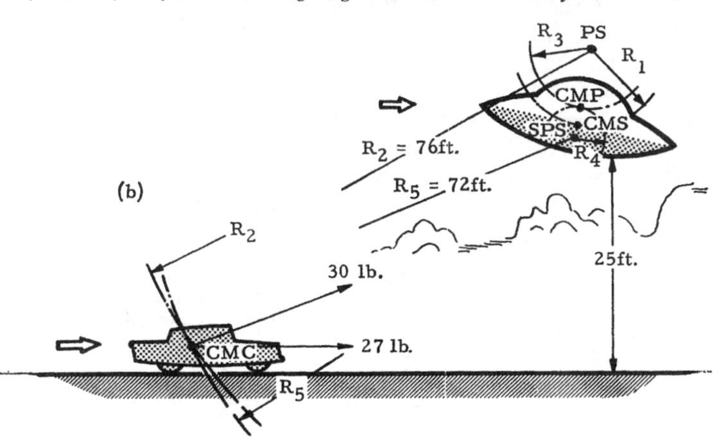

G field analysis of Mr Wildman's sighting on Ivinghoe road.
Fig 78.

vertical acceleration, with slight forward thrust, and that $R_1=7ft$, while $R_2=33ft$. From which we get upward primary field effects on Centre of Mass of the car:

$$G_c=\frac{R_1^2\,G_s}{R_2^2}=\frac{7\times7\times1}{33\times33}=.045g.$$

The distance from the primary field F.P. to the C.M. of the crew (R_3) is 5ft. Therefore the pilot's upward acceleration is:

$$G_p = \frac{R_1^2\ G_s}{R_3^2} = \frac{7\times7\times1}{5\times5} = 1.96g.$$

This is also the downward acceleration which must be exerted on the pilot to give him normal weight. The secondary field of F.L. to the centre of mass of the pilot (R_4) is 3ft, while the corresponding distance to the C.M. of the car (R_5) is 25ft, from which we can derive the additional upward aceleration on the car:

$$SG_c = \frac{R_4^2\ G_p}{R_5^2} = \frac{3\times3\times1.96}{25\times25} = 0.28g$$

∴ total upward acceleration on the car acting through saucer axis

$= .045 + .028 = .073g.$

Assuming 2,500lb weight for the car and two people, this represents an upward vector of about 182lb. From which it will be seen there is a small forward thrust of some 30lb imparted to the car. This combined effect therefore, is no more than would be experienced while negotiating a fairly sharp bump on the road, and in the circumstances it would probably go unnoticed by the occupants of the car.

Summarising the case we have:

(1) Small angle of tilt, low forward thrust component.
(2) Reduced focal length to minimise ground effects.
(3) Combined primary and secondary field lift components on car.
(4) Cockpit side thrust inducer vector on car as being negligible.
(5) Any increase in the forward thrust component of the saucer would tend to *pull* the car along.

In Fig 78(b) we analyse Mr Wildman's case also from the G field concept. In this I have assumed the UFO to be at a similar altitude to Mrs Jones' case, *i.e.* 25ft. Also the values for R_1, R_3 and R_4 are kept the same. A fair assessment for the distance from the primary P.S. and the C.M. of the car (R_2) would be about 76ft, while the corresponding distance from the secondary P.S. (R_5) would be 72ft. From which we get, primary field upward acceleration on car:

$$G_c = \frac{R_1^2\ G_s}{R_2^2} = \frac{7\times7\times1}{76\times76} = .0085g.$$

$$\text{And } SG_c = \frac{R_4^2\ G_p}{R_5^2} = \frac{3\times3\times1.96}{72\times72} = .0034g$$

$$= .012g \text{ total.}$$

Again assuming a car weight of 2,500lb this represents a force acting on the C.M. of the car of about 30lb, which would produce a forward thrust component of only 27lb. Again, as with Mrs Jones' case, such an effect would pass unnoticed.

Summarising this case we have:

(1) Small angle of tilt, low forward thrust component.

(2) Reduced focal length to minimise ground effects.

(3) Combined primary and secondary field lift components on car.

(4) Cockpit side thrust inducer vector on car as being negligible.

(5) Any increase in the forward thrust component of the saucer would tend to *pull* the car along.

(6) In this instance, the frost on the top branches of the trees would lose approximately 28% of its weight. But it is unlikely that the disturbance mentioned by Mr Wildman could be credited to G effects of this magnitude, though air displacement may have caused it. Quote, 'As it did so, it brushed particles of frost from the tree tops on to my windscreen'.

So much for the G field; now, let us examine both cases again from the repulsion field point of view. First Mrs Jones' case. In this we shall assume R_1 to equal 3ft, as in Fig 79(a). As we have already seen, in such a configuration, the pilot's C.M. must be 3.7ft from the ship's C.M., which in the hovering condition means the pilot loses about 34lb in weight. The distance from the P.S. to the C.M. of the car (R_2) equals 23ft, so the downward acceleration on the car normal to the axis of the saucer is:

$$G_c = \frac{R_1^2 \, G_s}{R_2^2} = \frac{3 \times 3 \times 1}{23 \times 23} = .017g.$$

As before, if the car weighs 2,500lb, then .017g=42lb.

Now it follows, that just as this vertical effect on the car is the reverse of the G field, *i.e.* downward instead of upward, so the horizontal component is also reversed. So, whereas we have a forward thrust component acting on the car in the case of the G field ship, we get a braking or rearward thrust acting on the car in the case of the R field ship. Which in this case amounts to no more than 8lb or so. And as before, it is very doubtful if the car's occupants would notice any change.

Summarising the case we have:

(1) Small angle of tilt, low forward thrust component.

(2) Reduced focal length to minimise ground effects.

(3) No secondary field lift.

(4) Cockpit side thrust inducer vector on car as being negligible.

(5) Any increase in the forward thrust component of the saucer would tend to *push* the car back.

Fig 79(b) reconstructs Mr Wildman's case, and as in the previous instance R_1=3ft, while R_2, the distance from the P.S. to the C.M. of the car is 71ft. And we get downward acceleration on the car:

$$G_c = \frac{3 \times 3 \times 1}{71 \times 71} = .002g.$$

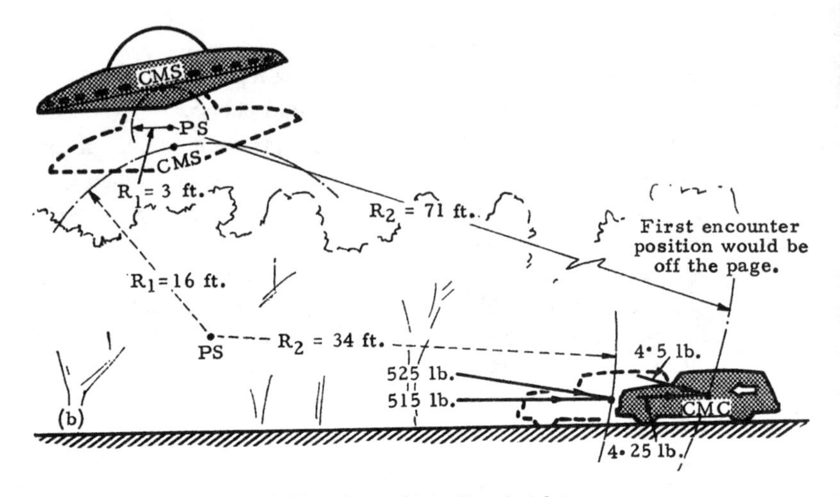

R field analysis of Mrs Jones' sighting.

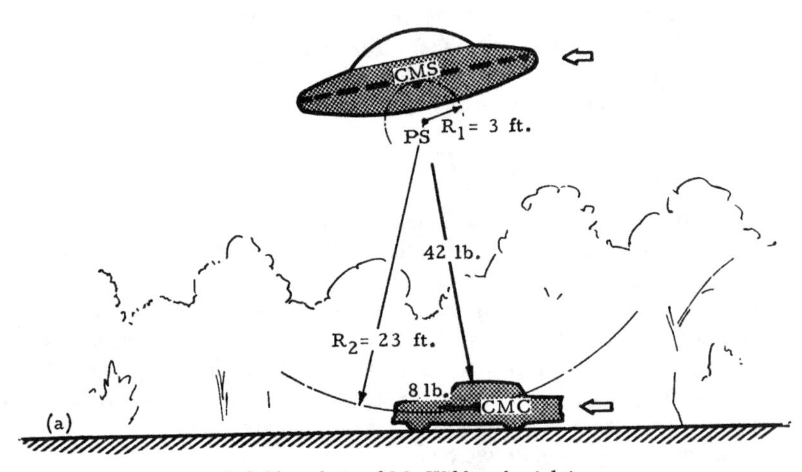

R field analysis of Mr Wildman's sighting.
Fig 79.

Which in the case of a 2,500lb car would amount to as little as 4.5lb thrust from the saucer's P.S., while the car experiences a rearward 'drag' force of approximately 4.25lb. Clearly hardly enough to be noticed by the driver.

Summarising this case we have:

(1) Small angle of tilt, low forward thrust component.

(2) Reduced focal length to minimise ground effects.

(3) No secondary field lift.

(4) Cockpit side thrust inducer vector on car as being negligible.

(5) Any increase in the forward thrust component of the saucer would tend to *push* the car back.

So far then, we find that Mrs Jones' case, due to the close proximity and matched speeds of the two vehicles, supports both the G field and the R field concepts, for, as we have seen, the resulting acceleration or deceleration forces on the car of both types of fields, would have been of small magnitude.

But in Mr Wildman's case the G field explanation breaks down, for small as the forces may be, we would expect the car to be accelerated if anything, not slowed down. We may be tempted to explain the deceleration in this case as having an electro-magnetic origin, as we considered earlier on, but no examination such as this can represent a fair assessment without due consideration of all the facts. One of the more obvious facts known to UFO researchers is that UFOs do *not* always interfere with the electrics of automobiles; far from it; for instance, Mrs Jones' case in question. Therefore a close analysis of *both* field effects is called for in Mr Wildman's case.

To begin with, the G field interpretation can be dismissed fairly easily by first assuming no E.M. effect and pointing out the fact that on a fairly flat road, an approaching car would be moving into progressively stronger regions of the saucer's field and therefore while it is true to say that 25lb thrust quoted would have little effect on the car, nevertheless this would rapidly increase due to the inverse square law effect, to the point where the car and saucer would be approaching the same relationship as in Mrs Jones' case. Certainly we would not expect the oncoming car to be decelerated and in this respect we have the inconsistency in the two above cases.

With regard to the possibility of the failure of the car's ignition system due to an E.M. interference, imagine a car being driven at perhaps 35-40 m.p.h. on a fairly flat road, when the ignition system becomes unstable. Instantly the state of affairs can change from one in which the engine is driving the car, to another where the momentum of the car is driving the engine, unless it is disengaged by the clutch. The latter was apparently not so in Mr Wildman's case, for he put his foot down to accelerate his slowing vehicle.

Now it is true that the momentum of a car travelling at this speed can be very quickly absorbed by a *stalled* engine, which in effect becomes a pump, but the engine of Mr Wildman's car was not stalled. Therefore, while we could expect his car to slow up a little due to an E.M. effect, we must also remember his car rapidly closed the distance between him and the hovering disc, in this instance, we must presuppose, aided by the increasing power of the G field. Compensating as these two opposing forces would be, we would rather expect the car to carry on with very little slowing down, according to the G field interpretation.

Similarly the R field point of view would appear to support the above argument, for if we allow loss of engine power, plus the calculated small, but increasing, backward thrust of the R field, then surely we need look no further? But I have pointed out the ignitions of car engines are not *always* affected by saucers. What if Mr Wildman's case belongs to this category, suppose his engine was maintaining full power, suppose the UFO was employing an R field, should we analyse the case further in these terms? These were the questions I asked myself. The following were my conclusions; which may not necessarily be the correct ones; you, the reader, will be the judge.

First, we assume no E.M. interference and begin where we left off the analysis in Fig 79(b); that is, the car in this case is approximately 60ft from the saucer, for Mr Wildman said: 'As soon as I came within 20 yards of it, the power of my car changed, dropped right down to 20 m.p.h. I changed right down into second and put my foot flat on the accelerator —nothing happened. The object kept ahead of me by approximately 20ft for 200 yards, then started to come lower.'

In calculating the R field force acting on Mr Wildman's car, we assumed the P.S. of the craft to be at very close focus, to minimise ground effects and to help create the maximum differential for the crew. And from this we found R_2 to be approximately 71ft, from which we obtained 4.5lb of force acting against the car. But the short focal length is also the one in which greater differential is set up within the structure and anything the ship contains, especially when high power is used as on a quick take off Yet again, that particular craft may not have been crewed, it just might have been using a long focal length, and even if the ship was crewed, then at the most they would be feeling very light footed, certainly not weightless anyhow, and for the hovering conditions this may not be very inconvenient, though we shall see later there are ways to obviate even this.

But there is another possibility. What if the occupants of the UFO *wanted* to slow that car down, to observe it operating at close quarters? Let us see, for instance, what might happen if it was using an R field at long focus, say down to R_1=25ft. Well, in the first place R_2 is also correspondingly shortened; in fact, it now becomes 63ft, from which we can find the alternative R field component acting against the car as:

$$G_c = \frac{25 \times 25 \times 1}{63 \times 63} = .16g.$$

which amounts to a drag or retarding force of some 400lb, quite sufficient to slow the car up, in fact, but not before a certain amount of the car's momentum has been absorbed; therefore it would be carried nearer to the 60ft range, being subjected to an increasing field strength as it did so. This would take mere seconds and the first phase is established. Mr Wildman was slowed down to 20 miles per hour and he got as close as 20ft or so from the object. Meanwhile the UFO was moving *away* from the oncoming car at the same speed. So relatively they were at rest with respect to each other. 'The object kept ahead of me by approximately 20ft for 200 yards, then started to come lower.' So begins the second phase, in which the relative positions of the vehicles are shown dotted in Fig 79(b).

Remember, Mr Wildman estimated the saucer as being down to within 20-30ft altitude, and in the diagram I have given this the benefit of the doubt and called it a mean of 25ft. But note, in the second phase, the disc begins to descend; therefore, if the R theory is correct, two things must be happening; that is, the vertical acceleration acting from the P.S. on the ship, already slightly less than 1g, must now be even less in order for the vehicle to sink, and without knowing the rate of descent there is little else we can do than roughly assess this on basic factors as approximately .95g. Also, as the machine descends, the F.L. of the field must be reduced so as to avoid dangerous ground effects.

By keeping the P.S. a reasonable distance above ground level, and assuming a minimum clearance height above the hedgerows and trees to be some 20ft, we can logically estimate the F.L. as being 16ft (R_1), while the distance from the F.P. to the centre of the mass of the car (based on Mr Wildman's estimate of approximately 20ft) is found to be 34ft. From which we get the repulsion on the car as:

$$G_c = \frac{R_1^2 G_s}{R_2^2} = \frac{16 \times 16 \times .95}{34 \times 34} = .21g$$

or some 21% of the weight of the car, which being 2,500lb means we could expect a force of something like 525lb. to be exerted on the car, which would yield a total horizontal retarding force of 515lb.

Now in perhaps no other incident are we in a more favourable position to do some profitable checking and although the following figures are easily calculable, I thought it would be more convincing to obtain the impartial opinion of experts. The following is a copy of a letter I had in reply to my inquiry from Vauxhall Motors concerning Mr Wildman's case, and for which I am greatly indebted.

Vauxhall Motors Ltd.,
Luton, Beds.
19 March, 1965.

Dear Mr Cramp,

I have been asked to reply to your letter of 7 March, which refers to the incident which occurred in 1962 affecting our delivery driver, Mr Ronald Wildman, while driving a new Vauxhall estate car.

The tractive effort required to propel such a vehicle at 20 m.p.h. on a level road with no wind would be approximately 40lb. The available tractive effort at 20 m.p.h. at full throttle and in second gear with a three-speed transmission would be approximately 470lb. The available tractive effort at 20 m.p.h. at full throttle in second gear with a four-speed transmission would be approximately 612lb.

Unfortunately, we have no record as to whether this particular estate car was fitted with a three-speed or four-speed transmission, but the figure you require can be arrived at by subtracting 40lb from the available tractive effort at full throttle.

We trust this is the information you require.

Yours faithfully,
for VAUXHALL MOTORS LIMITED,
(Signed) A. E. COOKE, Manager
Research and Development.

I could, of course, have pursued the case even further to establish exactly what gearbox was fitted to this car. But as the R field predictions are only 20% in error in the case of the three-speed gearbox, and 10% in error with the four-speed gearbox, and making allowance for the witnesses' judgment of distances, I could not help feeling the result was encouraging to an unexpected degree. I would like to add one more little observation with which to close this fascinating case.

In his report, Ronald Wildman also said: 'It continued like this till it came to the end of the stretch—then a white haze appeared around it, like a halo round the moon. Then it veered off to the right at terrific speed and vanished. As it did so it brushed particles of frost from the tree tops on to my windscreen.'

I must ask indulgence of the reader as I quote these words again, but they are vitally important to our case, for note: we are about to enter phase three of this epic sighting, Fig 80.

The saucer, now travelling at approximately 20 m.p.h., with slight forward tilt, and practically 1g contra gravitational acceleration, is tilted sharply to one side, whilst simultaneously the power and F.L. are greatly increased. At that instant some of the powdered frost lying on the trees' upper branches receives a violent upward impulse from the point source, which, like the debris from a crater, catches up with and surrounds the craft 'like a halo round the moon'. Also at that same instant, the frost on the lower branches is violently showered backward and downward into the path of the pursuing car, 'brushed particles of frost from the tree tops on to my windscreen'.

I pen these words in the early hours of the morning. Outside, also in the tree tops, a little owl calls. I imagine the Wildman case all over again.

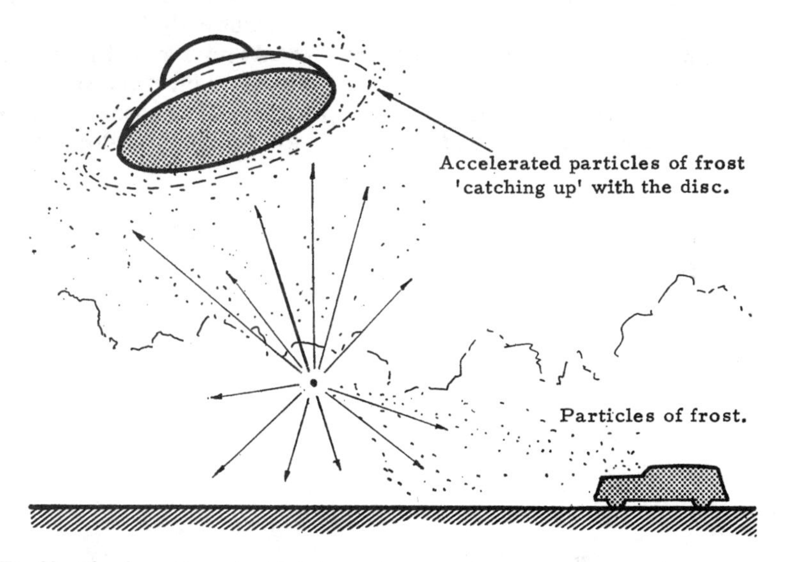

Fig 80. The formation of a 'halo' round saucers has been seen on other occasions. This may in part be due to loose debris and/or the formation of 'Angel hair'. Even so the phenomenon is consistent with the R field theory.

In the call of the little owl I can almost hear the echo of my thoughts: 'I wonder, I wonder if this is right.' And it just about sums up my honest feelings. I do not know if, in fact, my deductions are right; I only know I have found a pattern, a fascinating pattern in a fascinating story, which continues to fit. As yet I have found merely a few pieces. Others will no doubt add many more. But of one thing there is absolutely no doubt in my mind. There are too many facts making a good engineering story for this to be nonsense. There is more to come, but even at this stage I feel justified in saying, there is to be found in these pages more than a little proof of the saucers' existence.

Now I have spoken of these things at lectures, and I know how difficult it can be to illustrate some technical facts to untrained people. All too often, many of them are ready to give the arguments the benefit of the doubt and accept the case out of hand. This is not really satisfactory, and it is with some reluctance one has to give up trying. In a book it is different, and writing for a fairly large cross-section of the public one can feel more justified in examining as many facts as possible, for even those who accept the UFOs and are impatient to arrive at final conclusions, will be interested to learn that there exists so much technical evidence which appears to be completely ignored the world over. Not that I am discontent with my lot in taking part in revealing some of these facts, far from it, I feel it is a great privilege. Even so, I am regretfully aware that many will continue to ignore these facts, and not even therefore appreciating their worth, will still smile indulgently from lofty places. Without any sense of bigotry, to these I extend my sympathy, for I know the loss to be theirs. Perhaps for their sake, if no other, we should finish examining the corroborative evidence for the R field theory, I make no idle claim in saying, the reader's patience will be well rewarded later on.

19

Analysis Nine

Yet More Aerial Confirmation

IN chapter 9, I related the Alex Birch case to the G field theory and we saw how we could predict that a gaseous or liquid substance emitted from a hovering saucer would immediately coalesce into a spherical shape due to the fact that, acted on by the field of the ship, the mass would also be weightless. And we saw how, moving outward through progressively weakening regions of the supporting field, the mass would begin to change shape and fall.

A full analysis of the above aspect alone is beyond the scope of this book, but we can look back at it fundamentally from both the G field and the R field points of view.

Consider first the G field technique in which a disc is hovering, both the primary field and crew's secondary field being in operation, Fig 81(a). The diagram is almost self-explanatory, but observe how the crew are subjected to the combined inward components of *both fields;* therefore the side thrust inducers may be operative in order to counteract the total force. Thus, on being ejected from the craft, a liquid substance would be acted upon simultaneously by several forces, the resultant of which reveals clearly that an *ejection force* must be employed to expel the material away from the ship. If we assume this to be applied horizontally, then it follows that first the substance will coalesce, then begin to rise a little with the up currents of air forming round the ship. Then as the mass moves further out, both the aerodynamic and combined field intensities will get less; therefore, the aotual path taken will be slightly curved. Otherwise the phenomenon will follow the general assumptions made in chapter 9, where we first correlated the Alex Birch case with the G field theory.

A similar appraisal of the R field craft reveals once more a different picture.

This is shown in Fig 81(b) where a disc is hovering at 1g 'lift', and the crew something less than this. But here again they are also being subjected to the outward component of the main lift field; therefore, the side thrust inducers may be in operation. From which it becomes immediately obvious there is a combined side thrust tending to push any loose substance *away from the craft*. The chief difference here being that the material will not be subjected to a full 1g upward acceleration as it was

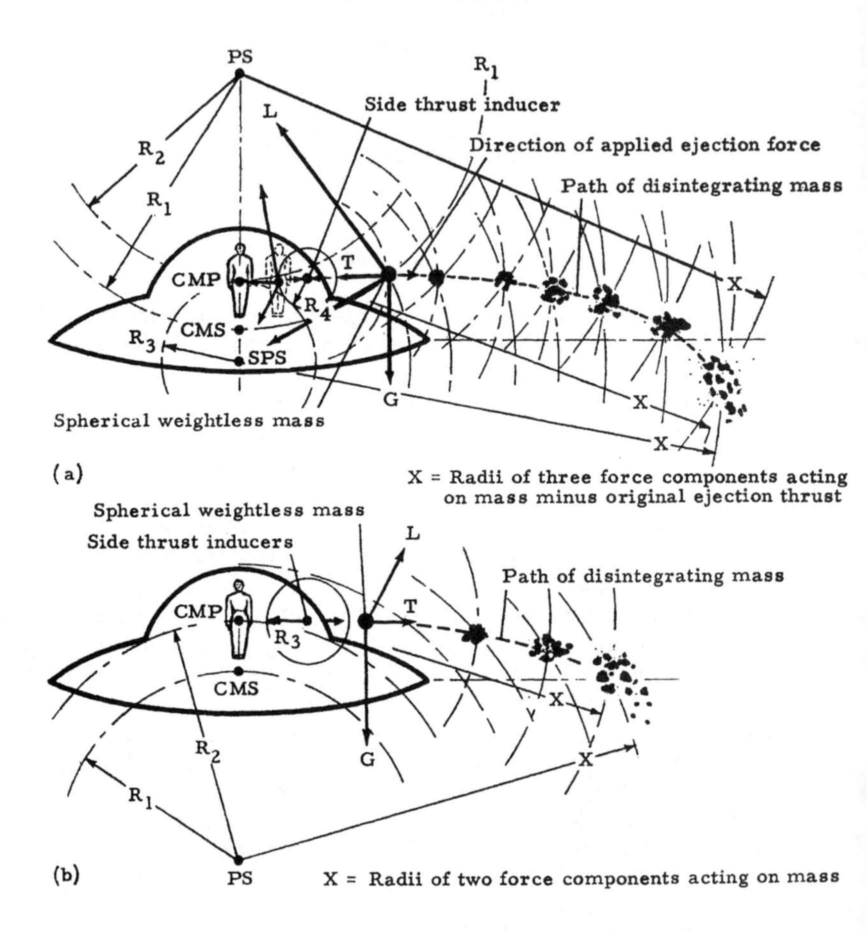

Fig 81. *A liquid or gaseous substance emitted from a hovering G field or R field saucer, would initially be weightless and therefore would coalesce into a sphere before disintegrating. This precisely confirms the sighting made by* 14-*year-old Alex Birch.*

in the G field instance; therefore, it will not be fully weightless; even so, it would still coalesce into a near spherical shape. But, as in Mr Wildman's case, the effect would be more marked if a long F.L. was being employed, and although the phenomenon is one we could expect of *either* lift devices, the tendency for secondary bodies to move away from a saucer is greater in the R field concept.

Now, concerning aerial effects of the R field, the reader might ask: 'If the point source can be responsible for a crater when it is formed below ground level, should there not be a similar effect caused in clouds?' And he would indeed be right; certainly we should expect such effects and we do not have to search long through the files to find possible evidence. Here are several cases.

On 3 May, 1966 a night nurse E. M. Doak of Colwyn Bay, N. Wales said: 'I was just getting up when my attention was drawn to a strange but beautiful celestial object directly facing my window. It came in rather slowly as though testing the layout, from a westerly direction, I would use the expression edge on. . . . The craft was low down quite close to the earth, it turned in a manoeuvring manner till it appeared to be tilted with the dome uppermost. I watched quite close . . . fascinated. The diameter would be around 130ft. The outer rim consisted of what looked like a ring of golden discs probably brilliantly lighted. There was a centre cabin or control room or section and what seemed to be three protuberances or propellers which were of a greenish purple hue, something like you find on a nearly ripe plum. The space which surrounded the inner cabin or control section seemed to be translucent.

I watched the craft for a space of half an hour or so, by which time the sky was beginning to brighten into a lovely morning. The craft hovered for some time a mere few feet above the ground. I am certain that it was intelligently controlled, whether remote or otherwise I don't know. But I did have the compulsive feeling that whatever I was looking at was being operated in a dimension unknown to mankind. Anyway, it stayed in one position for a while—then it made a terrific sharp turn into the flat state, then . . . almost violently it shot into an angular V turn and was soaring into the heavens in a south-easterly direction, becoming redder and redder as it blended into space. *There was no sound.*

What struck me as being odd was the terrific *whirling of the cloud bank* formation as the thing rose to the heavens. *There was no wind.* All was still again in a matter of split seconds.

The following report was obtained by Rear-Admiral Delmer S. Fahrney, U.S.N., Retd. It happened over Virginia in 1955 to a Navy Commander stationed at Anacostia Naval Air Station. He had looked back over his shoulder while flying to find a huge disc-shaped craft formating on him, no more than 75ft away. The Commander described the object as 'two saucers, face to face', apparently metallic and about 100ft in diameter. He said it was thick at the centre with a domed top through which was shining an amber light. The Commander, who also happens to be a Navy missile expert, said when he tried to ease his plane in for a closer look, 'the disc tilted and shot away, leaving the clouds swirling behind it'.

Earlier in August 1951, Central New Mexico, Alford Roos, senior mining project engineer for the U.S. Bureau of Mines and other government agencies, also a member of the American Institute of Mining and Metal Engineers, observed the performance of two lens-shaped saucers in interesting detail.

He was at his ranch 10 miles east of Silver City, when he heard a 'swishing' noise. Here is an extract from his official report:

'I saw an object swooping down at an angle of about 45 degrees, from a southerly direction, travelling at immense speed, coming quite close to the earth over Ft. Bayard, two miles to the N.W. Reaching the bottom of the swoop it hovered for several moments, then it darted up at an angle of about 70 degrees from vertical, in a north-westerly direction, directly over Ft. Bayard. . . . I neglected to state there were two objects that converged at the point of hovering, at which time they were in close proximity. . . . Over Ft. Bayard there was an isolated cloud island covering perhaps three degrees of arc and perhaps a mile across. The two objects shot up at this steep angle at incredible speed, both entering the cloud, and neither appeared beyond, and there was no trace after entering the cloud.

'Their track was as straight as a ruled line, no zig-zagging. The astonishing thing was that the cloud *immediately split* into three segments, ever widening, where the objects entered. . . . Each object left a pencil-thin vapour trail.'

On 23 July, 1948, in the vicinity of Montgomery, Alabama, two Eastern Airline pilots, G. S. Chiles and J. B. Whitted, had a near miss by an approaching UFO. It sported a brilliant light and hurtled on collision course with them. At the last moment the object veered to the right of the airliner, emitting a red 'exhaust', and shot straight up into the clouds. Captain Chiles described the UFO as appearing torpedo shaped, about 100ft long, with two rows of brightly-lit windows or ports along the side. He said his aircraft was rocked as the UFO shot upward, by a *strong blast*.

If an airliner could be rocked by the blast of the repulsion point source, then it follows that if for some reason a disc was having some kind of trouble which seriously restricted its speed, then an investigating aircraft could be in grave danger of approaching too close to the field Fig 82.

UFO researchers will know only too well of the tragic death of Captain Mantell, and once again I must ask them to bear with me, for the layman's sake, while we consider this case in terms of the R field theory.

It was getting on for three in the afternoon on 7 January, 1948 when a huge object shining brightly was observed travelling through the sky at very high velocity. As it happened it was travelling towards the big Air Force field, the Goodman Base. Earlier at two-thirty p.m. the State Police had given the warning. Hundreds of people had seen the thing and the Air Force was at the ready. The commanding officer, Colonel Hix, had three fighter planes up and climbing to meet the 'intruder'.

Through a gap in the clouds the control tower personnel observed the object through binoculars. If the reports had said the thing was large it was no exaggeration; it was colossal. It must have been at least, to quote

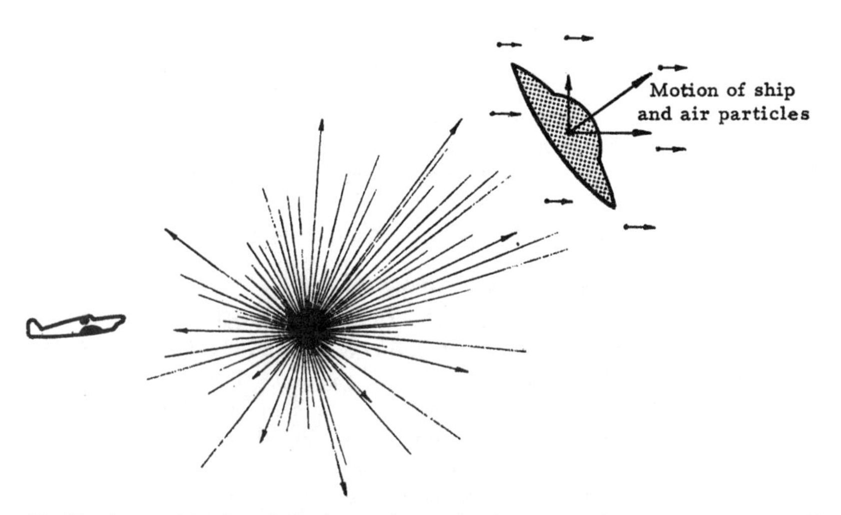

Fig 82. *A pursuing aircraft inadvertently running into a repulsion point source could be disintegrated.*

a conservative estimate, 500ft in diameter. Captain Mantell, leading the group of fighters, was in communication with the control tower. He confirmed that it was of fantastic size and certainly looked metallic. By now the thing was climbing with only half the speed of the pursuing aircraft. As they closed in, the object increased its climb and speed to something like four hundred miles per hour as if trying to evade them. When the control tower heard from the aircraft again, Mantell's companions had lost contact and had last seen him following the thing through a cloud gap. Eventually when Mantell came through at a quarter past three, he still had the thing in sight and was following it up to try for a better view; he would do so up to twenty thousand feet.

Captain Mantell's aircraft was scattered over a very large area in tiny bits; the exact cause of the disintegration is unknown. The possibility of his blacking out and the aircraft diving out of control was denied by his colleagues, for they claimed Mantell's aircraft, a P51, would fly itself 'hands off'. There were of course all kinds of notions as to why and how Captain Mantell lost his life, but surely the most ridiculous suggestion was that in fact the Captain was chasing the planet Venus, visible at that time. How anyone in their right mind could ask us to believe that a pilot with Mantell's experience, corroborated by his colleagues, could ever make a mistake like that, even allowing that there had been no original ground alarm in the first place, is difficult to understand. Most researchers are convinced that Captain Mantell *was* pursuing a UFO and some are inclined to believe that his death held sinister possibilities. But I have tried to show here at least one possible explanation; in this case, as in others, the deductions may be wrong.

If here and there throughout this work I have introduced a personal note, I have done so somewhat reluctantly and purely because I felt a certain personal experience, was at least authentic, or to stress some particular circumstance where it is difficult to bridge the understandably wide chasm between acceptance and downright rejection of UFO phenomena. As I attempt to develop the present hypothesis a little further, I am mindful of the fantastic realms into which this is taking us. I have long since acquainted myself with this line of thought and I am therefore somewhat prepared for further developments which it may take. But it may not be so acceptable to those who listen. Therefore, as we begin to outline the next stage in the evidence for the R field theory, I am reminded of an occasion when giving a lecture on UFOs to the otherwise sincerely rapt audience of a scientific group. I had warned them that what I was about to say next might in fact sound more like science fiction, etc., when from somewhere in the darkened hall I heard the whispered remark, '. . . is science fiction'. Trying to overlook the disconcerting rejoinder, I carried on to receive a warmly demonstrated appreciation of the talk. Later in the evening I discovered that the remark had been made by a technician of not really consequential standing, while other more responsible people had been very interested indeed. The whole point of this little narration being to emphasise to the reader that I am only too well aware of the so-called science fiction realm into which our journey is taking us, and would ask that we keep strictly to our examination of the clues presented, and not to be distracted by the voice from the dark, while endeavouring to remember that the science fiction of today is often the science of tomorrow. It will be for the reader to interpret the following and its possible relationship to the R field theory as he thinks fit, but I for one ran out of excuses for coincidences long ago!

On 23 July, 1947 John Janssen, of Morristown, New Jersey State, editor of an American aviation journal, reported a strange aerial encounter; he was flying his aircraft at 6,000ft when, he says:

'While my eyes played over the horizon, I became aware of a shaft of light that seemed like that of a photographer's flash-bulb. It came from aloft, very high up. It was above that position which, over a plane's nose, fliers call 11 o'clock. I at first thought it was merely the reflected sun, bouncing off the sides of an exceedingly high-flying aircraft. I gave it no further thought. Now, the engine of my plane *began to perform peculiarly*. It coughed and sputtered spasmodically. I pulled on the carburettor heat and gave it full throttle. This was to blast out accumulated ice from the carburettor at that height. The engine emitted one final wheezing cough and then quit. Now, the nose of my plane, instead of drooping to a normal glide, remained . . . rigidly . . . fixed on the horizon, in its *normal, level flight attitude*. Abruptly, I became aware that my plane was now defying the basic law of gravity. I became frightened, and close to panic, at so weird a predicament. I saw the air speed indicator was at zero! There was now an odd prickling, electric-like sensation coursing through my body. I had an eerie sixth sense feeling that I was being watched and examined by something that minutely studied my features, my clothing, and my airplane . . . with tenacity. I flicked a cold bead of perspiration from my eyes. *Then I saw it*! Above, and slightly beyond my left wing-tip, was a strange wraith-like craft. It looked like a flying saucer. Its flanged and projecting rim was dotted on either side with steamer-like portholes. It seemed to radiate in a dull metallic hue that conveyed an impression of structural strength, and a super-intelligence not of this planet. It was motionless. Perhaps a quarter of a mile away . . . beyond, and slightly higher, I could see another disc, seemingly fixed in the sky. I assumed that the second strange craft was but waiting for the one nearest to me to complete its examination. Then I had the most unaccountable urge to reach up and snap on the magneto switch, which I had turned off when the engine quit. I switched on both magnetos to the 'on' position. Slowly the propeller began to turn . . . then the engine burst into its steady rhythmic roar. She nosed into a stall, picked up air-speed and steadied under control'.

This is the kind of report which tests the credulity of even the most ardent UFOlogist, but let me hasten to add that in trying to correlate the technical evidence for the UFOs I have had similar experiences to Aimé Michell, who when developing the straight-line theory, now known as Orthoteny, found that he could only make sense of his investigation when he included *all* the sightings. I can best explain the sort of thing which occurs by adding the following:

I had read the Janssen case years ago when it was first published and, in the course of preparing this present chapter, decided to do some investigation into a possible relationship with the R field theory. But I was convinced that Mr Janssen had seen only one disc. In developing the analysis, however, it soon became obvious that it would be difficult to correlate the case with either the R field or the G field hypothesis. About to give up, I came to the conclusion that it *could* be explained by the R field, provided there was a second ship, according to my calculations, beyond the first and higher up. 'Perhaps out of Mr Janssen's sight', I thought, but this would be begging the question and of little factual use and, disappointed, I put the case aside.

Later the Janssen sighting did come to light and even before re-reading it I was convinced there had been only one saucer. I was of course delighted to find that there had been, in fact, two. The theory required *two* ships, Mr Janssen *had* seen them! The pictorial representation in Fig 83 is drawn to scale, the distance from the aircraft to the first disc

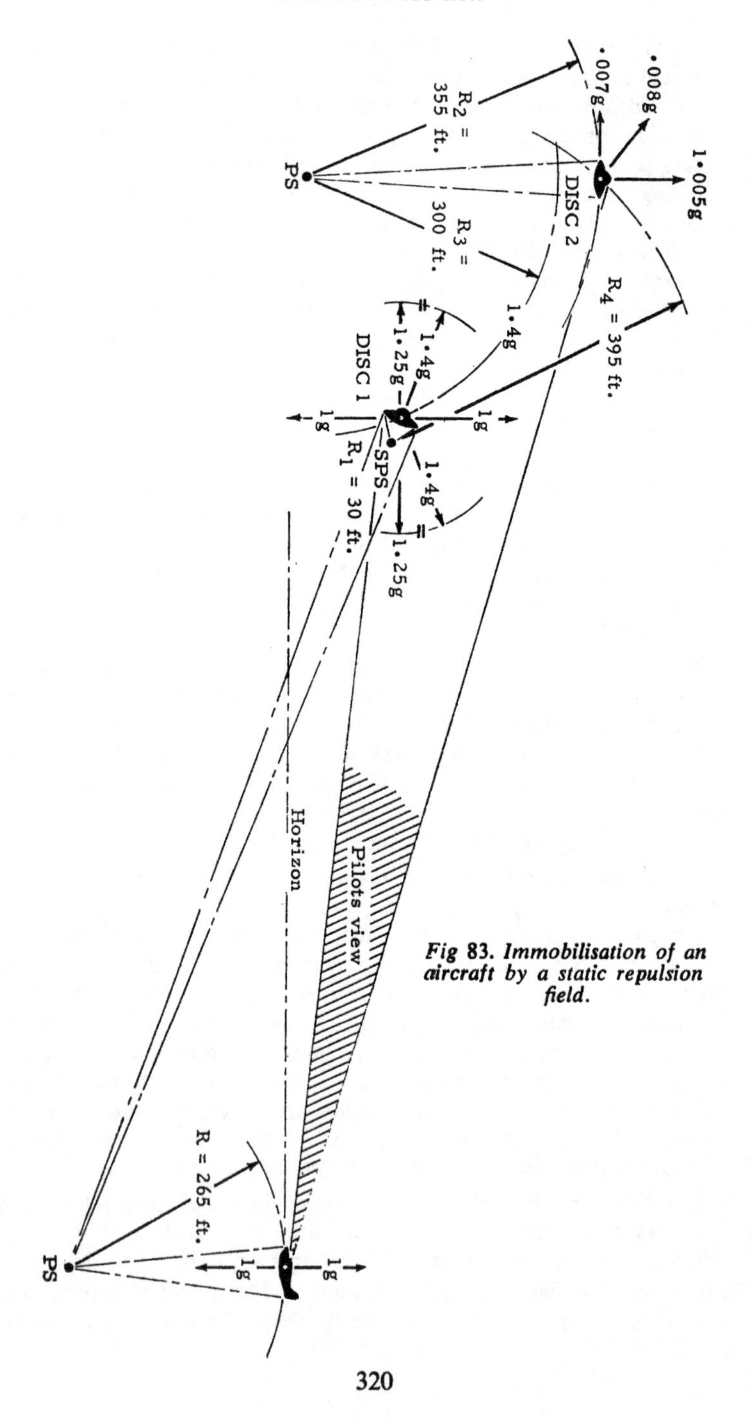

Fig 83. Immobilisation of an aircraft by a static repulsion field.

Whidby Island contact case considered in terms of repulsion theory. In this experiment the model is situated 3ft from the camera.

Plate 33.
The same model 3ft from the camera as above but photographed through a tilted reducing glass. Note the 'shrinking' and 'digging in' effect. This optical illusion and pressed down grass beneath the saucer described by the witness, are entirely consistent with the Repulsion field theory discussed in the text.

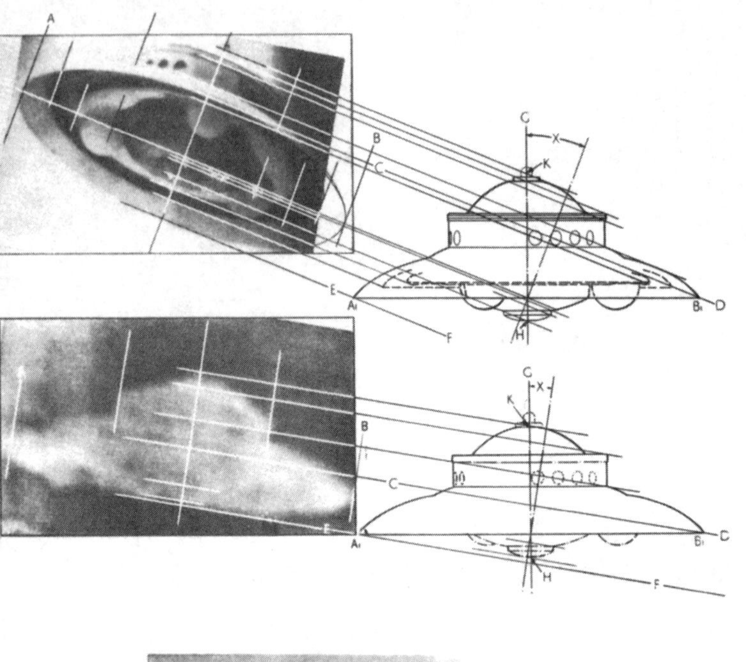

Adamski-Coniston orthographic test drawing reproduce from 'Space, Gravi and the Flying Saucer'.

Plate 34.

This photograph completed by scali reveals the striki similarity to the Coniston photogra below.

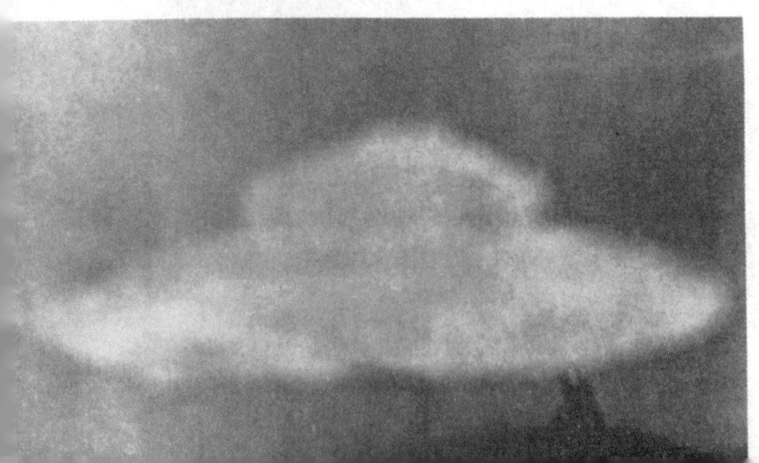

The Coniston photograph. Reade should note that t principal witness w 6 years old at the time of the sightir

being reduced to a thousand feet to save space, while the true relationships are unaffected by this difference. The reader will appreciate that herein is an additional advantage of the variable F.L. point source. The values of these lengths are of course hypothetical, as are the vertical distances separating the three craft. Before we check the arithmetical effects of these, we can logically arrive at the arrangement by the following assumption. First, the UFOs in this case wanted to isolate Mr Janssen's aircraft in space, and then to examine it from a *distance*, we can conjecture many reasons why the operators should so behave. A little reflection will show that to operate a G field type ship with the point source placed above and at a distance from the aircraft, we could expect the saucers to take up the reverse position, *i.e.* the first ship would be *lower* than the aircraft and the second ship would in turn be below the first.

But the positions shown in the sketch are the relative positions we could *expect* from the R field theory. In this, the first craft is shown tilted with the secondary point source R_1 focused at 30ft, and a power intensity suitable to produce a total repulsion vertical acceleration of 1.4g normal to the axis of the ship. Which means, the true vertical vector, acting on it normal to earth's gravity, is only .5g, thereby compensating half the weight of the craft. It will be seen there is also a horizontal 'thrust' component of 1.25g brought to bear on the machine, and normally it would be accelerating in that direction, due to the tilt.

If now a second craft is positioned as shown in the diagram, supported at 1g by its field set at long F.L., say R_2=355ft. Then the shorter radius of its field R_3=300ft will pass through the centre of mass of the first machine at exactly the same angle as, but in the opposite direction to, that machine's axis of tilt. And this compensating acceleration on the second craft can be shown as:

$$Gs1=\frac{R_2^2\ Gs2}{R_3^2}=\frac{355\times355\times1}{300\times300}=1.4g.$$

From this it will be seen all the forces on the first craft cancel out. That is, the sideways vectors of 1.25g are now compensated, so that despite the tilt this machine is immobilised. Whereas the combined effect of the two 1.4g vectors is to produce a total vertical acceleration of 1g, thereby immobilising the vehicle in this sense, *i.e.* it will hover.

But in turn the field of the first disc will influence the second machine, but due to the short F.L. of R_1 this effect will be small, as shown by:

$$Gs2=\frac{R_1^2\ Gs1}{R_4^2}=\frac{30\times30\times1.4}{395\times395}=0.008g.$$

Which means that if the second UFO weighed, say, 30 tons, then .008g represents about 537lb, giving vertical and horizontal components of 336lb and 470lb respectively. Now this comparatively small force will nevertheless still be sufficient to move the second disc at increasing speed, upward and away from the first machine, but a little thought will show that a very small tilt on this craft would be sufficient to cancel out this secondary effect. The whole arrangement is quite easily calculable, but has been omitted here for simplicity. With the saucers no doubt such 'field phasing' is fully automatic; we shall review other evidence for this later on.

The two discs now being 'locked' in space, the assessment of the Janssen case is quite straightforward. The crew of the first ship would focus the primary field P.S. to a position immediately below the vehicle they wanted to isolate, and a distance at which the electrical circuit of the aircraft is saturated. The engine having stopped, the aircraft would begin to lose speed and go into a dive, but the point source would be adjusted to 'support' the machine at just 1g in exactly the same manner in which the discs themselves are supported. By focusing the P.S. initially forward of the aircraft, it would be braked to an immediate stop, bringing the A.S.I. instrument reading down to zero without the pilot having the least sense of deceleration. There would of course be a small 'thrust' from this point source on the saucer which beamed it, but this would only mean that the whole trio of craft would move very slightly indeed.

The examination over, the P.S. would be moved a little toward the rear of the aircraft to give it a little helping hand, and it would be on its way at perhaps a slight climbing speed. By the time the pilot realised this and switched on again, the slipstream would start the motor! One final point. The UFOs could have still been moving, maintaining the aircraft's original speed, but as the air in the vicinity of the machine would be going along *with* it, as it does with the saucers, then the pilot might indeed think he was stationary. In any event, the Janssen case can be explained by motovating fields. Certainly it could never be considered in terms of plasma propulsion, or perhaps some UFOlogists as well as sceptics find it more convenient to dismiss such cases as outrageous. If such mechanical side effects of the R field can be entertained by ordinary folks, then the very force fields we are here examining may not be so remote from our ken either.

Naturally there are many, many aspects of these craft we could take up. We are, only too reluctantly, dealing with a few; but from the previous case, no doubt one of the more obvious questions which may come to the reader's mind is: 'There are many sightings of saucers flying in pairs, in tandem, line abreast, and in formation, will they not affect each other in flight?' Indeed, they will. Consider the case of two saucers first.

Fig 84 (a) shows two R field saucers hovering level at 1g contra gravitational acceleration. They are 20ft apart and the F.L. are set at 30ft. As in the Janssen case, it is apparent that the field of each ship will produce an acceleration on the other, and in this instance it would amount to .26g, of which the vehicles would experience an acceleration tending to separate them, while each would experience an increased vertical

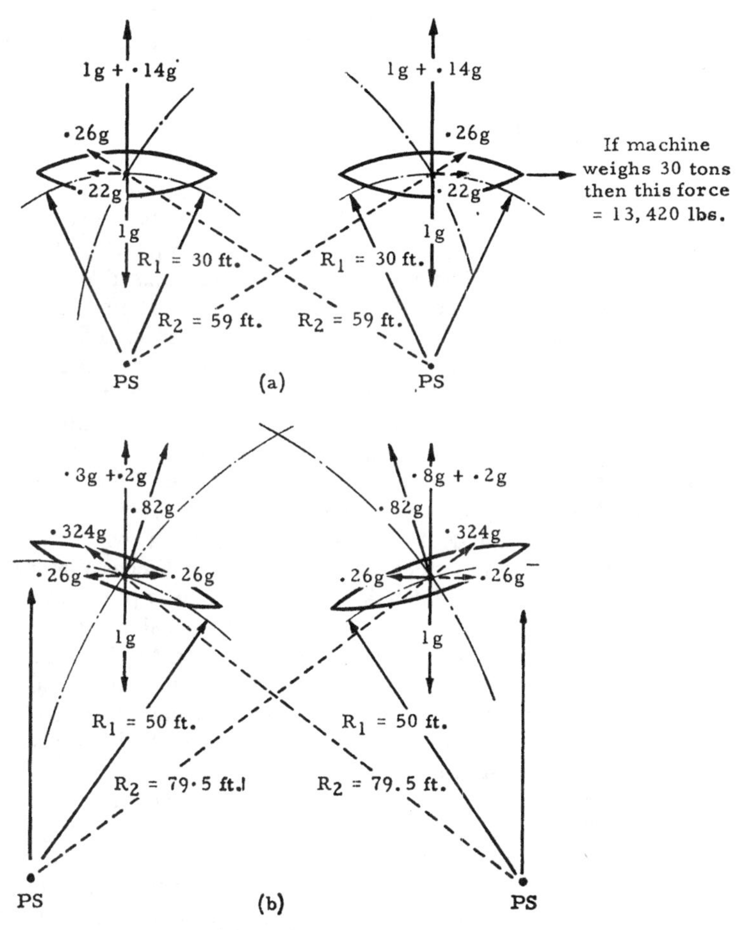

Fig 84. Two R field craft in hovering attitude, each craft producing 20 per cent of the lift force of the other.

acceleration complement from the other of some .14g. Again taking a total weight of 30 tons, this means that each craft would receive a separating thrust of no less than 13,420 lb. and a lift force of some 9,400 lb. From this we can surmise that two such ships approaching each other slowly in the hovering condition would have to do two things; decrease the power proportionally and increase the amount of tilt.

Fig 84(b) shows the craft in the hovering condition, again 20ft apart, but this time with sixteen degrees of inward tilt. If the F.L. were R_1= 50ft, the total vertical acceleration produced by each craft normal to its axis must be .82g, which in the tilted position gives the machine .8g vertical acceleration and an inward acceleration of .26g. But, meanwhile, each craft is receiving the 'thrust' from the other due to $R_2=79.5$ft, and this is due to:

$$\frac{50\times 50\times 0.82}{79.5\times 79.5}=0.324g$$

of which a vertical component of an additional .2g is gained. In other words, each machine produces 20% of the lift of the other. While the inward side thrust component is cancelled out completely. Incidentally, here again we see another design reason for the adjustable F.L. For in the rôle being considered, the rule must apply: shorter the F.L. the greater must be the tilt. From this it follows that should two 'R field' craft approach each other, they will both receive a retarding component of the other's field, tending to bring them to a halt. But their momentum will carry them on beyond this point a little until the repulsion component is sufficiently strong to overcome the momentum, when the discs will come to rest, and then be thrust away from each other again. Such behaviour has been witnessed. Here is an example:

In May 1964 an incident occurred near Canberra, which was investigated by the Scientific Attaché of the American Embassy at Canberra, Dr Paul Siple, and two N.A.S.A. engineers. It was just before daylight when observers spotted a large glowing object travelling across the sky in a north-easterly direction. It wobbled and appeared to be out of control. Hovering nearby was a smaller UFO emitting a red light. Suddenly the larger object accelerated towards the smaller one and appeared to collide with it. Whereupon it seemed to the observers that 'the impact' caused both UFOs 'to bounce a little'.

The larger object, now no longer wobbling, turned and moved *slowly* away from the smaller UFO *before* accelerating at high speed.

Now, from the theory, this latter behaviour is exactly what we would expect. For just as a swift take off from the ground will cause a local disturbance, so the smaller craft would have been affected by the R field of the big ship, had the power been intensified too near. Once again the behaviour of the discs in this case cannot be reconciled within the G field type of craft, unless separate R field inducers were incorporated.

It is conceded that such a faculty inherent in the R field ship may be frowned upon by the world's aerobatic teams, for they may feel it could take a lot of the skill out of formation flying. Indeed, this to some extent may be so, but there would be many other factors to reckon with. For instance, we have considered only two close-flying vehicles. Fig 85(a) shows that in the case of three, the centre craft must stay horizontal, while the other ones are tilted inward as before. Add to this the forward tilt as in forward motion, plus the effects of the secondary fields, *i.e.* side thrust inducers, etc., and the manoeuvre takes on additional complicating factors. In the case of several machines flying line abreast, then we could

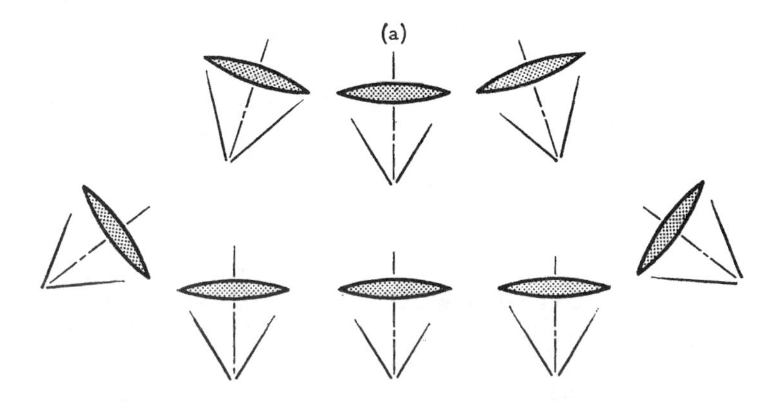

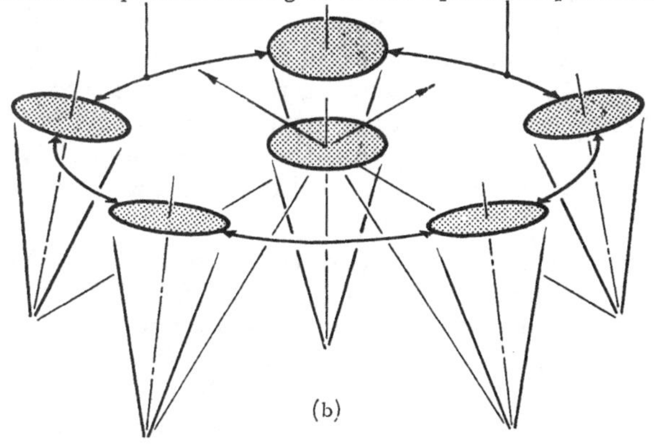

Fig 85. R field discs supporting crippled ship.

expect some of them in the middle to maintain a normal flight pattern while the outer ones took on an even greater compensating tilt. Is there confirmation of this in the following case?

On 3 October, 1958, a Monon Railroad freight train was making its way through Clinton County, Central Indiana, U.S.A., when a formation of four bright lights crossed ahead of it. As if interested, the objects turned and traversed the length of the train, which measured about half a mile, while the entire crew looked on.

After passing the rear end of the train, the UFOs turned east, swung back and proceeded to follow it. The exact shape of the objects was concealed by the bright glow they emitted, but some of the crew said 'they appeared to be flattened, operating part of the time in line abreast, with co-ordinated motions. *Several of them appeared to fly on edge.*' The UFOs turned away when one of the train's crew shone a bright light on them, but they returned and continued to follow the train. Finally they appeared to lose interest, turned away to the north-east and vanished.

At this juncture it is important that I point out that everything we have said thus far concerning formation flying will apply equally to the G field idea, save for the fact that the procedure would be reversed, *i.e.* when in pairs, the UFOs would be tilted *outwards* for the compensating side thrust to act against the inherent inward 'pull' of the G field. But here again each ship would automatically supply part of the lift force for its neighbour. Now in case you have not spotted it already, the R field technique would provide a very convenient way for a rescue team to transport a disabled saucer. Several discs would position themselves around the stricken craft and, tilted inwardly, they could gently bear it away, Fig 85(b).

Yes, indeed, we *have* seen this behaviour before. In chapter 12, as a matter of fact, when Captain H. A. Dahl and crew of the coastguard boat had their amazing encounter. Quote, 'six very large doughnut-shaped machines . . . stationary and silent . . . until the others began to circle round one machine which began descending rapidly as if in trouble . . . keeping about 200ft above it, the others followed it down. . . . All the time, the five were circling round the one which was stationary . . . then the rain of metal stopped. The strange craft silently lifted and went westward towards the Pacific. All the time, the centre one remained in the formation.'

A little earlier on, you, the reader, joined me on an imaginary trip in an R field machine, and I promised we would continue the flight later on. This might be an appropriate time, for you now have the opportunity of an unusual experience as far as piloting a saucer goes, for this sort of thing doesn't happen very often. With the aid of diagram Fig 86 we are going to re-enact the Dahl case with you at the controls of the stricken ship.

After your earlier take off, you were joined by five other discs, and as this was your first mission, the formation leader had placed you well inside the group. On the extreme flanks of the formation, No. 1 and No. 2 ships are tilted inwards at quite steep angles and as you increase power the forward tilt of your craft increases in unison with the formation. Very soon the instruments record your inclination at well over forty-five degrees, but you do not physically detect it. The porthole view of the lazy panorama just 180 miles below belies the clock reading of just under 10,000 knots, not much faster and your craft will require no lift component at all, for you will be at orbital speed! Your leader's voice over the intercom tells you to stand by for rotation, and you automatically check your lock-on phasing control and sit back comfortably to watch the dials unwinding at an alarming rate; you are decelerating at over 38g!

Outside the ship, the horizon swings crazily back to normal, but you feel no sense of motion whatever as the craft comes to rest at a gentle

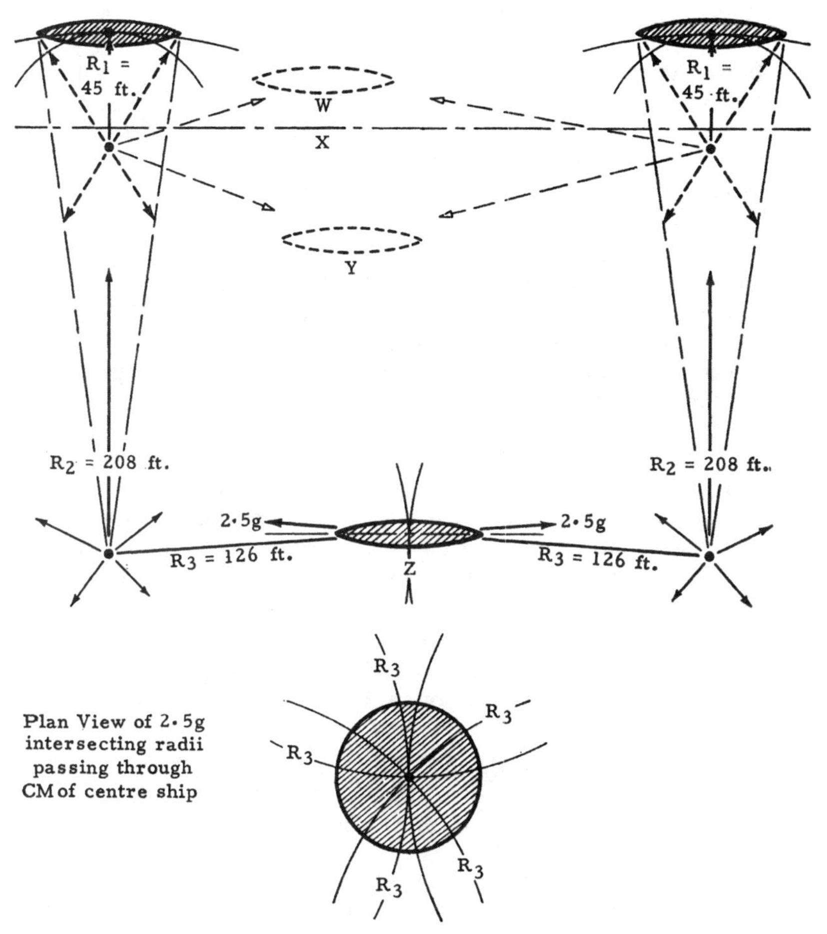

Fig 86. R field analysis of coastguard Dahl's sighting.

hover. Your group leader is talking again and you begin a rapid formation descent. Glancing over to the next ship, you receive a reassuring thumbs-up sign from your pal John. You feel good, grin back at him and admire the breathtaking view below. You are not prepared for the shrill note of the warning klaxon, but you instantly scan the instrument panel for the tell-tale red indicator light.

It takes a split second to find it, but already you have lost primary field inducer power and you are swiftly losing more altitude. You are down to point W, Fig 86. But you have your secondary reserve and you checked the ship's weight in the log book before take off so reassure yourself you have no worry as you snap in the secondary override toggles.

Blessed relief, as the craft's descent eases; but not for long, the secondary field is not functioning either! The slowing up was due to your craft falling through the maximum combined point source level of your formation at point X, just above R_1=45ft. Remember the other craft are flying in a compensating field and are therefore trimmed to less than 1g. Their combined lift components are not sufficient to support your craft; instead, like a shot from a gun, it receives the full downward blast and you plummet on through point Y.

You are alarmed and puzzled by the loss of both primary and secondary field power, but react instantly to the command of the group leader. 'Over to manual control gyro stabilisers.' All this has taken mere seconds and already your team mates are over 160ft above you. But the last instruction was a reassuring one, for it meant the emergency support system would bring your craft to a stop long before it hit the lapping waves below. Already the rate of descent is easing and you know the other ships have immediately formed a circle above your crippled craft and increased the F.L. of their primary field inducers to a point somewhere beneath you, R_2=208ft. So rapid was this switchover that the descending point sources barely rocked your ship, and you would not have felt it anyway, for it was momentarily switched on and off quicker than you could think.

As the descent of your vehicle is arrested, it automatically centres on the circling craft above due to the inward P.S. components, you have the comforting knowledge that your ship could not drift out of that circle even if it tried.

Soon you are drifting down to rest gently just 200ft below the formation and you know that the combined vectors at R_3 = 126ft are passing through the centre of mass of you and your ship. Your instrument readings tell you that at this stage the other craft have a compensated vertical acceleration of about .9g each, with very slight inward tilt. While R_3 is producing an inward vector of 2.5g on your own vehicle. The vertical component of this is only .2g, but you remind yourself there are five ships up there, each contributing this amount, which means you and your crippled craft are weightless at 1g. You are safe!

Over the intercom, the group leader tells you he is going to break formation to investigate the trouble, and you watch as his craft descends to formate on your vehicle. In order to do this he must cut back on the power of his craft until it, too, is being supported by the four remaining discs above. At this stage both your machines will be pressed firmly together by a force of more than twice their weight. There is no fear of them separating and the structure is enormously strong; it can take the load. You both open the access hatches and you go over to join the skipper as, weightless and therefore free from the fear of falling, he pulls himself over to join you.

A five-minute inspection reveals a major fault in the regeneration system, from which it is obvious that both the primary and secondary fields cannot be operated simultaneously. Also there has been a quantity of ballast left on board from a previous underwater mission; according to the log this should have been jettisoned. So the first thing to do is give the skipper a hand getting rid of the unwanted load. This, of course, must be done by explosive jettison, for the craft is held in a powerful surrounding R field. After the ejection you watch the stuff tumbling seaward for a moment, then together make your way over to the skipper's ship.

Talking into the intercom, he issues the next flight phase orders, and thankfully you prepare to leave the area. 'Hmmm . . . there'll be hell to pay for the chap who made that log-book entry. . . .'

Aerodynamics of the R Field

'They seemed to manoeuvre as one, as if tied together . . . in an uncanny manner.' How many times have these and similar descriptions of UFOs in formation been filed by witnesses? The researcher will be fully acquainted with them, and I need hardly state that this technique is just what we could expect of such machines, for with their controls 'locked on' to the leader, he could virtually manoeuvre them 'as one'. Slipstream, turbulence and other aerodynamic effects known to our pilots would be non-existent to the pilots of the discs, but as with the G field theory, we must allow certain aerodynamic phenomena to accompany an R field ship. Here again the effects will be similar, for when the craft is in motion, air particles will receive the same repulsion as the ship itself, according to the extension or power of the field and the distance from the P.S.

The diagrams in Fig 87 are the exact counterpart of those for the G field shown in Fig 34, chapter 11. As in that case, Fig 87(a) illustrates an R field craft in forward thrust attitude, the vertical component being equal to 1g sustaining horizontal flight. At R_1=30ft, a total acceleration of 1.8g is acting on the C.M. of the ship, while at R_2=34ft, a lesser acceleration of 1.4g is brought to bear on the air particles at the periphery of the disc. Therefore, in this respect there is a slight aerodynamic difference between this and the G field craft, in that there would exist a straight flow of air over the machine due to the g differential. But, as we have seen, as the F.L. increases in length, so the differential becomes less, and the relative air flow would be negligible. Fig 87(b) shows the disc in the same tilted attitude from above, from which an idea of the shape of the moving air belt can be formed. Here again it would resemble a tear drop trailing off to a point as in Fig 87(c).

Fig 88(a) analyses the condition a little further. In this the machine is tilted at forty-five degrees. R_1=38ft, the field producing 1.41g on the C.M. of the disc. Therefore, both the vertical and horizontal components

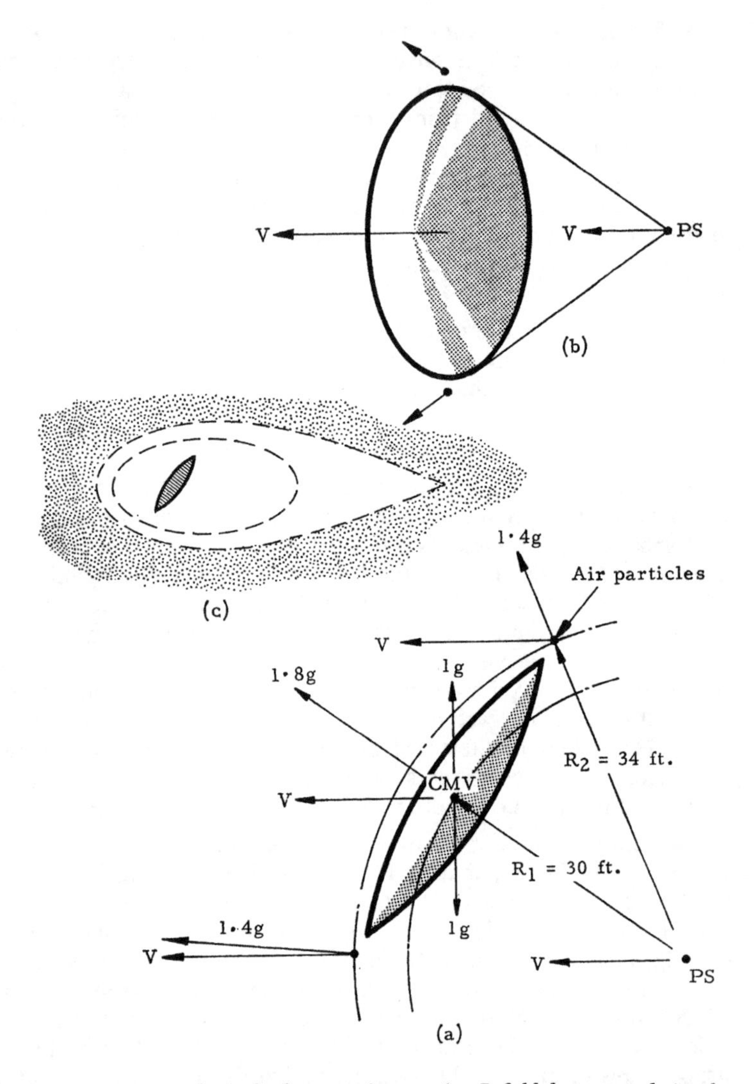

Fig 87. ***Atmospheric displacement in a moving R field due to translational velocity and acceleration from point source.***

are equal to 1g sustaining the level flight attitude. Neglecting the drag value of the moving air belt, we can see that after one second the vehicle will be moving at 32.2ft per second. While a particle of air situated immediately ahead of the P.S. on R_1, will experience an acceleration of 1.41g, which means that it will have attained a velocity of 45.5ft per second after one second. This means that particles in this region will be accelerating faster than the machine itself. The P.S. will also take on V, that is the same velocity as the machine itself, while particles of air situated immediately to the rear of the P.S. will also receive an initial acceleration of 1.41g backwards, but as from then on the P.S. itself is separating from it, the rearward velocity of the particle will be in the region of 13.3ft per second. Corresponding values for the velocities after ten seconds at 1g are also shown, but the position becomes a little clearer when we consider the case in Fig 88(b).

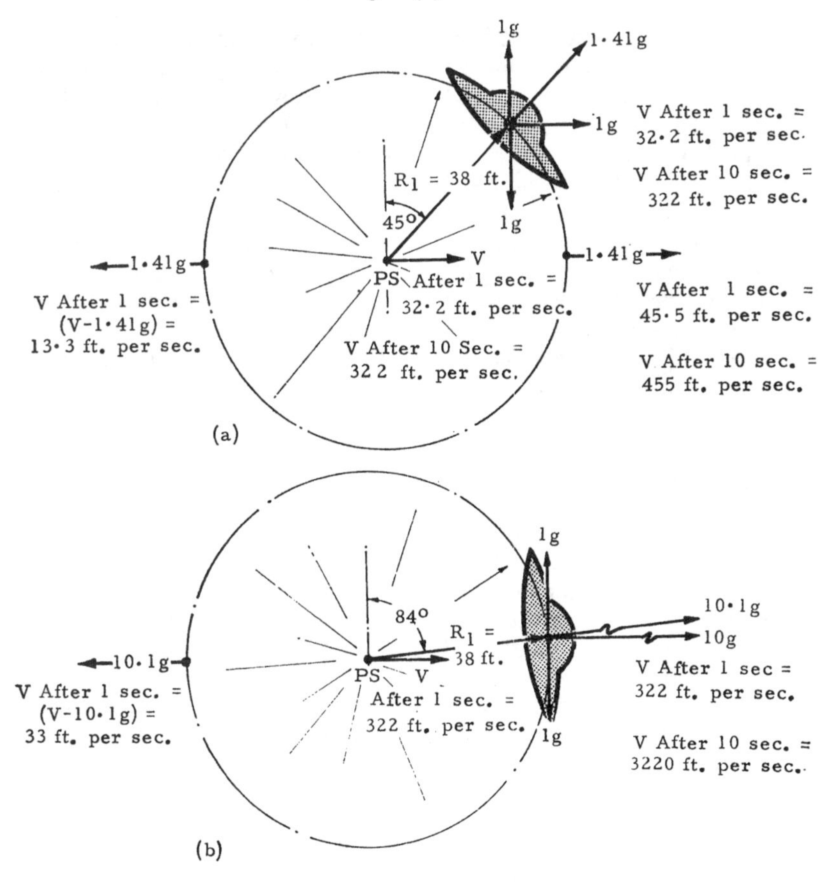

Fig 88. *Shortly after* (*a*) *the P.S. is leaving the area faster than the acceleration it will produce on a particle at this distance, therefore there will be a suction effect.*

Here a vehicle in a 10g propulsive field is shown. The first thing which becomes apparent is the angle of tilt, which in this case is some eighty-four degrees in order to furnish the machine with a 1g lift component. Cases for one-second and ten-second duration are quoted, which together with the vector diagram will show that the pear-shaped air belt will indeed tend to follow the craft; in fact, situated behind a fast-moving R field ship one would experience suction effects. In chapter 8 we saw how M Güitta of Casablanca nearly had his car pulled off the road by the strong suction following a UFO. Note, he said it was going 'at a terrific speed'. A violent gust of cold air followed the passage of the object.

In chapter 11, I used iron filings in a magnetic field to illustrate more graphically the air particles going along with a G field craft. This is more difficult to do with the R field counterpart, therefore I have chosen water as the medium representing the air in the analogy. Plate 27 shows a surface of water on to which is being blown a light jet of air. This forms an indentation which represents the repulsion P.S., which in reality would be forcing the air *i.e.* water radially away from it in all directions. In the experiment the air source is moved steadily along, which causes the water ahead of the indentation to be moved along with it and at about the same speed. If now, just ahead of the indentation we place a small piece of cork to represent the disc, it will be carried along with the water. The analogy may help to correlate the similarity between the two otherwise diverse effects of the R field and the G field.

As with the G field, we could expect local clouds to be formed in the vicinity of the craft due to the rapidly upward accelerated air particles near the P.S. The mass conversion may not always be one hundred per cent and, as predicted by Burkhard Heim's work, there might be field side effects, even the formation of large quantities of filamentous mass, which would be formed at the F.P. and, having been formed, repelled to fall earthwards, 'like long filmy cobwebs drifting down'.

As a fitting tribute to close this section, I am reminded again, as indeed I have been constantly reminded throughout my work, of the appropriate words by Desmond Leslie, whose inspiring maxim in *Flying Saucers Have Landed*, will remain with me always. Of the spaceman some time in the future he says:

'He will use gentle, harmonious forces that do not push and shove and heave and rend, but part the air smoothly, with courtesy and scientific good manners.'

20

Analysis Ten

Radar and Light Effects of the R Field

IN chapter 13, several UFO cases involving radar were dealt with and little more can be said from the R field point of view, save for the fact that a long focal length which might bring the point source nearer to ground level would be capable of local interferences other than ground effects, for the radiations accompanying such an intense field would be proportionally more powerful. If the point source moved across the countryside, as for instance if the UFO tilted, we can imagine it approaching close to a radar aerial and we would not be wholly surprised to hear that a radar image suddenly intensified. Such as the case at Goose Bay A.F.B., Labrador, in 1952. 'A red light which changed to white, hovered for a little, oscillated and sped away. The strange thing about it was, when it oscillated, the image on the radarscope *flared up*, returned to its original size and disappeared.'

I suggest that the original size image was the UFO, while the flaring up was the local effect of the beamed point source arcing across the sky. Indeed the repulsion field might offer an explanation for many of the otherwise unaccountable radar phenomena well known to operators.

We have seen how the direct conversion of mass into energy would be accompanied by other radiations, this being the 'exhaust' of the process and this would be equally true of both the G field and the R field principles; we could expect physical effects of light and heat as well as electro-magnetic effects in general. While a change in wavelength of the emitted light towards the blue and the red end is also likely to occur. In *Space, Gravity and the Flying Saucer* I mentioned the well-known cone effect associated with some of the saucers, and no doubt as we developed this present chapter, the thought will have occurred to some readers that here there may be visual evidence for the repulsion point source. Indeed this may be so. From the Clinton County Army Air Base, Wilmington, Ohio, comes the report: 'A flaming red cone trailing a gaseous green mist tore through the sky.'

June 1964. Mr R. W. Lambert, of Wallasey, Cheshire, stationed at Harrison Drive Coastguard Station, reported: 'A very bright cone-shaped object hovering over the sea for five minutes. Then the cone became inverted, moved out to sea and vanished.' The object was observed through binoculars.

9 February, 1950. Lt. Commander J. L. Krakev and five other residents of San Leandro, California, observed a UFO flying over the

Alameda Naval Station at about 5,000ft. It left a vapour trail and the speed was fairly low, about seventy to ninety miles per hour. They described the object as appearing 'like a thirty-foot ice-cream cone'.

There are many of them, the description much the same 'and for about five minutes it hovered, while a flood of blinding light poured from it, in the shape of a cone.'

But one of the most interesting incidents involving what might be visual evidence of the R field occurred to Mrs W. Felton Barrett, of Great Barrington, Mass., U.S.A., who in July, 1957, was on a cruise in Norwegian waters with the *Stella Maris*. While Mrs Barrett was ashore in Norway, she had photographed, among other things, a fjord, and much later, when the colour film was developed, she was amazed to find an object displayed on the background view of the fjord. The original colour print of this is far more impressive than the black and white copy, Plate 29, and cannot be satisfactorily dismissed as a lens flare. Although the field effect is distorted, there in the middle can be seen the conical beam effect. Mrs Barrett does not recall seeing or hearing anything unusual while taking the photo. It may well be of course that had an object been passing at high speed the instant Mrs Barrett was looking into the view finder, then in all probability she would not have seen it. To verify this we have only to superimpose the vectorial diagram on to the print to find that even if the object was travelling horizontally, then the angle of tilt over 45° implies a fairly high 'thrust' component, which of course implies high speed.

Had the object been climbing, as I am inclined to suspect, and the vertical component of repulsion greater than 1g, then at the angle shown the forward velocity component would be even higher. On the other hand, to reduce the forward velocity component to a low value would also require reduction of the vertical lift component in proportion; therefore, the object would be dropping at a pretty fast rate, which would seem to make the angle of tilt superfluous anyhow. In other words, the object shown in the photograph is in the very attitude required if we are to presuppose that Mrs Barrett failed to see it, because A it was silent and B it was too fast. Remember it was taken on a fast colour film; therefore, it is not required that the image should be any more blurred than it is.

We are indebted to the American organisation NICAP for the use of the following case which appeared in the August-September 1965 issue of 'The UFO Investigator', carrying with it a UFO photograph which might well bear visual evidence of the R field.

'Pictures of the UFO (see Plate 32) taken 8 August, in Beaver County, Pa., were declared genuine by three professional photographers who examined the negatives. (Signed evaluations given to NICAP in a five-day on-the-spot investigation.) Because of the close range (one-fourth mile) these photos have unusual detail.

'In the first, the bright spot at the left is the moon. The white mass under the UFO, like a double exhaust, was invisible to witnesses. (A NICAP technical adviser suggests the film may have recorded a force-field emanation, a clue to the propulsion.) The streak behind and below the disc is believed to be a trail the UFO left in manoeuvring over the trees.

'On the night of 8 August, James Lucci, son of a professional photographer for the Air National Guard, was making time exposures of the moon near his home in Brighton Township, Beaver County, Pa. With James, a high school senior, was his brother John, 20. A third witness, Michael Grove, saw the UFO from his home across the road.

'About 11.30, as James started another exposure, a round, thick object, glowing brighter than the moon, approached above the trees at the left. Realising the camera must have caught it, James closed the shutter, quickly wound the film for another shot—the slow moving UFO was then about a fourth of a mile away. Before he could get a third picture the strange device climbed steeply out of sight—its swift ascent also seen by Michael Grove.

(Later, the AF admitted to the press that a UFO was reported about this time, near Pittsburgh. Whether it was the same one, seen before or after the 11.30 sighting, it is not known, as the AF refused to give details.)

At first, fearing ridicule, the Luccis avoided publicity. After two days friends persuaded them to show the pictures to the Beaver County Times. The newspaper's three photographers superimposed the negatives and made other tests which showed the UFO had slowly moved closer, left to right, as described. After a full evaluation, they labelled the photos genuine. This was later confirmed by NICAP Photographic Adviser Ralph Rankow.

'According to the Luccis, the disc appeared 'larger than a Piper Cub aeroplane'.'

When the Times story broke, the negatives were quickly obtained for NICAP analysis by William B. Weitzel, Chairman, Pittsburgh-NICAP Subcommittee. Aided by the Times staff, Weitzel and Subcommittee member Robert Brown spent five days cross-examining the witnesses, checking the photos against aerial and contour maps, and recording technical photographic evidence. The Times gave them this statement.

'To the best of our knowledge, and in our considered judgement, the negatives are not the result of photographic or physical faking . . . nor is the UFO image thereon the result of any photographic accident such as reflections . . . lens flare . . . developer bubbles and the like . . . and said image appears to be the result of photographing a self-luminous object positioned in the field of view of the camera.'

'These conclusions were signed by City Editor Jack Mitchell, Photographer Peter D. Sabella, Reporter Thomas W. Schley, Managing Editor F. N. Hollendenner and Photographers Harry K. Frye and J. C. Gardner. In addition, the witnesses' high characters were vouched for by the Chief of Police, Brighton Township, the high school principal, and Beaver County Police.

'Up to our press time, the AF has kept silent, though James Lucci offered them the photos for evaluation. *The Times*, calling the AF 'elusive and mysterious', said one spokesman 'bluntly described the official position when he said, 'My mouth is zippered'.' Commenting on this 'official brick wall', the newspaper added:

'Where UFOs are concerned, the AF conception of co-operation is a one-way street . . . IN ONLY.'

Lucci camera data: Yashika 635 with Altipan 120 film (ASA 100) Lens opening f3.5, set at infinity; exposed 6 seconds. Film developed 12 minutes, fresh D76, 70 degrees, with agitation. Prints shown specially exposed to bring out detail. UFO slightly out of focus due to motion. A TV tower (not shown in prints) is in focus.

A fuller analysis of this photograph would be interesting but our space is limited; even so the reader will not find it difficult to relate the cone shape beneath the Lucci UFO with the repulsion theory. If we vertically bisect the saucer, an imaginary focal point can be found at the centre of the white haze, from this can quite easily be traced a ball-shaped formation from which are projected two filaments *thrusting up towards the disc*. We cannot help being reminded of the 'halo' of the Wildman case and the frost being thrust down etc.

Here and there throughout this work I have stressed my earnest wish that in pursuing this subject UFOlogists should themselves avoid the pitfall of bias with which they are often apt to tarnish UFO sceptics at large. By this I mean the tendency of some researchers to accept one particular group of sightings without question, but simply because the remainder fail to fit their particular interpretation of the phenomena, it is rejected out of hand. I need hardly say that this attitude, apart from being unscientific, may be very misleading in the long run. It has been my experience to discover that many sightings which have made even very open-minded UFOlogists curl an eyebrow have something of extreme technical merit, which have been otherwise discourteously overlooked. In fact, only by including *all* the more notable sightings is the story within these pages beginning to make technical sense. In this regard I feel honoured to walk in the footsteps of the late Waveney Girvan, Aimé Michel and his good friend Jean Cocteau, Arthur Constance, Desmond Leslie and many others. It is due to this line of enquiry that I have detected technical consistency even amongst contact or near contact with saucer occupant reports. We might just be able to establish the validity of some of these claims by technical substantiation; it should at least be one of our primary concerns.

Light and R field propelled saucers; let us take a look at the following bizarre account published in the November-December 1964 edition of the *Flying Saucer Review*.

Whidby Island Contact

We are indebted to Marvin W. Smith, a member and active researcher of the Everett Flying Saucer Club, in the state of Washington, for sending us a full report of a new contact story.

It seems that Mr Smith came to hear of this tale during a casual conversation with a friend. There was nothing casual, however, about his subsequent investigation. A diligent check through directories, a series of telephone calls, an exchange of letters, and finally an uninvited visit to Whidby Island, across the Sound from Everett, brought him face to face with the reluctant contactee.

The description 'reluctant' is used advisedly. Mr Smith points out that the contactee is a retiring, middle-aged woman who shuns publicity. Furthermore, he stresses the fact that she did not invite him to hear her story: it was he who sought her out. He describes her as 'a matter-of-fact, reasoning individual who has had quite a bit of experience in public life', and he goes on to reveal how, when the overall UFO picture was discussed with her, she was suspicious of the validity of many other contact claims. She suspected that some claimants had invented stories, hoping subsequently to 'cash-in' on them through publication.

In accordance with this lady's wish, Mr Smith has withheld her true name; instead he calls her 'Mrs Ruth Brown'.

Mr Smith's first visit to Whidby Island took place early in February 1964. He found Ruth Brown living in a detached house, with her nearest neighbours living more than a quarter of a mile away.

Enquiries elicited that there had been two visitations by unidentifiable craft, the first in July 1963, the second the following October.

The first 'contact', which proved to be rather a tenuous one, occurred when the good lady was on the point of retiring for the night after an evening's televiewing. She stated that throughout the evening she had experienced an uneasy, prickly feeling that she was being 'watched', but it was not until she looked by chance from her bedroom window that she was aware that anything unusual was happening. It was then that she saw a strange craft hovering near the house. Her immediate reaction was not one of fear, as one might have expected, but one of inquisitiveness instead, for she hurried outside the house to get a better view. In this she was unlucky: when she reached the yard the UFO had gone.

The second visitation was quite sensational, however, and more than compensated for the disappointment of the July sighting. It happened at breakfast time, in broad daylight.

Mrs Brown saw a machine of similar appearance to the July visitor. It was about 10ft long, coloured grey, and rather like an aeroplane to look at, shorn of wings, tail, fins and undercarriage. There was no visible means of propulsion.

The UFO approached from the north, and as it halted it hovered some 5ft from the ground, quite close to the eastern side of the house. The witness reports that the front part was transparent, and that she could see at least three figures inside. Whilst the craft hovered, *the grass underneath was flattened.* The italics are mine (author).

Quite suddenly one of the occupants, a being of human shape and size, was seen to be standing on the ground, having seemingly emerged through the side of the craft! 'He' was clothed in 'asbestos textured coveralls . . . which concealed the feet, hands and face, except for openings where the eyes should be, but where no eyes were visible. . . .'

The craft, led by the being on the ground, now began to move slowly round the house. Mrs Brown hastened to the door and cried out to the solitary figure: 'What do you want?' Whereupon he conferred with his companions in the machine, before turning and saying quite clearly: 'One of our party knows you, we will return!' He then re-entered the craft in the same disturbing manner that he had left it.

The craft now continued its movement round the house, with Mrs Brown following, and then a second strange thing happened. Without warning, it suddenly *shrank considerably in size* and tilted so that its rear portion *dipped into the ground.* The startled witness at first concluded that it was taking in water, but subsequent speculation favours the view that it was seeking fuel of a kind.

The occupants signalled to Mrs Brown that she was not to touch or go near their craft. She claims that she had an 'awareness' of this, and suggests that it would indicate some form of thought transference, or telepathy.

When the 'digging' operation was over, the UFO swelled up to its previous size and then moved off in an easterly direction, with '. . . much steam, smoke, a flash and a noise, rapidly disappearing from sight'.

Some little while after this visit, a strange fungus appeared on the ground at the spot where the 'digging' had taken place.

Marvin Smith had suggested that the object of the visit was not to study the locality, but to gauge the effect of such contact on human beings. He feels that this theory is supported by the space visitor's statement, that they knew of Mrs Brown and that they would return.

The first more obvious technical observations concerning this case, are that the witness claimed the grass beneath the hovering machine was *flattened*, also that the machine was possibly a disc viewed on edge, *i.e.* 'it hovered some 5ft from the ground'. The witness estimated the craft as being 10ft 'long' and one of the crew walked a little ahead of it. We can assume the machine to be at least 5ft thick and the centre of mass somewhere near half this, which if the estimated hover height of 5ft is correct, places the c.m. about 7.5ft above ground level. Now the flattened grass suggests the use of an R field and we can assume the point source of this to be stationed at some convenient point between the craft and the ground, and not too near to either. A fair judge might be 1.5ft above ground level, which, assuming 1g at the centre of mass of the vehicle, gives us the downward acceleration on the grass as being:

$$G_g = \frac{R_1^2 \; Gs}{R_2^2} = \frac{6 \times 6 \times 1}{1.5 \times 1.5} = 16g.$$

As we have seen, this could quite easily account for the grass being 'flattened' immediately beneath the ship, but could a member of the crew alight and stand near to the hovering device?

Well, the witness claimed the vehicle was about 10ft long and looked like an aeroplane minus wings, etc., and that she saw several people in it. Now if the craft was about 5ft thick, as it must have been in order to have human beings, albeit sitting down, then the proportions 5ft × 10ft would hardly remind one of an 'aeroplane'; therefore, I take the liberty to suggest the craft was much longer than this, and without begging the question I would think it more conservative to suggest a minimum of 12ft 'long'. If this were so, then a human being standing about 9ft from the F.P., *i.e.* 3ft outside the perimeter, would be subjected to just under $\frac{1}{2}$g radial acceleration. In other words, if a person were expecting this, then by leaning slightly 'into' the force, they could no doubt counteract it quite easily, as one would in a centrifuge.

Regarding the curious technique of exit and entry by the being, here again, once practised, a person might well find in the R field effect a handy 'leg up' into the craft and, in this case, the entrance may well have been hidden from the view of the witness.

I would add that the explanation for the latter may well be far more fantastic than the very down-to-earth one I have offered, but there is no point in complicating the issue further at this stage.

So far, from the above consideration, the Whidby Island case is supported by the R field theory, but this is the more obvious example; are there other clues and, if so, are they also in keeping with the R field theory? Let us see.

Earlier, when discussing light in terms of the G field theory, we saw how we might expect such a field to form an atmospheric lens, producing optical effects which might be further augmented by other field effects as well as the gravitational bending of light. Also how these phenomena might be obscured by the fog produced by the increased air density dropping below the dew point. Now it follows that if there would be a local *increase* in atmospheric pressure due to a powerful G field, then similarly we could expect a *decrease* in atmospheric pressure to accompany a powerful R field, and again we would not be surprised to find optical effects.

Most of us while driving at some time or another, chiefly in summertime when hot air is rising off the road surfaces, have seen a sheet of 'water' mysteriously flood our approach, only to disappear before we reach it. This is caused by a local decrease in atmospheric density, with an accompanying change in the refractive index; it is in fact a kind of mirage, to which, of course, nearly all flying saucer stories are accredited by our good Dr Menzel. I might digress for a moment to say here that I shall be intrigued to see how the good doctor will explain the technical corroboration among sightings I have attempted to set out in these pages, in terms of his rather outdated temperatures inversion, mock suns and mirages, as he has done in the past, or it may well be his only comment will be as argumentatively stimulating as an earlier remark he made concerning *Space, Gravity and the Flying Saucer*, 'Rubbish!' I put this on record in all goodwill, for if the gentleman is really sincere in his oft-questionable arguments, then I know this to be his loss, not mine, and with whatever shortcomings I may have garnished myself in this world, one of them is not a desire to fool myself or others. If I find a pebble on the beach I can marvel at its exquisite being; it is an added joy to show it to comradely souls. One can only feel a sense of regret and sorrow that some of them cannot even see it in your hand!

Now, if the 'sheet of water' phenomenon can be caused by even a *slight* drop in atmosphere density, then we would not be surprised to find similar optical effects near the R field, but what kind of effects might these be?

Well, to begin with, as there would be a *drop* in pressure instead of an *increase* in pressure as with the G field, then we might expect the very opposite effect to occur.

This is shown in Fig 89(a), where a bi-concave lens is interrupting the cone of vision producing a small image, the effect becoming more marked as the lens is brought nearer to the eye.

Now let us correlate this with the R field, as in Fig. 89(b). First, we note that the atmospheric lens formed is less dense, from which we could expect deviation of the light rays, *i.e.* rays which we would normally see are deflected past the eye. Next, light rays which would normally be focused before they reached the eye will now be deflected to meet it, therefore making the object appear smaller. To this we must add the effect of the R field on light rays, Fig 89(c), which would be to augment the *deflection* in exactly the same manner in which the G field augments the *contraction* of the light rays.

In a word, then, we can now say, while a G field might produce optical magnifying properties, an R field could produce optical *reducing* properties. And if the grass being pressed down was indicative of an R field in Mrs Brown's case, then her claim to have seen the UFO 'shrink' is technically corroborative. That this was in fact an optical effect caused

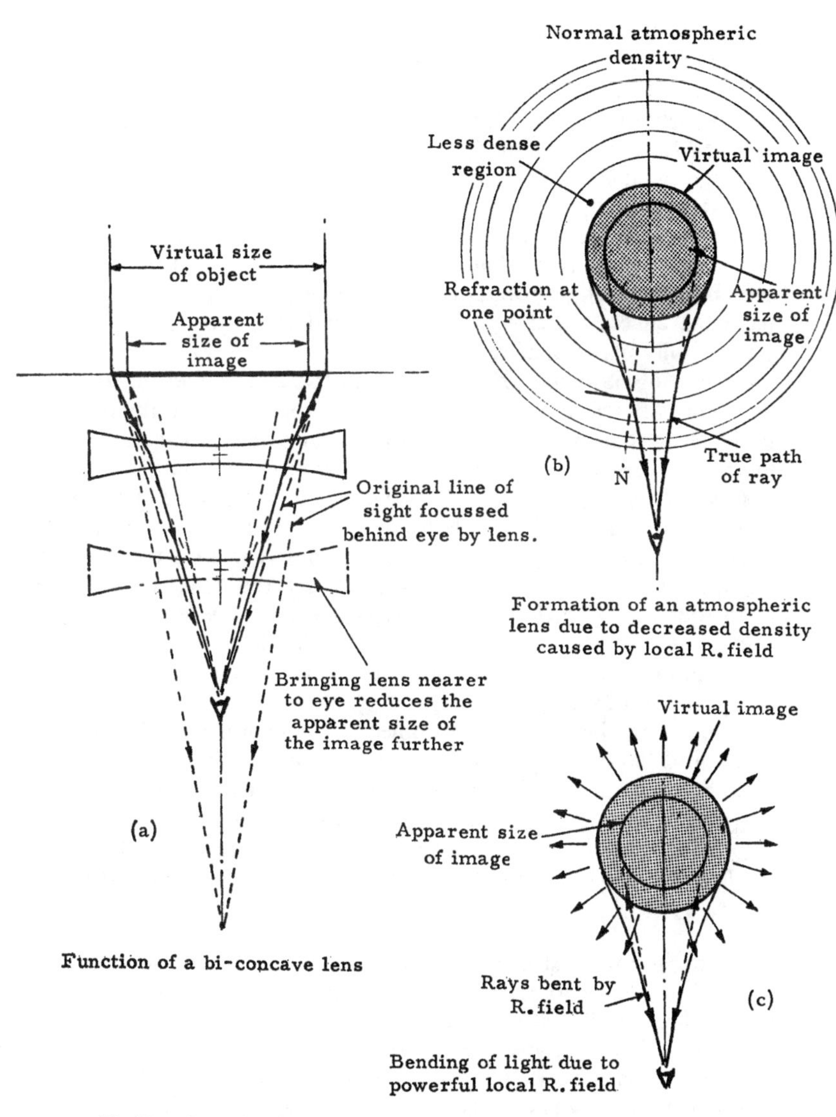

Fig 89. Optical reducing properties of the R field are the opposite to the magnifying properties of the G field.

by a change in the field is, I think, borne out by the fact that the occupants warned the witness to keep away.

In addition, it is of perhaps only passing interest to note that if an object is viewed through a reducing glass which is tilted to or from the viewer, then an additional vertical 'shrinking' takes place. If, on the other hand, the glass is stood vertical, but with one edge swung further away from the viewer, then horizontal 'shrinking' of the object takes place. But even more interesting, if *both* these planes of tilt are produced simultaneously, then the object not only shrinks in *all* dimensions, but immediately appears to 'dig itself into the ground'. To illustrate this, Plate 33 is a photograph taken of a model disc suspended straight and level when viewed both with and without a reducing glass tilted in this manner. Naturally, if there were other markings or objects behind the disc, they too would appear tilted, but as Mrs Brown's home is in the country, it is possible that such surroundings, grass, trees, etc., might tend to camouflage the effect; therefore the background in the experiment has been purposefully made nondescript.

By way of endorsing the link between the R field and the atmospheric lens due to the accompanying drop in pressure, I would like to point out to the reader that we could also expect a change in temperature. Therefore the rule: wherever there is a *repulsion field* it should be accompanied by a *drop in temperature*. Now I should like to recall your attention to the bizarre Romand affair which took place at the village of Prêmanou, France. Note: young Raymond had almost run into the queer figure outside the barn. 'Plucking up his courage the boy had advanced so as to touch it, when he was instantly *knocked to the ground* by ' an *ice-cold invisible force*'.' Not only is this additional fact corroborative, but it also substantiates the use of small secondary R field inducers. We shall hear more of this later on.

I would not be true to the subject if I did not add that I am very mindful of the fact that these are but my conclusions. They continue to fit, and as long as they do I can do little else but to accept them. But they may not be the right conclusions; though the corroborative value is there, throughout our entire journey, we may be examining only the interrelated shadows, albeit correct in their sequence, of pieces of a jig-saw puzzle yet unseen by us.

As we near the last stages of our journey, there, over the distant hills, the end of the road is in sight. Familiar voices echo just ahead, as turning into the last stretch we come across a clearing formed by a road junction. Waiting there are the rest of our party who took the other branch way back down the trail. Soon we join them and rest while we compare notes. We find that at first the signposts were diverse, the terrain somewhat different. But then the pathways began to bend towards each other again, until now, looking back at them, an interesting symmetry blends the scene. Viewed separately they would indeed appear as different settings, perhaps one wild and uninviting, while the other is more placid and serene. Yet the entire picture we now observe would be incomplete, one without the other, the diverse pathways have led both parties on to common ground.

PART FOUR

Familiar Scenery at the end of the journey

21

The Bi-Field Theory

WE have seen how electric, magnetic and gravitic fields are probably synonymous in the sense that attraction and repulsion phenomena have a common cause. Similarly, gravitational attraction or repulsion might be different effects of that common cause, in that two bodies are 'attracted to' or 'repelled from' the surrounding space, rather than to or from the bodies themselves. So that the condition is either gravitative or repulsive space, depending on how you look at it. It would be equally true to say that both conditions were a state of gravitative space. As a fair simile, engineers get quite used to thinking in terms of a vacuum exerting a 'suction' on a body, when what we really mean is that a 'pressure' is exerted on the body. On this same premise, gravitational phenomena is still 'gravitational' whether the effect on matter be acting inwardly or outwardly. Therefore, G field or R field, we are still dealing with phenomena in the gravitational domain.

In correlating and analysing some UFO cases at this threshold, we have allowed that such local conditions can be generated by a machine to cause it to move, or more correctly, suffer spacial displacement. In a word then, we could have been examining a phenomenon from two different points of view. Of the G field-propelled craft discussed earlier, we could say that it was gravitationally moved *towards* its point source, or, due to a *decreased* gravitational field strength above it, it is repelled by the 'denser' space beneath, and either could be equally true; how are we to know? Similarly of the R field-propelled vehicle, we might say that it was repulsively moved *away* from its point source or, due to the *increased* gravitational field strength below it, it is attracted to the less 'dense' space above, and again either could be true, we could not know. In other words, the phenomena we describe as gravitative in terms of matter, may in reality be repulsive when considered in terms of space. In either instance it would be more convenient to speak in terms of unbalanced inertia, which in actual fact is more strictly true. The argument becomes even more rational when we remember that matter is probably a conglomeration of waves, which will move along carrier waves in any direction, despite our efforts to categorise such movement into brackets of different natural phenomena such as 'attraction' and 'repulsion'.

But it is mentally convenient for us to think of phenomena in this way, therefore this we shall continue to do by saying: if unbalance of a kind can be created above or below a machine by the use of a G field and an R field, so as to cause it to move in a chosen direction, then it is surely logical to use not one but *both* techniques, though this may not always be convenient. By studying some of the predicted and noted effects of both fields, the student will be able to visualise the more obvious combined effects for himself, but this analysis would be incomplete without a précis of some of them.

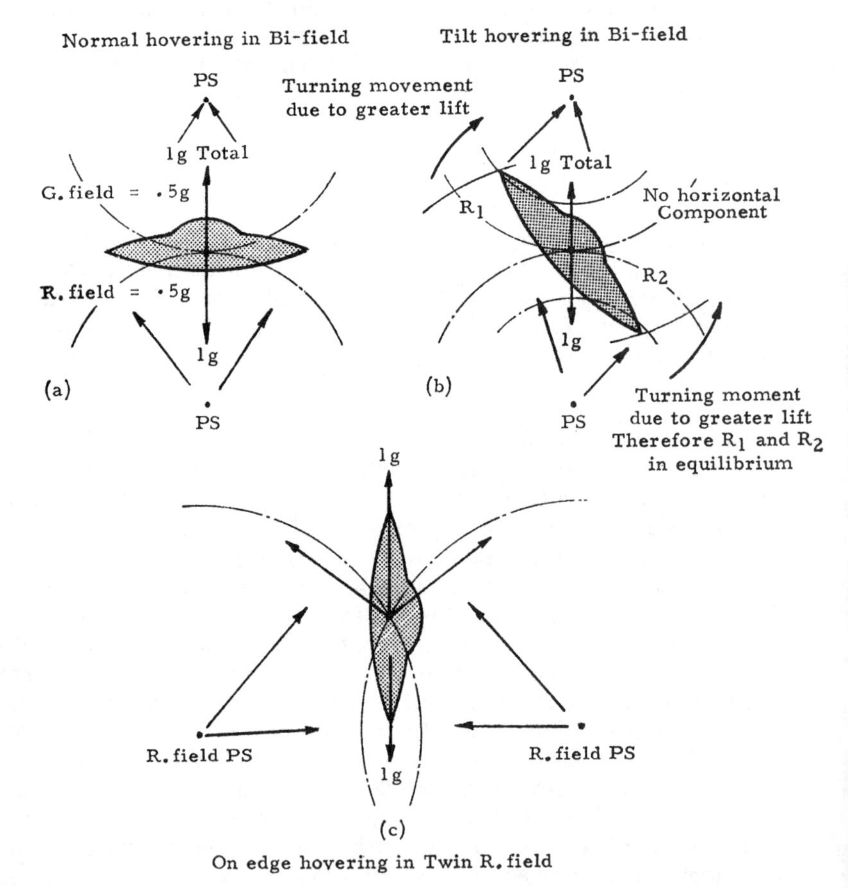

Fig 90. Some inherent advantages of the Bi-polar field.

In the first place we note that if both fields are employed, then they will share the lift power required, as in Fig 90(a). This shows a Bi-field craft hovering at 1g with both fields contributing .5g each. One of the more obvious advantages of this is, that the dual arrangement permits the use of much shorter focal length, which requires less power and therefore is more efficient. Earlier we considered the possibility that the P.S. could be shifted to cause an asymmetrical force to be brought to bear about the C.M. of the ship, therefore facilitating orientation. But I also pointed out that, if a disc was hovering and to be tilted, then it would move in the direction of the tilt due to the horizontal component. Nevertheless, saucers do frequently hang motionless in the sky while tilted, and tilted acutely at that. This could be explained logically by the use of secondary field inducers; but note, if both P.S. are displaced as in Fig 90(b), then a tilted machine can hang suspended at 1g vertical acceleration without being subjected to a horizontal component, because the increased force of the G field at the short radius R_1 is exactly offset by the increased force of the R field acting at the short radius R_2. It is also logical to suppose that the capability of producing either types of field could be used for other purposes. For instance, the conversion to twin R fields or twin G fields would permit the craft to 'roll along' through the sky on edge. Note, such phenomenon has been recorded, Fig 90(c).

But of the more obvious physical requirements demanded by the gravitationally propelled machine, there is one which we have not yet discussed. Neither could we have discussed it until now, though it is anticipated it has already occurred to the more mechanically minded reader.

It is a problem whose answer also advocates the employment of *both* G field and R field and, further, it requires that both fields *must* be *convertible* one into the other. Indeed, without such an arrangement, neither the G field nor R field craft could function at high speed in the atmosphere! The following will explain this in broader terms.

First let us consider the case for the singular-operated G field and R field machines. Beginning with the take off, everything of substance that we have predicted still holds true, such as the tilt and forward flight. With increase of tilt and field intensity we get increase in forward velocity, air particles will move along with the craft, etc. But we have seen that if, while the craft is at high velocity, the field should collapse, then the vehicle would run into near stationary air with dire consequences, depending on the magnitude of the velocity, as in Fig 91(a) and (b). Prompted by this, the reader might now ask the question, 'Just how do we arrest high-speed atmospheric flight of either the G field or the R field type craft?'

Well, let us look at the more obvious answers first. We could, of course, simply rotate the machine so that there is a rearwardly acting thrust component as in Fig 92(a) and (b). But while this is possible *outside* the atmosphere, certainly it would be impossible *in* it, for it is apparent that during the manoeuvre the hurtling machine will be subjected to the same

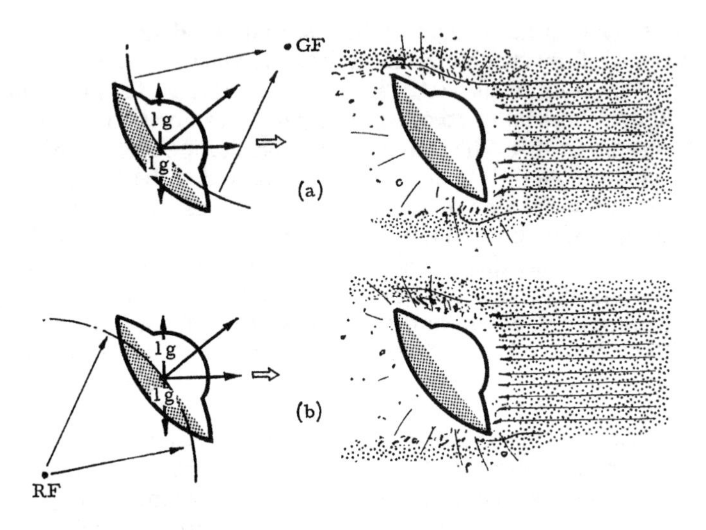

Fig 91. Both types of vehicles would be subjected to aerodynamic forces if the propulsive fields collapsed.

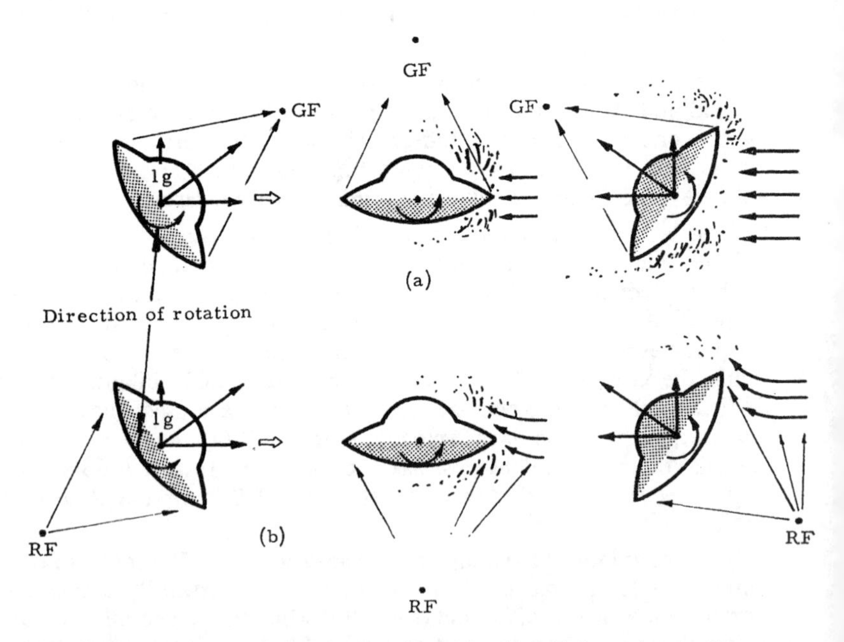

Fig 92. Rotating the machine would reverse the field thrust, but would not obviate aerodynamic forces.

intense blast of the onrushing airflow, almost as surely as if the propulsive field had collapsed, as in Fig 91(a) and (b). While it will be seen that in this respect the R field ship comes off slightly better.

Therefore, in order to decelerate the vehicle it must be rotated, but not *backwards* as in Fig 92, but *forwards* as in Fig 93(a), while the 'attractive' G field must be converted into a 'braking' R field.

Now this suggests that such craft *must* be capable of converting from G field to R field. Therefore, the most logical arrangement must surely be the convertible Bi-field, as in Fig 93(b), in which the machine with polar axis aligned with the direction of flight would be retarded by the converted fields, the R field point source now acting like a dispersing spearhead or umbrella to the oncoming airflow, which is both decelerated and shed round the slowing machine. But due to its attitude, the craft would momentarily lose height, until after shedding forward velocity it would be brought back to an even keel again. The temporary loss of vertical acceleration could be compensated in several ways, such as displacing the point sources or completely inverting the machine, as in Fig 93(b).

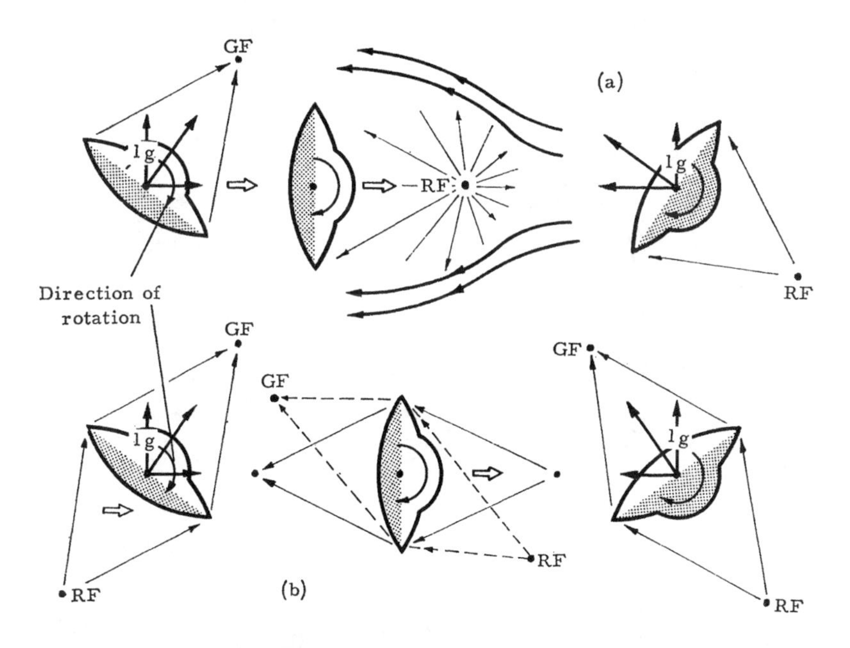

Fig 93. Converting an R field to a G field would not stop air blast, but converting a G field into an R field would, and if fields can be mutually converted, then the Bi-field is the logical combination. During the deceleration-turn about stage both focal points can be displaced so as to maintain a lift component.

But if the need to be aerodynamically shielded during deceleration is the chief advocate for the convertible Bi-polar character of the saucer's lift propulsion system, then the need for controlled vertical descent must surely run a close second. In Chapter 11, we considered this in terms of the pure G field craft and the same can be said of the R field saucer; that is, in order for either craft to descend, the vertical acceleration must be reduced to something less than 1g. But this means that the machine will be subjected to aerodynamic forces, for clearly it is falling, swiftly or slowly, depending on the magnitude of the field lift. Which also means that, allowing for the craft to be stabilised by other means, and the field lift to be zero, the fastest sinking speed attainable will be that governed by the aerodynamic support. What then if the pilot of the saucer required to increase the rate of descent as might the pilot of a jet? The answer is simple, he must convert the sign of the field, and 'fly' the machine down!

As in the deceleration rôle there are several alternatives, but the diagrams in Fig 94 illustrate all too clearly that the best control is offered

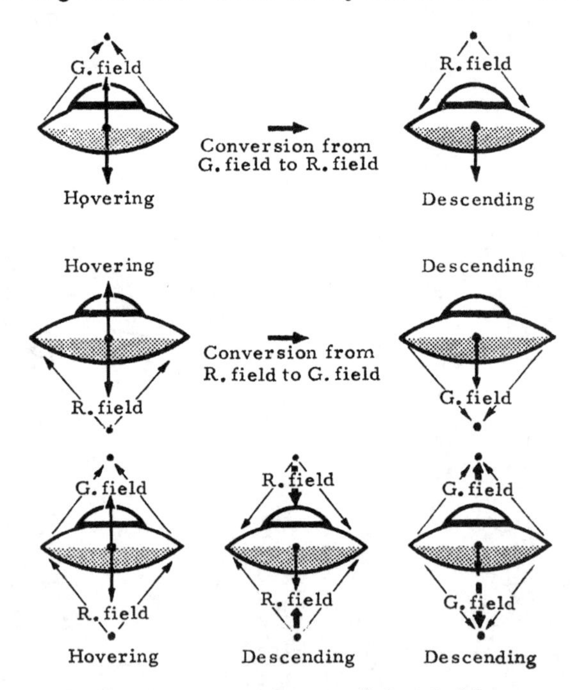

Fig 94. Arrangements for controlled vertical descent.

by either a twin G field or twin R field. In any event it is pretty obvious there is another need for the saucer's propulsion field to be controllable, both in sign, magnitude and direction. Nevertheless, saucers *do* 'nose dive' in the lower speed forward flight régime, in which case it will be seen that the controlled Bi-field arrangement offers the best solution, in which the vehicle would be tilted as in horizontal flight, but the lift component reduced to something less than 1g. In this particular phase, therefore, the machine would almost certainly be subjected to some complicated, though restricted airflow, but stabilised by the variable point sources and/or other means, this would not amount to any serious consequences, any more than a fast-moving conventional aircraft. And while talking of conventional aircraft, it is interesting to note that included among the previous observations should be the need for the gravitational machine to be banked on turning, and for exactly the same reasons as its more down-to-earth counterpart. Again, such behaviour with UFOs is common.

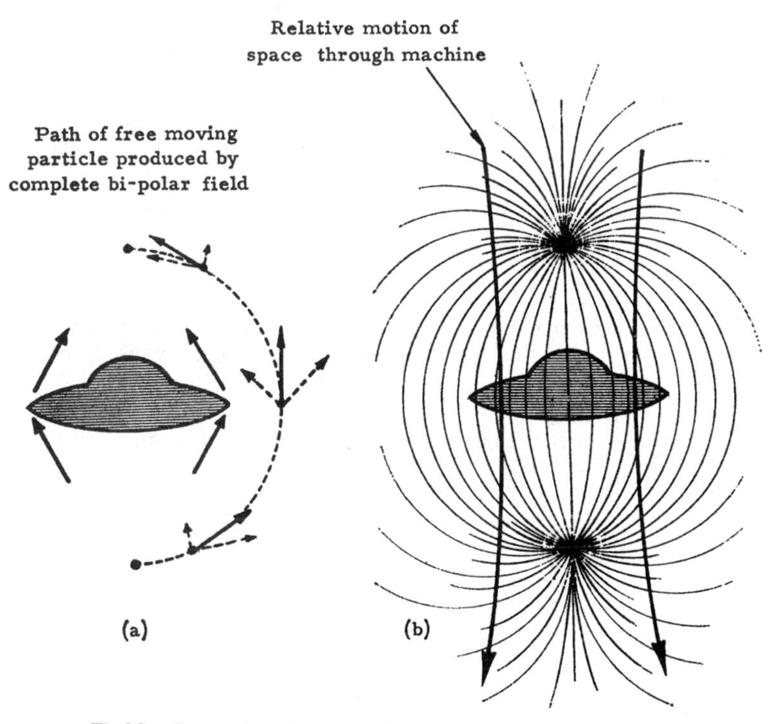

Fig 95. Interaction of space and machine due to Bi-polar G field.

Other flight phases might require the use of convertible fields, as suggested by some of the cases reviewed; in fact, there is reason to consider the possibility that, due to certain terrestrial conditions being encountered, the fields may not be completely stable and the employment of a duplicated system would largely offset this.

We are now in a position to draw up a picture of what the Bi-polar field may be like and begin by assuming the vehicle to be hovering at

1g and to trace the path of an imaginary particle repelled at the R field P.S., as in Fig 95(a). This shows the forces acting on the particle at three different positions and it will be seen it traces out a near circular passage. If we complete the diagram, as in Fig 95(b), we find a surprising likeness to the lines of force displayed by the common magnet; we have arrived at the Bi-polar G field.

From this we can see that in principle the flying saucer may still be very much a kind of rocket, in so far that it is drawn towards 'space' above it, and ejects 'space' below it. So that in this particular arrangement space is virtually passing through the system as a kind of reaction flux. In this respect it is interesting to note the remarkably similar conclusions arrived at by Lieut. Plantier. Writing on Plantier's theories, Aimé Michel says: 'The liberation of this cosmic energy makes it possible to create, at the scene of the liberation, a local field of force which can be varied at will. This local field of force is like the magnetic field which exists in a solenoid or between the two poles of a magnet, or the earth itself.'

Plantier himself goes on to say: 'It can be presumed that the device utilises a method of liberation analogous to that which in nature creates the primaries of cosmic radiation. The cosmic corpuscles thus generated would radiate in the form of corpuscular-undulatory fluid through the machine, in the direction in which it is being propelled, at a speed approximating to that of light. There would thus be a kind of continuous cosmic jet, pulsating right through the machine.'

As I said earlier, if there had been any communication between Lieut. Plantier and myself, or if one's work had influenced the other, the value of the independently similar conclusions would be void and I would not be placing these facts on record. As it is, my own findings are different in many respects, but none can deny the fundamental similarity is there.

Ground Effects of the Bi-Field

If a powerful G field or R field might cause a crater in the ground beneath a low-flying machine, this can also be said of the Bi-field. Indeed, if the R field P.S. is formed below ground level, the combined fields will serve to augment the effect even further. But as both fields will be sharing the lift of the machine, the R field effect will be reduced proportionally. To illustrate this further, Fig 96(a) depicts a machine taking off at 10g by a G field only. It is some 12.5ft above ground level and the P.S. is 20ft (R_1) above the ship's C.M. If R_2 equals 35ft, then the ground effect immediately beneath the craft will be

$$G_{gl}=\frac{20\times 20\times 10}{35\times 35}=3.26\text{g}.$$

But suppose we now consider the craft in the same conditions but powered by both R field and G field, Fig 96(b), which means 5g is

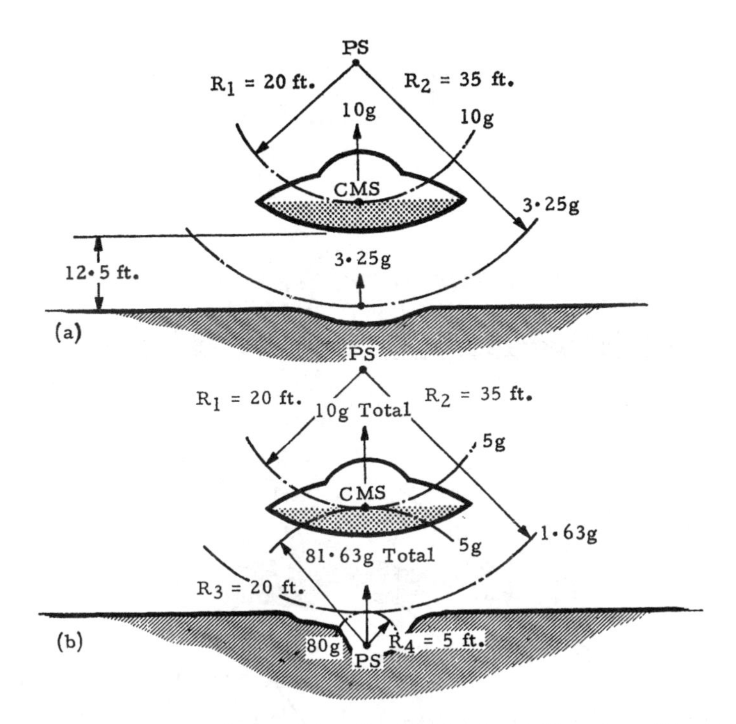

Fig 96. Ground crater effects of G field and Bi-polar field craft at lift-off from 12.5ft and 10g acting at centre of mass. Ground effect of R field craft only, at same 10g would be 160g.

provided by both fields to produce the original 10g. We shall see later that the F.L. can now be much shorter, so the ground effect will be less, but for the moment we shall assume R_1 remains at 20ft. But there is now only an acceleration of 5g supplied by the G field, therefore its contribution to the ground effect will be exactly half, viz:

$$G_{gl}=\frac{20\times20\times5}{35\times35}=1.63g.$$

If now the R field F.L. is also made 20ft, then the P.S. will be 5ft (R_1) below ground level. Therefore its vertical g contribution will be:

$$G_{gl}=\frac{20\times20\times5}{5\times5}=80g.$$

Therefore a total upward acceleration of 81.63g will be exerted on the ground, but as previously stated, with both fields' F.L. reduced, this effect would not be so pronounced; however, in either case, Fig 96(a) or (b), a crater will be formed. Fig 97(a) shows an R field vehicle borne aloft

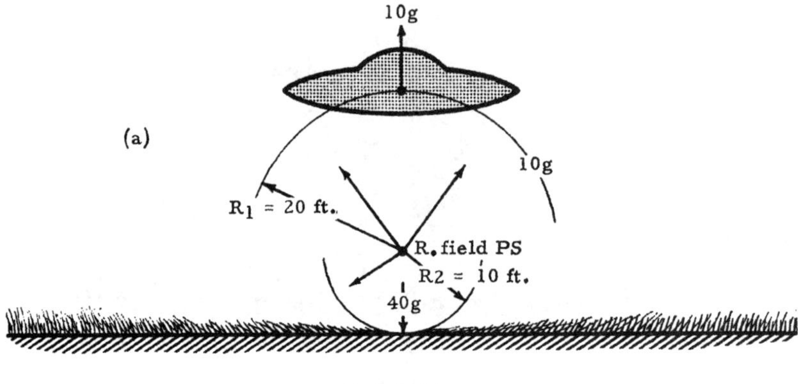

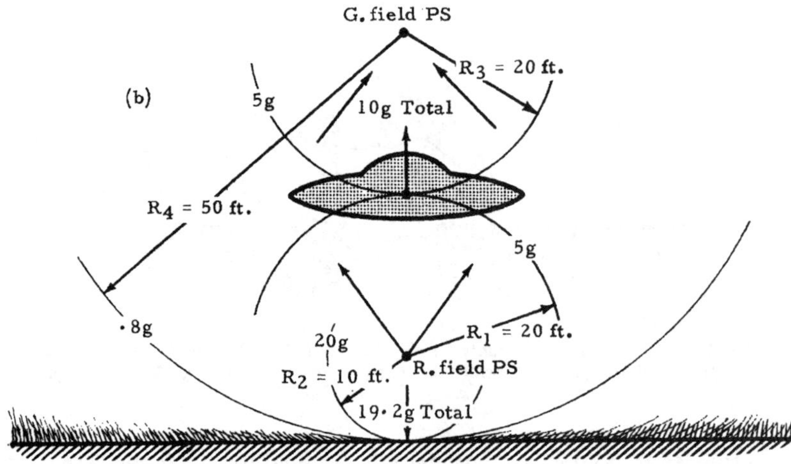

Fig 97. Ground effects of the R field and the Bi-polar field.

also at 10g, while the F.L. is still 20ft (R_1) and the distance to the ground (R_2) is 10ft, which gives the downward acceleration on the ground as being:

$$G_{gl}=\frac{20\times 20\times 10}{10\times 10}=40g.$$

By introducing a G field complement of 5g, as in Fig 97(b), and reducing the R field to an equal amount, the R field ground effect now becomes:

$$G_{gl}=\frac{20\times 20\times 5}{10\times 10}=20g.$$

But this will now be reduced a little by the upward component of the G field, whose P.S. we shall also place at 20ft (R_3), which makes R_4, the distance from the G field P.S. to the ground, 50ft. From which we get:

$$G_{gl}=\frac{20\times 20\times 5}{50\times 50}=.8g.$$

Therefore the total downward ground effect will be 19.2g, which again would be further reduced by the application of shorter focal lengths.

Bi-Field Effects on Crew and Ship

For the sake of simplicity, we made the focal length of both field components equal, but this does not necessarily infer that it has to be so, for there are conditions when *either* F.L. may be varied. Also it was pointed out that a much shorter F.L. can now be justified. This will be quite obvious when you think about it, but a word or two will not be amiss.

Fig 98 shows the interplay of forces about a random point Y in the ship. The full line represents the R field components, while the dotted line represents the G field components. It will be seen that the chain dotted resultant total lift is parallel to the axis of the ship, and the horizontal components about the point Y cancel each other out, as will be the case for any other point within the machine. By now the reader will no doubt appreciate why it has been desirable to approach our subject in two broad stages, unifying them later, rather than attempt to present the issues simultaneously.

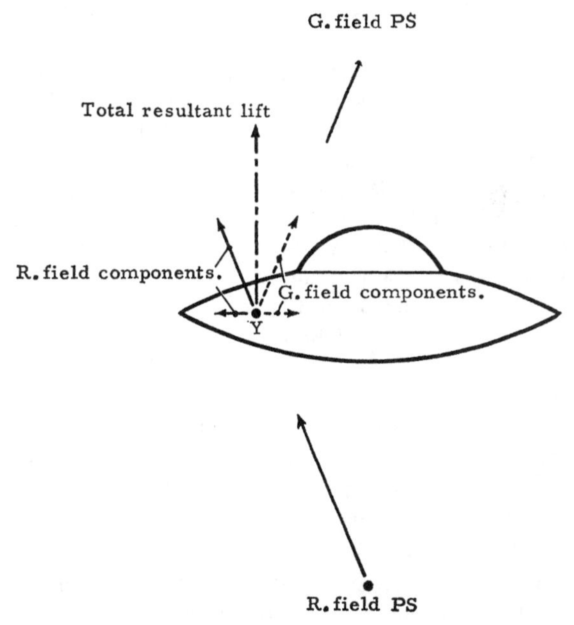

Fig 98. R field and G field components, in which the horizontal vectors cancel each other, leaving only the vertical vector operative.

From the diagram it will be clear that the focal lengths may now be kept very short indeed, thereby taking advantage of the fuller coverage and greater field area available. To do so with one-half of the field only, would subject the machine and crew to undesirable lateral stresses, either outward or inward, depending on the half of the field used. Should the focal lengths be kept equal, then it also follows that side field inducers would be unnecessary, but as already stated, there may be conditions prevailing which would require asymmetrical focal lengths, in which case there will be a small lateral force brought to bear on the crew, which will require correction, in which case side thrust field inducers would be employed. Sensing devices situated within the ship would phase these fields in or out automatically.

At this point we are in a position to consider in greater detail the G differential to which the human body may be subjected within a gravitationally propelled vehicle. You will remember that earlier, when dealing with the G field analysis, I pointed out that there would be a serious distribution of g differential over the pilot's body. At that stage we could not go into the matter further, because the problem could not be satisfactorily solved by the application of the G field type secondary field inducer only. But it can be reconciled by the introduction of the R field counterpart. Even so, we are still restricted to a somewhat basic analysis of this problem here, a more complete explanation being beyond the scope of this book.

First let us imagine a Bi-field vehicle with the field intensity set at 20g, *i.e.* 10g being provided by both parts of the field, as in Fig 99(a). The F.L. of both fields are equal and set at 10ft (R_1 and R_2), this passing through the C.M. of the machine. At this point it should be understood that although the vehicle may be subjected to 20g acting at its centre of mass, there still exists a g differential distribution on the structure. If only one part of the field is considered, then we assume the value of g at the C.M., which is approximately the average of the greater force on the craft *nearest* to the point source, and the lesser force acting on the craft *farthest* from the point source.

When both sections of the field are mutually considered, however, this differential takes on a different pattern, for the *two* extremes of the machine are subjected to a greater force than the centre of mass, Fig 99(a).

Now it follows that if the pilot is placed with his C.M. coincident with the ship's C.M., then he also is exposed to the foregoing differential. Let us consider the value of this at three main stations, *i.e.* the head, the trunk and the feet as in Fig 99(b).

First, we assume a 6ft pilot, his head and feet extending 3ft either side of the C.M. The total acceleration acting on the C.M. of the trunk we know to be 20g, no allowance being made at this stage for earth environment 1g differential.

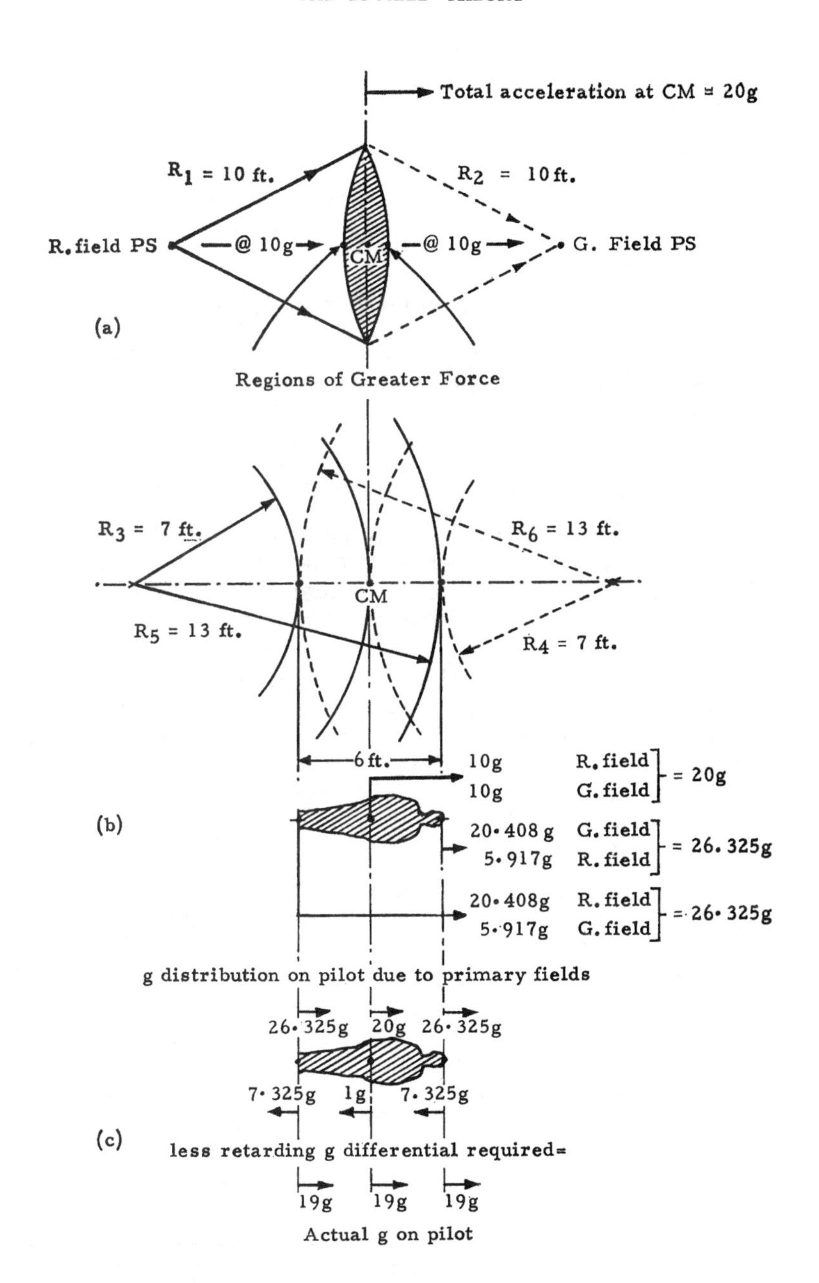

Fig 99. Required retarding acceleration distribution at three main stations on pilot to provide 1g ship-crew differential in a Bi-field craft.

Situated as they are at identical distances and closer to the two point sources, the head and feet stations will be subjected to the same accelerations, which are shown as:

$$\text{Primary G at head and feet stations} = \frac{10 \times 10 \times 10}{7 \times 7} = 20.408\text{g each.}$$

But both stations are also situated farthest from the respective point sources, providing a lesser force, but which will be additive to the general thrust on the pilot, viz.:

$$\text{Secondary G on head and feet stations} = \frac{10 \times 10 \times 10}{13 \times 13} = 5.917\text{g each.}$$

Therefore both the head and the feet will experience a total acceleration of no less than 26.325g each, and if free to move, the pilot would find himself shot towards the ceiling of the ship's compartment. As I have said before, the problem is made more complicated by the *actual* rate of acceleration of the vehicle, but the student will not be far out to assume these static values for g.

So the accelerations acting on the pilot are: head 26.325g, trunk 20g, feet 26.325g; in fact, not a very enviable situation! Assuming the original acceleration of 20g acting through the centre of mass of the machine, we must provide secondary field inducers, which will generate on the pilot a decelerating force which is so distributed through his body as to bring about a uniform 19g, or a 1g differential with the ship. The required correcting factors in this case would, of course, be: feet 7.325g, trunk 1g, head 7.325g, as in Fig 99(c). Now a secondary G field inducer placed suitably beneath the cabin floor, could produce 7.325g at the feet and only 1g at the trunk, but its effect on the head would prove negligible, leaving the pilot feeling very, very light-headed indeed, by about 7g in fact!

In a sense it may be inopportune to make fun at this particular stage, for we have now arrived at a threshold of revealing architecture. Its blending simplicity is there for all who will care to view it. The earnest traveller will notice there is nothing at all superfluous about the design. For if we require a secondary G field to produce a diminishing retarding gradient *upward* from the feet of our astronaut, then equally we might use a secondary R field above the roof to furnish the diminishing gradient downwards from his head. In other words, just as the employment of both fields are complementary to the needs of the ship, so the use of both types of secondary fields are complementary to the needs of the crew. The only difference being the polarity of the fields must be inverted, see Fig 100.

At which stage I would take time to emphasise the fact that the Bi-polar G field theory has been built up not only from technical analysis of UFO reports, but from a good deal of fundamental engineering reasoning. And I feel bound to add that it is really asking too much to suggest

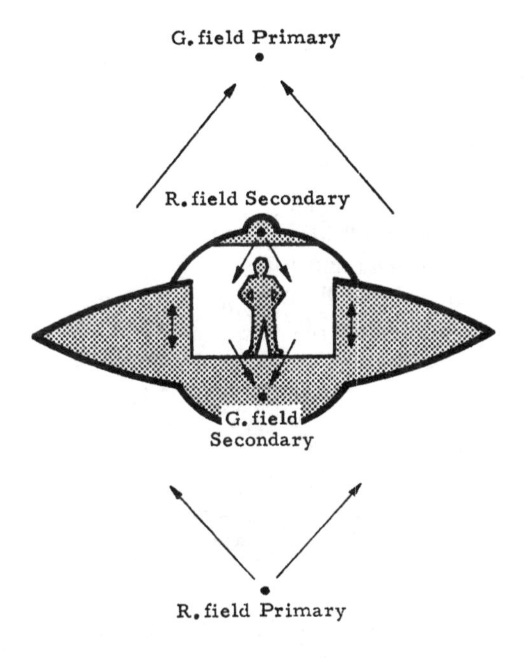

Fig 100. The employment of an inverted Bi-field to provide a 1g differential on the crew regardless of ship's rate of acceleration.

that by mere chance again, the Bi-field principle would also conveniently be the means of protecting the crew of such a vehicle from the hazards of enormous accelerating forces! Indeed, I would make so bold as to say, if the UFO sceptic having access to these pages still remains sceptical, then there can only be three reasons for such continuance. Either (a) he cannot understand the material presented, in which case he is not qualified to express an opinion; (b) he understands it, but is too proud to accept the evidence, or (c) he is too prejudiced to take the trouble to read the facts at all.

To have a theory is one thing, to see it borne out by calculated application is another. Let us now evaluate the protective Bi-field idea a little more fully.

Referring back to Fig 99(c), we saw that a retarding acceleration of 7.325g was required for the astronaut's feet and head respectively, while 1g was required for the trunk in order to provide an even 19g at these three stations. Fig 101 shows the interplay of the required forces in more detail.

First, we know the secondary G field must produce .5g in the solar plexus region, as its share of the 1g retarding acceleration required there, and furthermore it is necessary for this point source to be somewhere beneath the cabin floor, in fact calculation shows this to be just 1.08ft (R_2). Thus, by making the distance from the P.S. to the trunk 4.08ft (R_1), we get the retarding acceleration, on the feet due to the secondary G field inducer as being:

$$R_f = \frac{R_1^2 \, R_{sp}}{R_2^2} = \frac{4.08 \times 4.08 \times 0.5}{1.08 \times 1.08} = 7.15g.$$

But this same field will also produce a small acceleration at the head (R_3=7.08ft) which amounts to:

$$R_h = \frac{R_1^2 \, R_{sp}}{R_3^2} = \frac{4.08 \times 4.08 \times 0.5}{7.08 \times 7.08} = 0.167g.$$

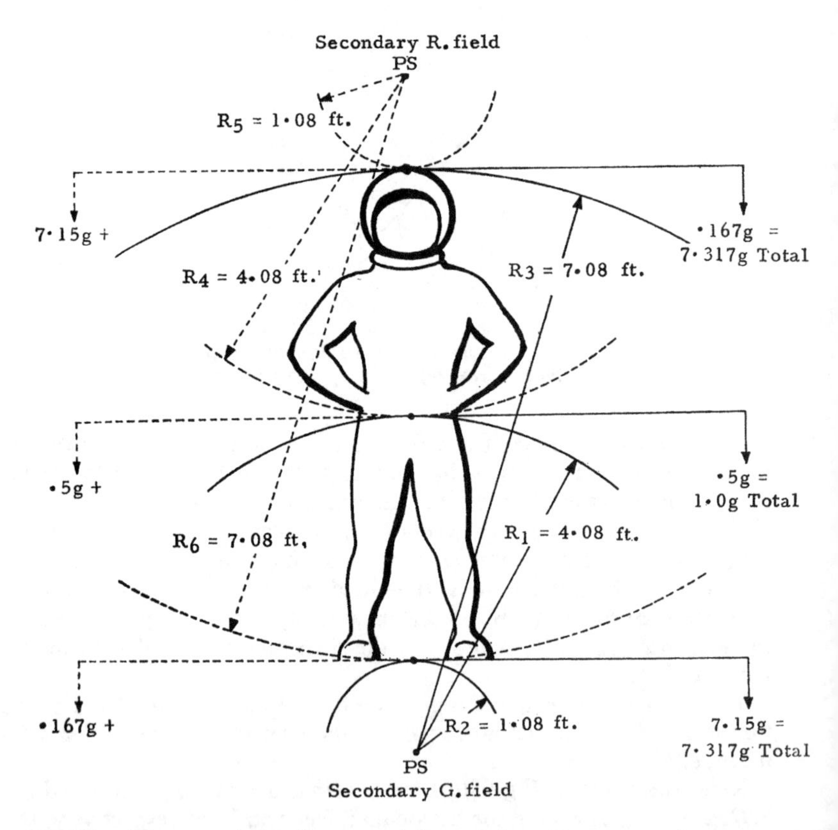

Fig 101. The approximate distances of the secondary point sources from astronaut's head and feet required to provide 1g ship-crew differential, when the primary fields' focal lengths are set at 10ft and a total acceleration of 20g is acting on the centre of mass of the ship and crew.

And as the secondary R field will also produce exactly these accelerations in reverse at the three stations, the additive effect will be: retarding acceleration at head and feet=7.15+.167=7.317g total, or very nearly the 7.325 required. While the combined effect of both fields at the solar plexus, of course, will be 1g retarding acceleration.

Now this is the state of affairs at the three main stations, *i.e.* a 1g differential produced on the pilot by the simple procedure of employing the identical ship's drive technique in reverse as secondary field inducers, the dual Bi-field.

So far so good, but further examination of the reciprocal action of both components through other stations of the pilot's body reveals complications. For instance, at the knees and chest there will be a discrepancy of some .7g and the enthusiastic reader will observe that while the outward and inwardly acting components of the fields cancel out as we saw earlier, nevertheless, the calculations have been assessed on the crew member's body standing co-axial with the ship's centre. What happens if he moves out *further* towards the cabin walls? Well, it is obvious that due to the short radius of the fields, his body will encounter different areas of varying field intensity, but other circumferential secondary fields would obviate this. In addition, the focal lengths of the primary fields may not necessarily be equal; in fact, all that we have said of both fields individually still applies. The reader will appreciate that by making the primary field lengths unequal and by adjusting the centre of mass of the cabin in relation to the centre of mass of the ship, also by adjusting the F.L. of the secondary inducers, any number of permutations of field strengths can be generated. One can imagine a beautiful ship where all this would be achieved automatically by built-in field sensors and computers, a more detailed study of which aspects alone could fill several volumes, and I feel it is fair enough to say that I have shown one more interesting piece of engineering evidence for the saucers. Of one thing, however, we can be sure, the above physical problems are soluble, while even a basic analysis such as this establishes among other things, that it is entirely logical for the cabin of such a vehicle to be circular!

We could now carry on re-examining in terms of the combined field theory, many of the conclusions arrived at elsewhere, such as effects on witnesses, UFO formation flying, aerodynamics effects, etc., but not only would this prove needlessly lengthy, but also it would deprive those who are interested enough to examine the theory for themselves, some fascinating research. One thing I would ask at this stage is that readers would take it for granted that many of the more obvious associations of the phenomenon have simply been neglected here because of lack of space. Our subject is vast and we have to leave something out. I could mention the lighting effect of the Bi-field and how angel hair might follow the field to form a ball, Fig 102, while some of this etheric 'candy floss' would

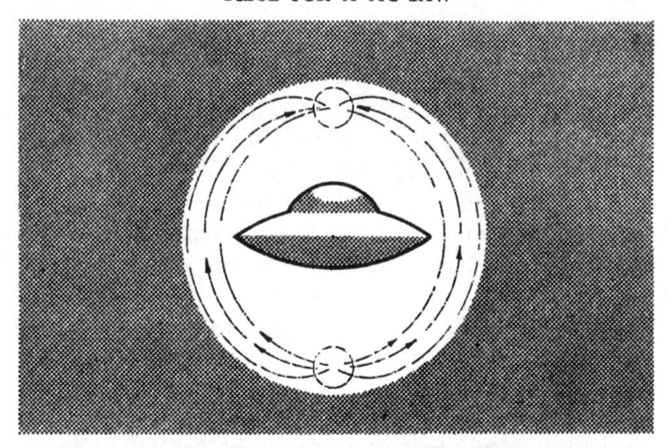

Formation of globe of 'Angel hair'.

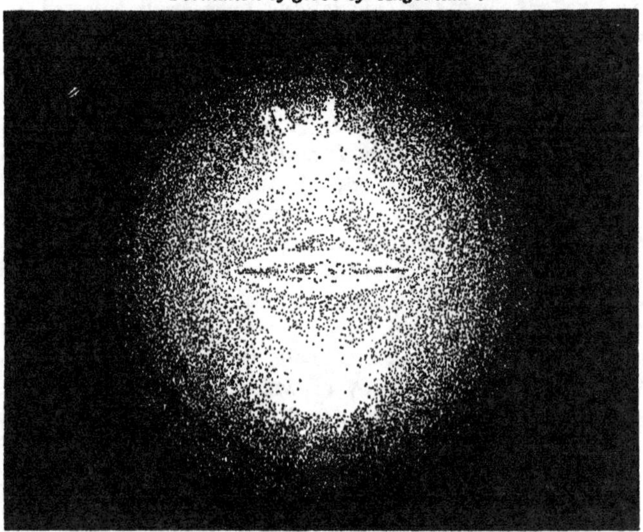

General luminescence effect of the Bi-field.
Fig 102.

be pressed upwards to hug the belly of the parent which created it, and a host of other side effects, but we must go on to one other important analysis.

Some of my readers will have been way ahead of me in these past assessments and no doubt will have anticipated my next intention. Yet others will be a trifle impatient that the following correlation should be made. But I must stress, I try very hard not to be biased, science is far more interesting left as it is. Therefore I am sincere in my next analysis. I trust the reader will be, too. I have purposely chosen the following title for the evidence which I hope may help shed some truer light on one who may be a very maligned man.

22

Vindication of a Scout Ship

SOME time ago I heard a learned gentleman proclaiming bitterly about a certain type of UFO. A more convinced and ardent researcher could be difficult indeed to find. Yet this same man, who would hotly denounce the prejudice sceptic, is himself so sadly beridden of this intellectual disease. Saucers? Of course he accepts them, only a fool could ignore the facts, but *that* certain type of saucer—well, anyone could see that it just doesn't *look* right; nobody in their right mind could accuse a technologically advanced race of producing a monstrosity like that! Pearls of 'wisdom' such as these belie the intellect; in fact not only does it herald from the rooftops its owner's very restricted technical vision, but it sadly reveals a mental block which seriously restricts him in making completely unbiased research. It is a pitfall which the author has seen too often; it is a pitfall which I have diligently tried to avoid. Therefore, I cannot stress too emphatically that I am not prejudiced about the facts I am about to set down. As an engineer I shall go on seeking the technically corroborative among UFO reports, no matter how much I might *like* this or that case. And I for one will certainly never reject a UFO sighting purely on the grounds that the thing does not look *aerodynamically* correct. If, after reading this work, any researcher still clings to the idea that saucers are fundamentally aerodynamic or electromagnetic, my advice is, take up gardening—it will save you a lot of disappointment in the long run.

On the other hand, I have often found it refreshing to talk to openly declared UFO sceptics, some of my best friends among them, on some technical assertion such as the above, and have actually had these fundamental blunders used as fuel against many UFOlogists on the grounds that 'How can you take the opinions of such unqualified people seriously?' To which I have had no satisfactory answer, except perhaps the fervent wish that technically untrained people occupying influential places in this subject should take heed not to mislead even lesser folk with often erroneous deductions.

The whole point in levelling this well-intended criticism is that such observations make it quite clear that the perpetrator is not acquainted with sufficient technical information to denounce this or that case, and he or she would be wise not to let silly statements betray their underlying prejudice. So from now on I want to carry out the rest of this analysis with the above facts firmly in our minds; remember we are travellers exploring an unknown land; if the road ahead bends a little to the left, there is no use trying to pretend it does not, simply because that particular direction doesn't appeal to us. The colour of the trees, the wild plants, the trickle of a mountain stream have revealed as it were a pattern in our minds; at the moment that pattern is in the form of the Bi-field propelled space ship; we are about to find another very valuable specimen.

I would like to begin by saying to you, let us carry on with the design of our space ship. We have attempted this here and there as far as we were able but, now we have more data, let us see how the craft is shaping up, while being mindful of the fact that we have arrived at these conclusions from consideration of fundamental laws. Also we should bear in mind that all mechanical contrivances are the result of compromise. Our task now is to find the one with the least number of disadvantages.

1. We have established that due to the employment of force fields which obey the inverse square law, the best shape for the vehicle plan form is circular.
2. The very nature of these propulsive fields suggests that at least two major parts of the vehicle are concave in section. The alternative arrangements of layout are shown in Fig 103(a), (b), (c), (d), (e), (f), the black squares representing the primary field inducers.
3. Due also to the nature of the force fields, the crew's quarters are best suited at the centre of the machine, and preferably above the centre of mass, this produces Fig 103(h) as being the best compromise, with Fig 103(g) following as second by providing the annular type field inducers.
4. Provision must be made for the cabin to raise or lower in relation to the centre of mass of the rest of the machine to help furnish the crew's g differential.
5. It must be possible for parts of the craft to rotate as a stabiliser cum generator apparatus.
6. A landing gear must be provided to keep the R field point source clear of the ground; this may be either a central pedestal type which would form part of the cabin and retract or extend with it, as in Fig 103(i) and (j), or alternatively it could be a conventional leg-type undercarriage.
7. The craft may hover a few feet above ground level with restricted ground effects, particularly if the G field component is being chiefly used.
8. There are several methods of flight attitude correction, two of which we have already considered. These were the offset focal point, and the utilisation of the couples available in the rotating masses.

PLANFORM

POSSIBLE ARRANGEMENTS OF FIELD INDUCERS AND DISCS

(a) (b) (c)

(d) (e) (f)

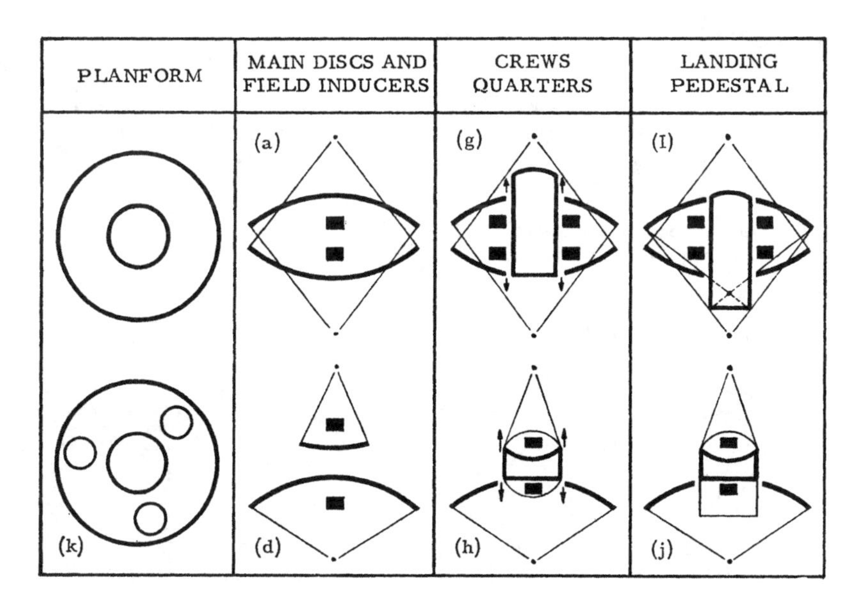

Fig 103. *Logical development of (a) and (d).*

Thus far we have reached and we can now take several steps further.

9. If we can use small field inducers to protect the crew, then we can also use them to stabilise the machine in exactly the same manner as our present-day VTOL aircraft employ their main jets to support the aircraft and secondary wing tip, nose and tail jets for orientating it in all planes. Due to their configuration, however, these aircraft require four such compensating jets, but on a circular-shaped vehicle which centre of mass coincides with the geometric centre, the minimum number of equally spaced stability field inducers required are three, Fig 103(k).
10. Again we can take the cue from conventional VTOL aircraft practice and swivel the stability field inducers so that they can be usefully employed in other flight conditions; moreover, by placing them in gimbals, so that they can be moved in any plane, they can be used to steer the craft as well, Fig 104(a) and (b), in addition

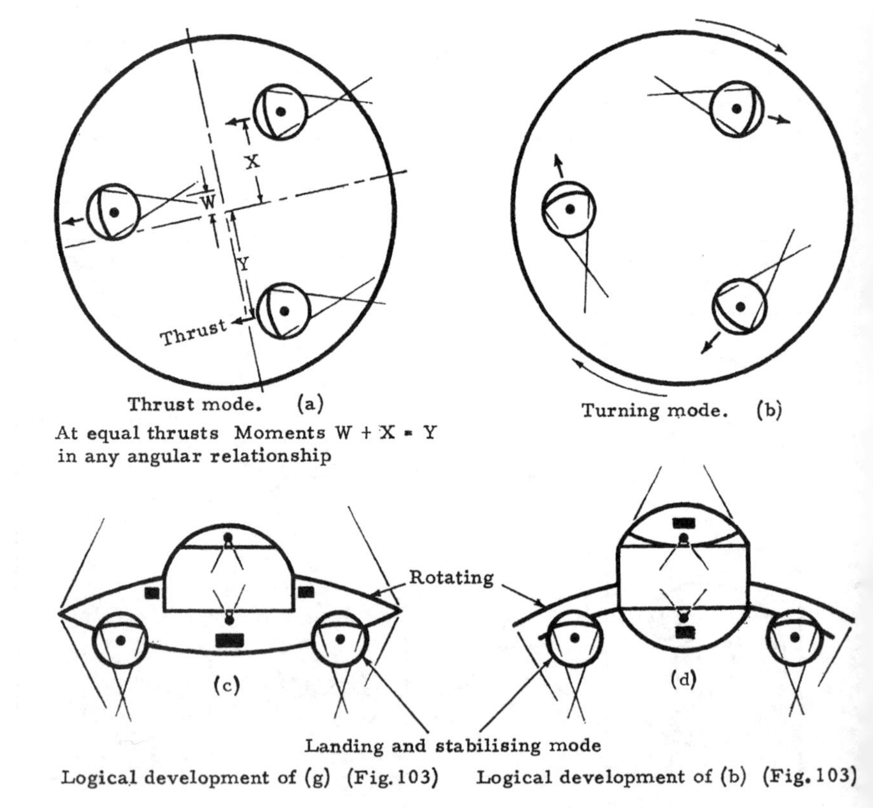

Fig 104. Logical arrangement of the stabiliser field inducers.

to augmenting lift for the machine during the deceleration turn-about stage in Fig 93(b).

11. As the gimbal is spherical in nature, we might as well give it a spherical protective shield, which if suitably suspended could function as a spherical wheel cum float type landing gear. Therefore, the logical position for the stabilising inducers is the lower half of the machine, and due to this the R field type inducer is automatically suggested as in Fig 104(c) and (d). It is interesting to note that in October 1932 the designer, B. von Loutzkoy, took out patents in Germany for a spherical wheel-float system for aircraft.
12. In order for these units to work at their maximum effectiveness they should be given as large a moment about the centre as possible, and preferably attached to a non-rotating member. This also automatically suggests either arrangement Fig 104(c) or (d). In operation we would assume the machine to rise on its main field while stabilised by the rotating masses until the stabilising units were sufficiently clear of the ground, where they could then be brought into action. On the other hand, the combined effort of all three units could support the craft while low-level hovering with a subdued main field. Should the field intensity suddenly be increased, then several small craters might be formed, together with local induction heating. The researcher will know that there are UFO cases on record bearing evidence of this.
13. We have seen how the 'exhaust' of a mass/energy conversion process would be in the form of radiation, some of it in the light and heat range. Therefore, the craft would be subjected to heat waves associated with the fields, the intensity varying as the efficiency of the system varied. The crew's quarters might be shielded from this by the secondary field system or other means, but an additional practical way to keep heating down on the rest of the machine might be to make the surface highly reflective, or even *translucent*.

Although some of the other configurations have been reported on numerous occasions, the two most common are Fig 104(c) and (d), the latter of course strongly resembling the late George Adamski's scout ship. Now the reader must understand I have not deliberately set myself the task to vindicate this man's claims, but I would be guilty of the same bias I have levelled at others and would not be true to the subject if I did not reveal these facts; it will be for the reader to explain them, either as purely circumstantial or supporting Adamski's claims, as the case may be.

14. In the last section it was mathematically shown that the crew could be protected from extremes of acceleration by the use of a secondary inverted Bi-field system installed within the ship, here indicated diagrammatically in Fig 104(c) and (d). Of course the strengths and

focal lengths of these fields must be variable, one set of conditions only being considered. When discussing the undercarriage in chapter 17 we saw that it might be a good idea to shroud the point source in an evacuated tube for at least part of the lift off phase. Now if in order to adjust centre of mass relationships the cabin of the ship must move up or down relative to the main disc mechanism, then there would be times when the secondary focal points come dangerously near to or even through the ceiling and floor structure. Therefore you, the designer, would immediately suggest placing them in an evacuated tube running through the centre of the ship, as we

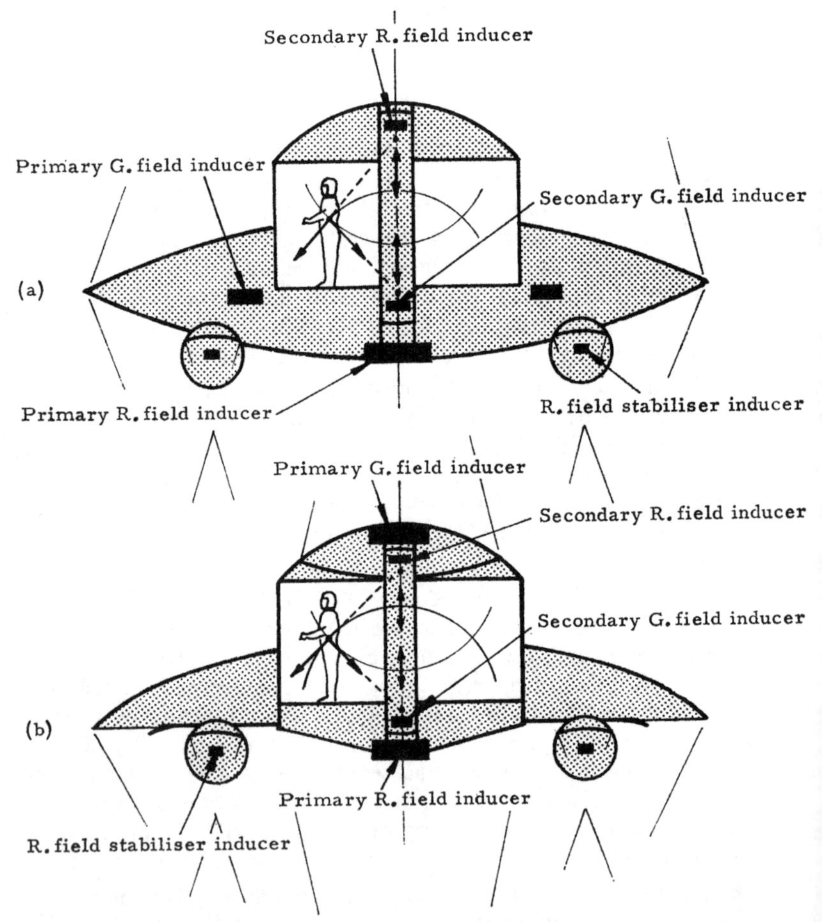

Fig 105. *Whether George Adamski sceptics like it or not the logical development of a gravitationally powered space craft, purely from engineering considerations, would almost certainly take the form of the well known 'scout ship'.*

did the primary focal point in the R field in chapter 17, and we arrive at Fig 105(a) and (b).

Now readers will remember G. Adamski described the same central column in *Inside the Space Ships* and, lest we forget, may I remind you that this is the design we have arrived at from engineering principles. You would have done so whether you had heard of George Adamski or not, whether you had seen his photographs or not, you and I have reached the end of our journey, we have designed a machine which looks remarkably like the well known 'scout ship'.

I would most earnestly stress that I have not deliberately set myself the task of supporting Adamski out of personal inclination, rather have I set myself the task of discovering the truth as an engineer. At this juncture I would like to take the opportunity to place on record an interesting fact about the central inducer shield or column, which some readers may remember I showed in the cut-away frontispiece of *Space, Gravity and the Flying Saucer* in 1954, which otherwise would not be known publicly.

The late Waveney Girvan introduced me to Desmond Leslie, author of *Flying Saucers Have Landed*, and we discussed some of George Adamski's photographs. I mentioned the likelihood of the central column to Mr Leslie, whereupon he gave me a sketch made by Adamski showing the same thing clearly. This sketch was also later printed in *Inside the Space Ships*. At that time it would have been natural, therefore, if readers of my book, in which I portrayed the central column, had harboured the thought that either Adamski or I had directly or indirectly influenced one another, but this was not so; I considered the idea independently, but Adamski had already placed it on record in the sketch. It is just one more little technical point in George Adamski's favour, unknown to the public, which I have to mention, for, together with the other data I shall offer, it looks as though many of us may have a lot of reproachful thinking to do.

Some time after the publication of *Space, Gravity and the Flying Saucer* I received a letter from Mr Adamski, in which he very warmly commended my sketch of the interior of a 'scout ship'. He asked me frankly if I had ever been inside one, and if not, he went on to say, 'it was the nearest representation an earthman could get by mere imagination alone.' I offer this with no sense of self aggrandizement, rather as a measure of his sincerity, after all, I never wrote to Adamski, therefore he was not obliged to write to me.

It would be natural if the reader suspected that some of my conclusions had been influenced by reading *Inside the Space Ships*, but this is not so, for strange as it may seem, only very recently have I read this completely, and have been rather astonished to recognise some of these facts myself. Indeed at this very moment I sit penning these words with two documents before me. One is the galley proofs of this present work just received from the printers for checking. The other is a copy of *Inside the Space Ships*. The page is open at forty-two. If I *had* been influenced by this narration, I would not now be adding to an already long overdue galley proof. But before I quote the following I would remind you that we arrived at the need for the convertible Bi-field and the optical effects of these fields in general, from purely engineering considerations. Now let us look at page forty-two of *Inside the Space Ships*, the third, fourth and fifth paragraphs read: 'It was all so exciting that I was obliged to take myself firmly in hand in order to concentrate on any one thing. I wanted to leave this ship with a clear picture of everything in order to give a lucid account of what I was seeing.

'I estimated the inside diameter of the cabin to be approximately eighteen feet. A pillar about two-feet thick extended downward from the very top of the dome to the centre of the floor. Later, I was told that this was the *magnetic pole of the ship*, by means of which they drew on Nature's forces *for propulsion purposes*, but they did not explain how this was done.

'The top of the pole,' Firkon pointed out, 'is normally *positive*, while the bottom, which you will notice goes down through the floor, is *negative*. But, *when necessary*, these poles *can be reversed* merely by pushing a button.' The italics are mine (author).

If witnesses might observe optical effects from immensely powerful fields of saucers, then logically we might expect such effects to be apparent to the crew of such craft also. And although we might be at a loss to see how such phenomena might be utilised, it is interesting to note that on page forty-six of the same book, Adamski says:—

'My attention was now called to the big lens at my feet. An amazing sight met my eyes! We appeared to be skimming the rooftops of a small town; I could identify objects as though we were no more than a hundred feet above the ground. It was explained to me that actually we were a good two miles up and still rising, but this optical device had such magnifying power that single persons could be picked out and studied, if so desired, even when the craft was many miles high and out of sight.

'The central pillar of magnetic pole serves a double purpose,' explained my bench companion. 'Besides providing most of the power for flight, it also serves as a powerful telescope with one end pointing up through the dome to view the sky, and the other pointing down through the floor to inspect the land below. Images are projected through it into the two big lenses in the floor and ceiling, as you can see.'

'He did not explain whether this was done electronically or by some other means. Its magnifications could be varied at will, and I suspect that there was more to it than a simple optical system such as we know on earth.'

At this juncture I can almost hear many an Adamski sceptic breathing a sigh of incredulity, hardly able to believe that I should dare to include such material in a semi-technical work of this nature. In which case I must remind you, that throughout this book it has been my policy to deliberately include *all* types of technically qualifying cases wherever I can, without prejudice or preference. Moreover, there are very few cases on record where a witness has claimed to have boarded a saucer. Therefore if George Adamski's claim to have done so is the only one available, then we are bound to include it in this analysis. It so happens, it appears not to be inconsistent with the rest of our findings.

One would like to think that if a contactee was to be given a ride in a saucer, then secondary field inducers would be in action to ensure he suffered no discomfort. You will remember earlier when discussing the R field we showed mathematically how at *take off* and *landing* there might be occasions when the 1g ship-crew differential could not be maintained, that is unless a secondary field were in use. There would be all manner of computations and different reasons why the machine's Bi-field might have to be modified. A more obvious instance being when the ship might be hovering at ground level and near to trees or other obstacles where it would be desirable to reduce or cancel completely the G field component of lift to avoid the ingestion of any foreign matter into the intense focal point. Now although I cannot understand why, for there were no trees or the like near the 'scout ship', it is none the less interesting to note that the resulting temporary field imbalance is exactly what George Adamski claimed, for he said: 'I was quite unaware we had taken off, although I did suddenly register *a slight feeling of movement*!'

The reader will appreciate that a loss of weight or an increase of it, would be registered by Adamski as *movement*, for as we have seen relative acceleration or deceleration is taking place.

He also claimed to have stood beside a hovering saucer; our calculations have shown that indeed he could. But by the same token, neither would the field be strong enough to snatch up his arm as he said it did, at best his arm would become weightless, and in any case there would not be an R field or G field component which would thrust his arm down again, as he claimed. But before I attempt an explanation it is interesting to consider the following:

A Stockton, Kansas, farmer said that just before an aerial explosion he saw a strange-looking thing which he described as a 'funnel-shaped saucer' wobbling over his house.

'It came within six feet of me,' he said, 'then it stopped in the air, level with my face, and again wobbled around for an instant, with fire *belching out* and then being *sucked back in*.' Sparks had suddenly showered from the thing, 'as if a fuse had been lit'. Then it took off in a north-west direction, very fast, gaining altitude as it went. 'My wife came out and watched it fly off, leaving a trail of smoke. On a sudden, a great cloud of smoke appeared in the sky and in a few seconds we heard a terrible explosion. I could still feel the heat from where the object came near the ground.'

There are other cases on record where witnesses have claimed to have observed this reciprocating characteristic of saucers, and you might be inclined to think of course that any similarity to George Adamski's claims are no more than circumstantial. But one thing we can be sure of is, that while a witness might stand near to a hovering saucer, he would be very foolish to deliberately touch it, unless he wore heavily insulated shoes, for what little our theoretical sifting has revealed tells us beyond much doubt the surface of such a machine would be very highly electro-statically charged. To see this effect in action does not require very much in the way of laboratory equipment. A sheet of perspex or celluloid and a few scraps of wood and graphite off your pencil will do. Preferably in a warm, dry atmosphere, the celluloid should be statically charged by rubbing, then held low over the particles. Instantly these will be seen to snatch up to the sheet and just as quickly shoot down again, so that a regular little 'snow storm' ensues. The effect, which is well known to most schoolboys, will go on until the celluloid gives up its charge, which it does by attracting the neutral particles and saturates them with a like charge, whereupon they are repelled to give up their charge to the earth, and the cycle is repeated. In any event, true or false, the statement made by Adamski in this respect figures absolutely technically corroborative.

There is another interesting point I take opportunity to record and it concerns the equally disputed Coniston photograph taken by Stephen Darbishire. Among the UFOlogist Adamski sceptics are those who equally deny the authenticity of the Coniston photographs, on the grounds that the Adamski photographs are fakes and the Coniston photos poor copies of them. Therefore, the orthographic test drawing made by me and later shown in *Space, Gravity and the Flying Saucer* proved no more than that, that one was a copy of the other, Plate 34.

Concerning this, the other fact I want to put on record is another behind-the-scenes affair which the majority of people are unaware of, but this small snippet of information, together with the others, may help to shed light on a very enigmatic story.

All the Adamski saucer pictures which appeared in magazines and national dailies were copies from those used in *Flying Saucers Have Landed* and, like everyone else at that time, Stephen Darbishire had access to them, which the boy never denied. None of these shots showed groups of more than *three* portholes, but Stephen Darbishire made sketches of the thing he had photographed, in which he showed a group of *four* portholes. Later, subsequent questioning did not deter the boy from this fact; he was adamant, there had been four of them, he was certain.

Now bearing in mind the exact duplication of all the dimensions in both Adamski's and Stephen Darbishire's photographs, as shown by the author's test drawing, is it not rather strange that they should differ in this one respect, that is, from the sceptic's point of view? But what is not known to the general public is something the late Waveney Girvan told me, and that is, George Adamski's photograph *did* show a group of *four* portholes, but in order to suit the format of the book, Waveney had trimmed the photograph of the scout ship to size, leaving only *three* portholes visible! No-one else knew about this, certainly not young Stephen, yet the boy could not be shaken, he *had* seen *four* portholes, in a row, close together.

As our space between these covers runs out, I cannot help thinking of the many other clues I would like to have added in this last section, and in association with G. Adamski I am reminded once more of one of the most sincere and level-headed flying saucer researchers the subject has ever known. Indeed, one of the earliest contributors the subject has ever known. I refer of course to the late Waveney Girvan, founder and editor of the top international UFO periodical *Flying Saucer Review*, publisher of *Flying Saucers Have Landed* and author of *Flying Saucers and Common Sense.* As I sit penning these words I cannot help thinking of the occasion when visiting Waveney at his office one day he solemnly handed me a manuscript and asked if I could go into the adjoining office and scan through the work for an hour, then to tell him what I honestly thought of the contents.

Alone in that office I did look through those papers and after half-an-hour I returned to Waveney. He looked straight at me and said, 'Well, what do you think?' I handed him back the manuscript and said, 'You and I might find the substance of the claim difficult to accept and concerning this I can only say I don't know, but technically it's logical enough.'

That paper bundle was the manuscript for George Adamski's *Inside the Space Ships.* What I told Waveney that morning I can only repeat to you now, I do not know the truth about that strange affair. I only know that Waveney Girvan chose not to publish it. I offer this small token as a measure of his sincerity of purpose. Even so, Waveney never hesitated to support Adamski whenever he felt justified, despite the growing differences of opinion among his readers.

In this respect I feel this book has been influenced indirectly by the common-sense outlook of Waveney and I like to feel the title was influenced directly by him. I also feel it appropriate to quote one of the first corroborated sightings which Waveney asked me to include in *Space,*

Gravity and the Flying Saucer—a title, incidentally, which he chose, but I never quite took to.

At the time this particular sighting puzzled Waveney, as it also puzzled me, and I have included it here so as to emphasise my own belief that we must continue to regard this subject from a completely unbiased point of view. Tying up facts where they tie up and not indulging in either wishful or pessimistic thinking. This was the way I felt when in 1954 I coined the term 'G field' in relationship to flying saucers. It is still the way I feel now.

The Observer, Sunday, 11 October, 1953

Seen over Norwich

Sir,—While observing the sky over Norwich at 7.15 p.m. on Tuesday last, 6 October, I noticed a bright and very large object appear from the south-west. It appeared to be a very large yellow star. I then noticed it was travelling on a level plane, and with the naked eye it now appeared oval-shaped, like a cluster of tiny stars. I waited for a favourable opportunity and focused the object in my 3½in reflecting telescope.

On bringing the object into focus, the apparent cluster of stars took on the appearance of a dome on a large flat disc. The dome had apertures placed at intervals around it, four of which were in my field of view. Light from these apertures made the disc visible. The top dome did not rotate. There was no noise to be heard from where I was observing, and the object kept a constant altitude. Under the disc a cavity could be seen, and this glowed a dull red colour. I saw no traces of gas or flame.

The object travelled south-west to north-east and remained in my view for three-and-a-half minutes. The sky was perfect with no cloud. All the constellations were visible, and this object was seen independently by at least seven other Norwich people—members (like myself) of the Norwich Astronomical Society and the British Astronomical Association.

Norwich. F. W. Potter.

In a later article which appeared in the *Daily Mail*, 11 February, 1954 written by J. Stubbs Walker, and headed 'Was it a 'Saucer' they saw over Norwich?' Mr Walker tells us:

'Now, here is a strange thing about what the Potters saw and what Mr Potter drew. The whole of his description is very much like the much-questioned photographs of a flying saucer supposedly taken at short range by Mr George Adamski and published in the book *Flying Saucers Have Landed*, except for the vital fact that Mr Potter drew what he saw in his *reflecting* telescope, *which reverses the image.*

'His flying saucer was not flying the same way up as those of Mr Adamski, and no amount of arguing will make him change his mind. Mr Potter had previously seen a representation of the Adamski saucer and was consequently aware that what he saw might be expected to be flying the other way up.'

Now we have already seen that one way we might consider to decelerate a fast-moving Bi-field craft would be to rotate it *into* the direction of flight as the field is simultaneously converted, *i.e.* R field component into G field and vice versa, at the end of which manoeuvre the machine would be *inverted.* This would be of little consequence to the crew, for the only real 'up' or 'down' to them will be that relative to the ship.

But there could be other reasons for this behaviour; here is one. From what has been said in the foregoing pages, it will be apparent that should the machine be employing G field lift in the atmosphere and particularly when it was raining, then water droplets will tend to collect about the point source and a little thought will show that it will form a pear-shaped mass due to the combined gravitational effects. Now there would be several ways of dispersing this mass, one of which is shown in Fig 106(a). In this we assume the craft to be hovering with the G field component closed down, while the R field employing a short focal length beam contributes the total lift for the machine.

The water, almost unsupported in that region by the diminished value of the R field, will now fall, and in falling will be thrust outwards over the disc, finally tumbling to earth at some distance from it.

The second method would be to invert the machine while simultaneously converting the fields, so that the G field, which a moment before hugged the watery mass to itself above the craft, now becomes a dissipating field beneath it, and here also the fluid would fall to earth as rain, Fig 106(b). Although not common, researchers will know that there have been falls of rain locally—too locally to a hovering saucer.

I would now like to place side by side with these, to me, priceless titbits the case of Mrs Smith and her daughter, Clare Taylor, of Ryde, Isle of Wight, whom I had the pleasure of interviewing and getting to know well enough to ascertain beyond doubt the validity of their claim to have seen a disc craft hovering low over trees one night in July 1961.

I took the precaution of taking along with me an independent witness, an impartial friend, Alan, and my collection of UFO photographs, together with tape recorder, etc. No-one listening to the subsequent recorded interview could doubt the sincerity in the voices of these two women, whom I must point out were very reluctant to attract attention over the experience.

Mrs Smith had told her husband of the sighting, but he and the painters decorating the house had laughed at her, and she resolved to keep quiet about the affair. While viewing the TV that evening she was surprised to see a Commander Mole interviewed, who together with his

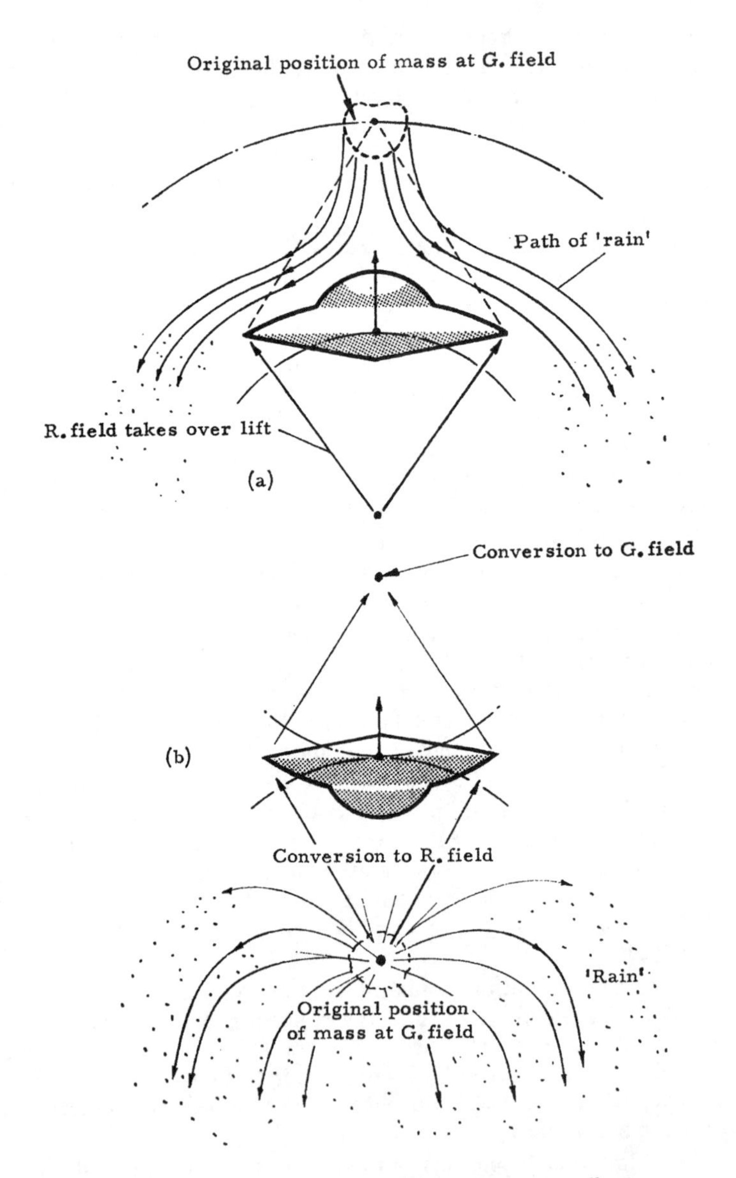

Fig 106. Two possible methods of dissipating water collection.

wife had viewed a red glowing disc through field glasses in the early hours. Puzzled, he had alerted the police. The following day the story got out and he was whisked off to the television studios at Southampton, where Mrs Smith saw him being interviewed concerning the affair.

Thinking the Commander had seen the same object, she had ventured to telephone him to see if she could corroborate his story. Despite the good lady's protests, the Commander took it upon himself to tell the police, thus the story of her sighting came to light.

Mrs Smith was staying overnight at her daughter's flat on the top floor of a large house which stands overlooking the whole of the seaside resort of Ryde, with splendid views over the Solent and surrounding countryside.

At about fifteen minutes after midnight, Clare decided to collect baby linen from the little roof garden above the flat, saying to her mother that they might see the lights of her husband's ship which she was expecting in from a trip abroad. Visibility was very good, they could see the mainland lights quite distinctly, yet there was cloud cover for none of the stars were visible.

Suddenly Mrs Smith pointed out to sea where there was a row of lights, apparently from a distant ship on the horizon, but this was no ship for, as Mrs Taylor pointed out, they were *above* the horizon and rapidly getting near. At first, thinking it might be an aircraft, the women were not unduly concerned and, even when they realised it was not, felt quite certain it was some kind of hovercraft, for at that time the Saunders Roe N1 was having its trials. Anyone having seen the N1 could easily understand how it might be mistaken for a saucer.

But this thing was uncommonly silent and less than a quarter of a mile away, where it hung, almost at eye level, over some trees. Mrs Smith said they could see the shape of the UFO distinctly, it being lit up by a row of *five* porthole lights. She remembers this, for when thinking it might be a ship she had counted them.

The rounded dome-like structure at the top was reflecting a brilliant light and the bowl-shaped lower body was surrounded by a soft glow. As the object tilted now and then, they could see that it was circular, the trees below being lit up by the furnace-like glow emanating from the underside.

By then the two witnesses were pretty scared and when the UFO started to come nearer, 'we backed down the stairs, thinking it was coming straight at us'. But then the UFO had *tilted*, showing its glowing red base for an instant, and was gone at vertiginous speed. Said Mrs Smith, 'One moment it was there and the next instant it was going away in the direction from which it had come, like a shooting star; it went so fast it made me feel dizzy at the speed of it, and in a split second it looked just like a bright star in the sky.'

Breathless with amazement they had stood there, open mouthed for an instant, before they realised something was happening at the spot the UFO had so dramatically vacated. There, slowly rising, was 'an expanding ring of smoke, *or leaves and debris;* it was luminous and hung there for several minutes after the object had gone, before it finally disappeared.' (Author's italics.)

I then asked the witnesses if they could identify the craft, were they to see it again, and then produced my photographs. They assured me with a laugh that they would have no difficulty doing this, for they would never forget the experience as long as they lived. Looking through the prints they said 'perhaps' yes to this and that, but when I produced George Adamski's scout ship, their reaction was spontaneous; there could be no doubt, they said, they were certain this was it. It so happened that young Mrs Taylor was a fine artist, and she now, half apologetically, produced a coloured sketch she had made of the UFO. There could be no mistake about the likeness; any UFO student would identify it immediately.

Both women were anxious not to attract attention to themselves; they were sincere and cultured people whose word I would unhesitatingly accept.

Now I went up on the roof and photographed the distant trees over which the witnesses claimed the object had hovered. These were easily located on an ordnance map, and accompanied by the same impartial friend, I went out to find them. During the interview with Mrs Smith and her daughter, Alan had said that he knew the area over which the disc hovered, as he had a friend living there, and now, sure enough, we found that these same trees were adjacent to this man's house.

Alan suggested we call there and ask if anything unusual had occurred on the night in question, which we did. After cordially inviting us in, Alan's friend listened patiently to our questioning. 'No,' he said, 'nothing unusual had happened, but I was disturbed in the night by a cat knocking the lid off the dustbin, which made a hell of a clatter.' The time? 'Oh, about a quarter past twelve; I noticed the time as I got back to bed. The strange thing though was, the following morning I found the dustbin lid about 15ft from the dustbin; odd that; it could have rolled, I suppose.' I did not bother to offer any other explanation. Several days later it was noticed that the leaves of the trees over which the UFO had hung, seemed to be turning brown.

This story is, of course, not unique. Adamski-type saucers have been seen and photographed. Indeed, many Adamski sceptics are prepared to accept the genuineness of his pictures, it being his subsequent claims to have spoken to the saucer's occupant which appears unacceptable. Did George Adamski have an experience like this? Did something happen which proved too much *for him,* or was he simply telling the truth? The truth which is too much *for us*!

Before we hasten to decide, I feel the following point warrants some attention. Throughout the whole of this work I have tried to show the evidence for the gravitational nature of saucers. We have briefly reviewed the work of one brilliant German scientist who is devoting his life to a theory which could make gravitational space craft a practical possibility. Now I ask you to consider very carefully the following: George Adamski claimed to have met a man from another planet and one of the questions he asked him was: 'How does your ship operate?' Adamski said he finally got the question over and 'He made me understand that it was being operated by the law of attraction and repulsion, by picking up a little pebble and dropping it, then picking it up again and showing motion.'

Like so many other saucer researchers, Adamski immediately interpreted this as meaning *magnetic.* But *because* he interpreted it as meaning magnetic, to me is a point in his favour. As I have said throughout this work, and I hope not over tiringly, I try not to be biased, but I could think of no better way of demonstrating a machine being thrust along by the force of a gravitational field than by 'picking up a little pebble and dropping it, then picking it up again and showing motion'. I should add that when I had the good fortune to meet Mr Adamski at Bournemouth in April 1959 he was still inclined to a magnetic interpretation.

Out of the confusion and the shadow surrounding this most enigmatic of flying saucer contact stories, emerges four very important facts.

(a) Adamski-type saucers are not uncommon among UFO reports.

(b) On purely engineering grounds the configuration of the 'scout ship' offers the best all round compromise which could embody the functions required by the gravitational propulsion theory.

(c) Adamski's description of inside a saucer is not inconsistent with the theory, on the contrary, theory seems to support *it.*

(d) Thus on these grounds alone I for one feel inclined to accept the first part of his experience as fact, and in my opinion any inconsistencies which *appear* in his subsequent claims are insufficient to invalidate the previous conclusions.

Summary

Our journey on the technological pathways of development has taken us through the domain of the aeroplane and that of the rocket. We have seen that such development is likely to take us into the realms of gravitational research and finally to the gravitationally propelled space ship which such research suggests. Indeed it would appear there can be little doubt that the earth is being visited by intelligences who may be employing similar means of transport themselves.

In general the conclusions may be summarised as follows:—

1. The technically corroborative evidence for the existence of flying saucers brought to light within these pages, supports the existence of humanoid saucer crews. Which cannot be denied simply because our science fails to explain the existence of such peoples.
2. We have seen that from present day technological considerations, one way of achieving gravitational fields would seem to reside in the domain of direct conversion of mass into energy.
3. In so far as gravitational waves are so relatively weak compared with electro-magnetic ones, we must expect that any conversion process involving the generation of gravitational waves would in turn liberate large quantities of accompanying radiation, which would represent an enormous loss. Therefore we could expect any really efficient technique to allow for this excess energy to be reconverted into mass, where it would be exchanged once more within the system, that is in accordance with Burkhard Heim's theory.
   Thus in a word at the press of a button, mass would be converted into a colossal quantity of radiant energy, among which would be found a correspondingly small amount of gravitational waves. But, although small by comparison, even this would represent a factor which if concentrated, would be capable of producing in its vicinity a herculean motivating field on any material device.
   A more down to earth comparison would be exemplified by considering a fuel burning engine which turns an electric generator, which produces a *magnetic field,* which in turn attracts and repels the *armature* in an electric motor, thereby turning wheels and producing motion on the carrying vehicle. Whereas in the case of direct conversion of a microscopic amount of the same fuel, a propulsive field is generated in the vicinity which acts directly on the *whole* mass of the machine, Fig 107(a).
4. Some of the radiations would be lost, depending on the efficiency of the system, thus in the vicinity of such a device we would expect a whole range of electro-magnetic effects to occur, which, although exceedingly powerful, would still be a microscopic representation of the whole radiant energy.
5. The possibilities beyond this stage are varied and many. For instance some might suggest that if according to Heim this excess energy might be reconverted into mass, then something like the mass-energy conversion regeneration cycle in Fig 107(b) might take place. In which an original mass at A would be converted to energy, represented by 1. Of course no attempt could be made to show this to scale. Accompanying this enormous amount of radiation, we would find a correspondingly small amount of gravitational waves, which would still be extremely powerful nevertheless. These would be

allowed to leave the system to produce the motivating field, while the remaining radiation would be reconverted to mass at B, though now minus the immeasurably small portion of the mass represented by the centre dot.

The process would now be repeated through condition 2 and on to C, then 3 and on to D, each time the original mass losing a little of its substance to create the gravitational waves which are inherent in all formations of matter. In this way the cycle repeats itself continually, until the original substance is expended, completely converted not into *all* kinds of radiations, but back into gravitational waves or the C rays of space, from which it was originally formed. Others might feel that there can be no loss-free system and that gravitational waves must be generated by nuclear means. And again, others might cling to the idea that saucers are inherently magnetic. Herein I have offered *my* conclusions, and as I have already said, they may not be the correct ones, but of one thing we need be in little doubt, strange space craft *are* landing on the planet earth.

In setting these facts down, I am reminded once more of my good friend's remark when he said 'You might be in the position of a man examining the steam from a kettle without having the slightest idea of the latent heat which produces it.' And again later, when reviewing this text another friend said 'You seem to have covered most everything, except how to push the button!' No, as I said in the foreword of this book, such know-how is very much beyond my ken, but I think I have kept my promise in so far that you the reader, have been shown many, many clues. Would you *really* care to have the responsibility of finding that button? I wonder. For my part I am content to have offered this little piece for an exciting jig-saw.

Now as we go to press it looks as if more parts of the jig-saw will be offered, for at long last there are signs that officialdom's policy of silence concerning flying saucers is changing. For the American Air Force has just announced its decision to sponsor the scientific investigation of UFO reports at University level with £100,000 in the kitty. While more and more eminent personalities are voicing the acceptance of extra-terrestrial visitors publicly. From the 9 October, 1966 issue of the *Sunday Mirror* we find the latest two.

'I feel the Air Force has not been giving out all available information on these unidentified flying objects. You cannot disregard so many unimpeachable sources.' John McCormack, Speaker of the U.S. House of Representatives, and . . .

'Reliable reports indicate that there are objects coming into our atmosphere at very high speeds. The way they change position would indicate that they are directed.' Rear Admiral D. S. Fahrney, U.S. Naval Missile Programme Chief.

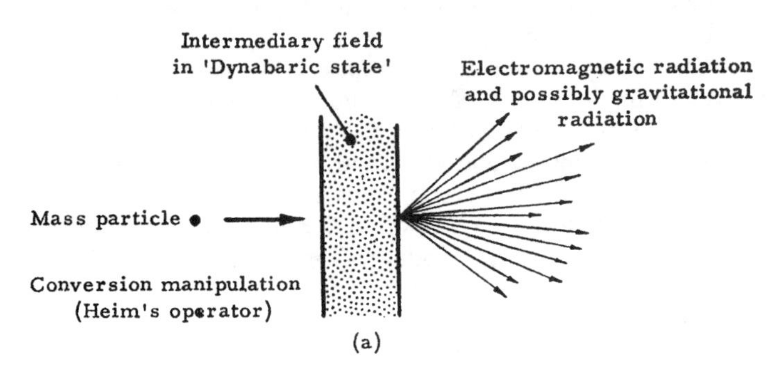

Loss-free or low-loss conversion of mass into energy in accordance with B. Heim's comprehensive field theory.

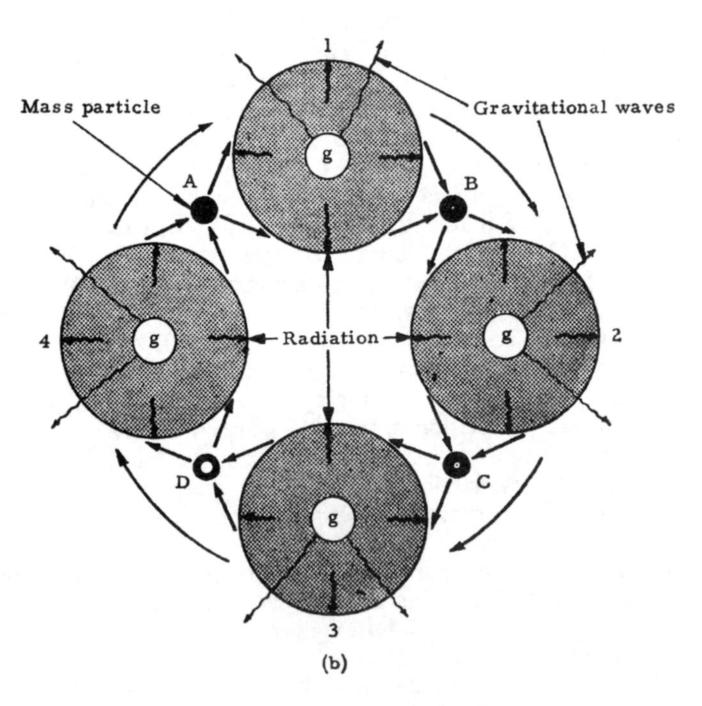

Mass-energy conversion, regeneration cycle.
Fig 107.

There is, of course, far more to the flying saucer story than the mere mechanical aspect I have chosen to demonstrate; in truth the story only begins here. It had been my hope to talk about a personal experience of thought-provoking nature, of the Mother ships, why I have not theorised about them in this text, more about the energy the discs use, the little signposts which indicate where those who want to, may look for it themselves. I wanted to offer my own findings on where the big carriers may be coming from; how they come; nay, why they bother to come at all. I had planned to offer this for those who would care to listen at the end of this book; indeed, some of it has already been written. But in presenting the technically corroborative evidence for the flying saucers, I have found too much to say, too little space to say it in, and thinking about it now, I feel it right and proper to let it stand for what it is, so that my carefully planned contents sheet has had to be cut right through the middle. Reflecting on it with some regret, I am partially reconciled into thinking that whereas what has been offered in these pages may at least make a few more people think, what was to come after would almost certainly have put some of them off. For, to some extent, it represents a different order. Its proper place must be between separate covers. In making this decision, I realise that many folks who might have borne the semi-technical nature of this work in order to read what was to follow, will not now do so. To them I can only extend my apologies with my feeling of deep privilege, and renew my efforts to complete the second half—their half—of a fascinating story.

INDEX

E—cont.

F

G

H

I

J

K

L

M

UFOS AND ANTI-GRAVITY

Appendix

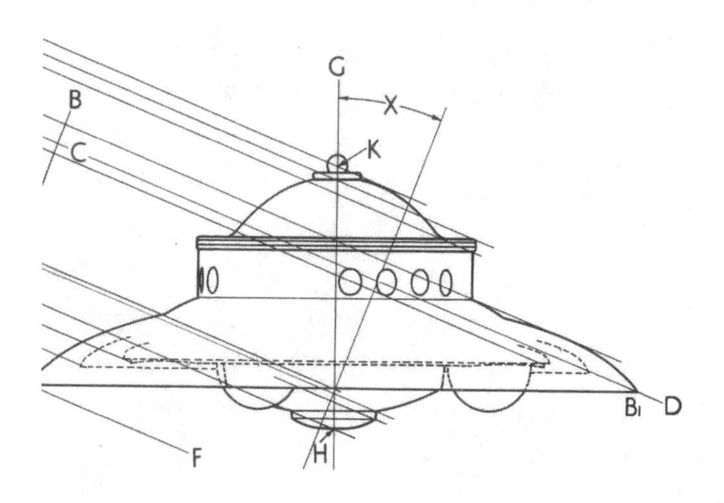

From Leonard Cramp's book *Space, Gravity and the Flying Saucer* (1955).

Space, Gravity & THE FLYING SAUCER

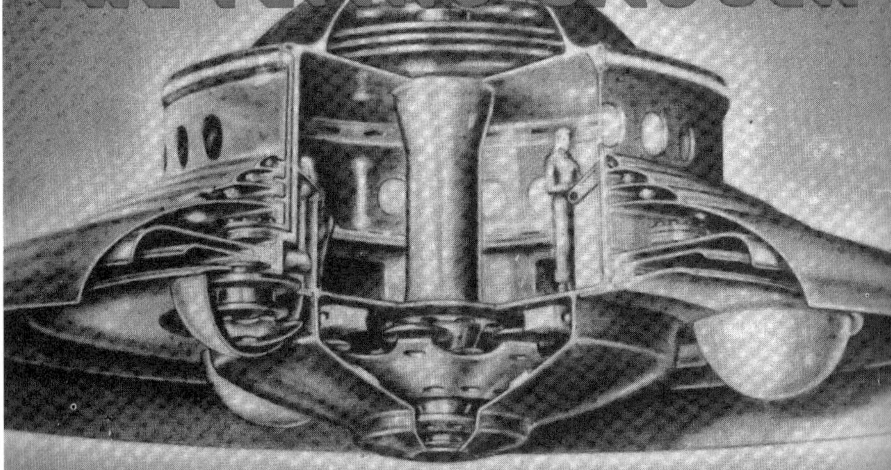

Introduction by Desmond Leslie

LEONARD G. CRAMP

Back cover of Leonard Cramp's book *Space, Gravity and the Flying Saucer* (1955) showing the flying saucers and cigar-shaped motherships with windows that we are so familiar with today.

Plate 4 Electromagnetic Repulsion of a Conductor

Permission of Rider and Company

Plate 5 Levitation of a heavy table

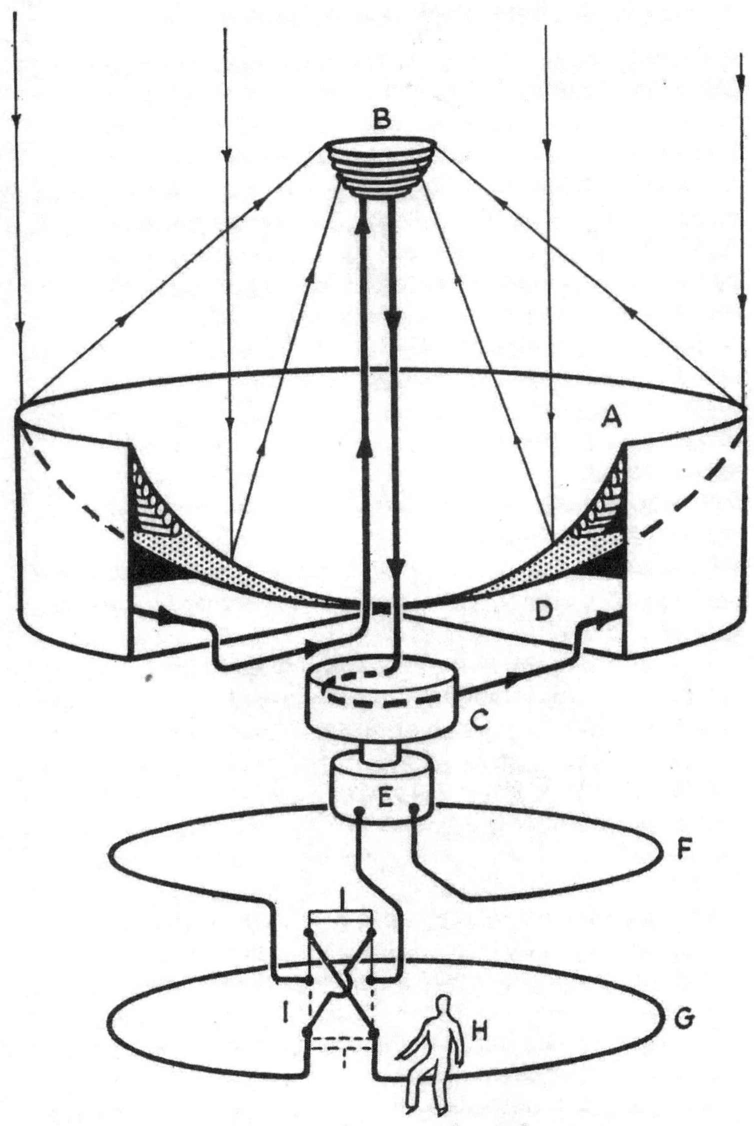

Fig. 6 Thermo-electric analogy of gravity

From Leonard Cramp's book *Space, Gravity and the Flying Saucer* (1955).

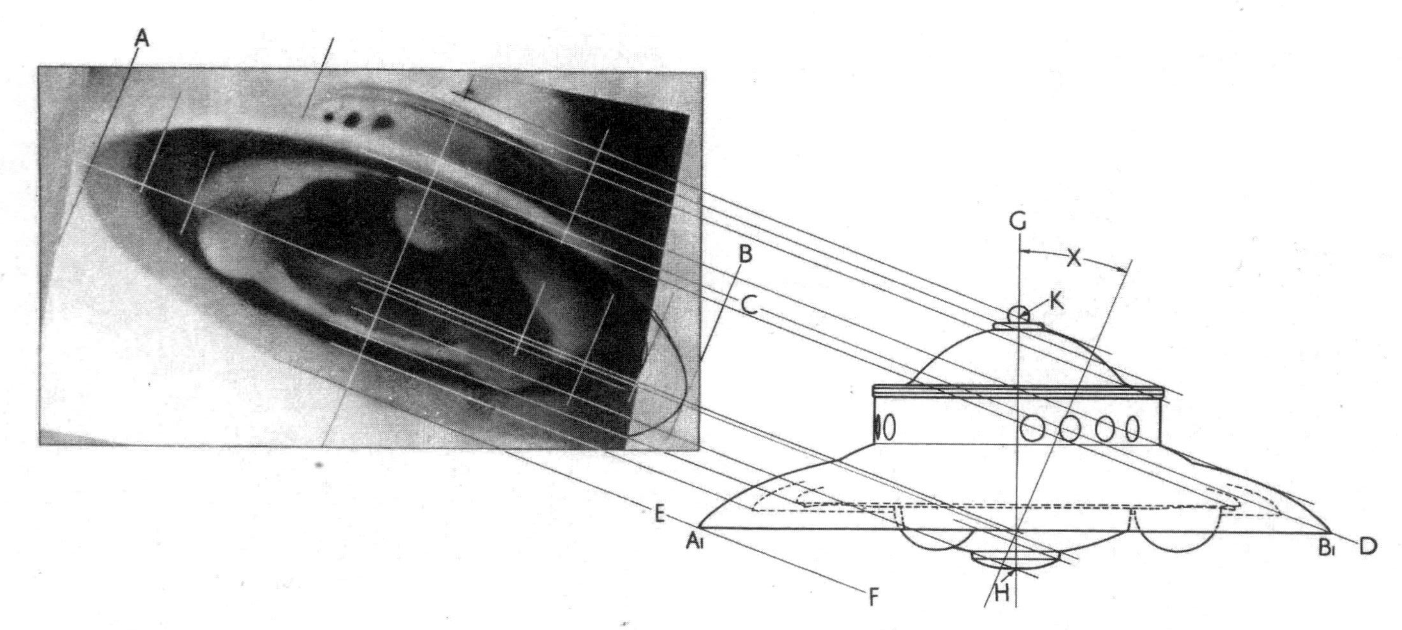

Fig 15a Comparison by orthographic projection of the Adamski (top) and Coniston (lower) photographs

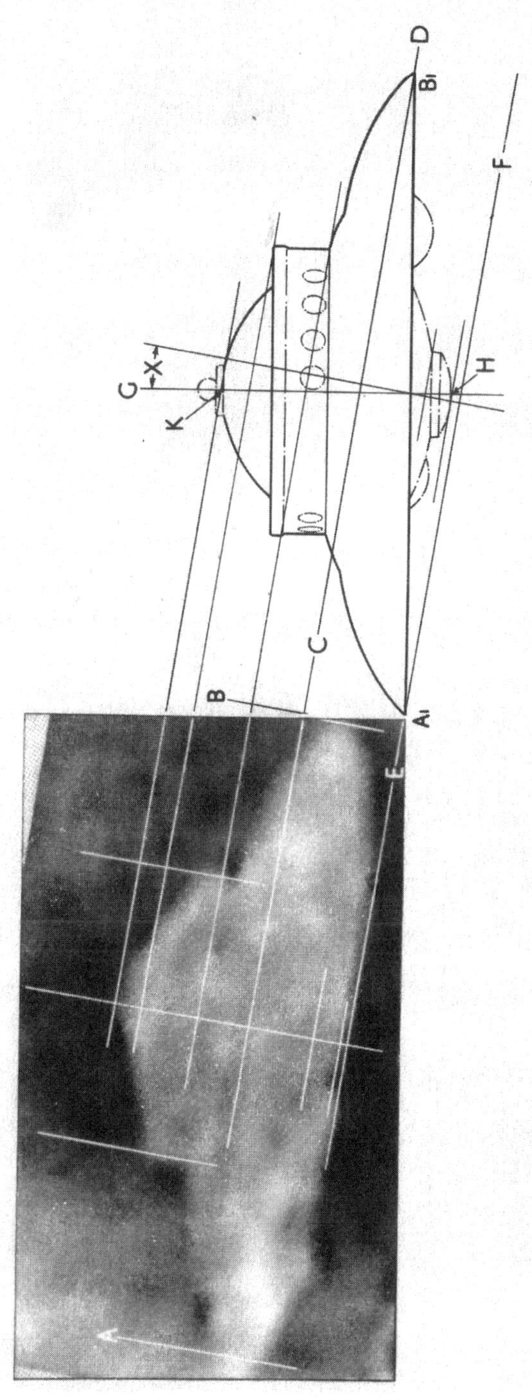

From Leonard Cramp's book *Space, Gravity and the Flying Saucer* (1955).

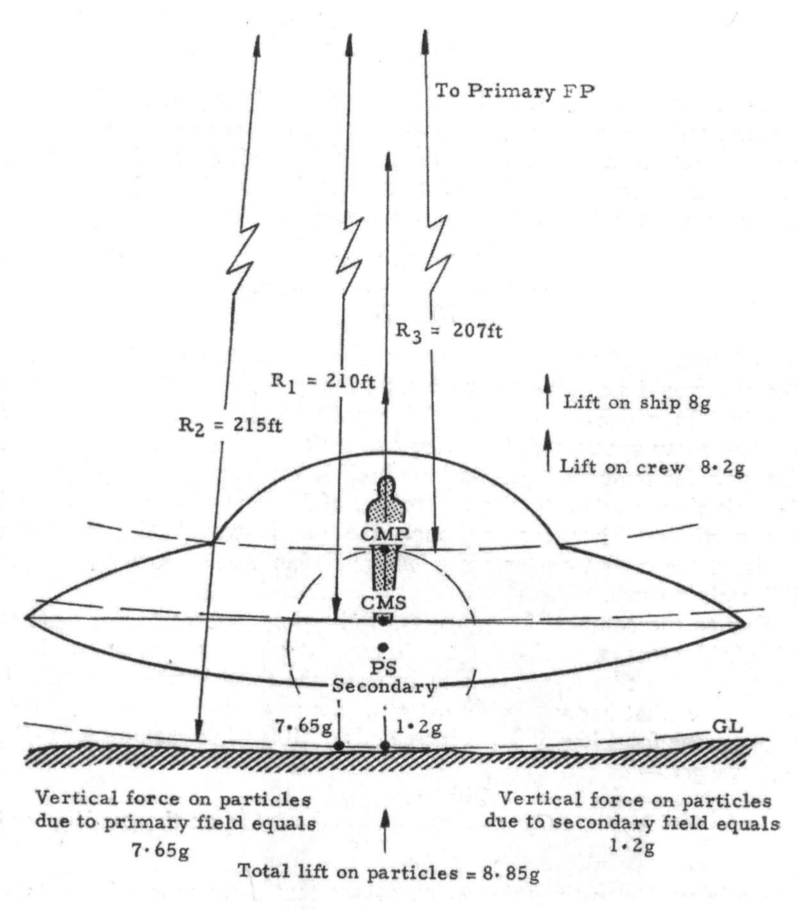

From Leonard Cramp's book *Space, Gravity and the Flying Saucer* (1955).

www.AdventuresUnlimitedPress.com

VIMANA:
Flying Machines of the Ancients
by David Hatcher Childress

According to early Sanskrit texts the ancients had several types of airships called vimanas. Like aircraft of today, vimanas were used to fly through the air from city to city; to conduct aerial surveys of uncharted lands; and as delivery vehicles for awesome weapons. David Hatcher Childress, popular *Lost Cities* author and star of the History Channel's long-running show Ancient Aliens, takes us on an astounding investigation into tales of ancient flying machines. In his new book, packed with photos and diagrams, he consults ancient texts and modern stories and presents astonishing evidence that aircraft, similar to the ones we use today, were used thousands of years ago in India, Sumeria, China and other countries. Includes a 24-page color section.

408 Pages. 6x9 Paperback. Illustrated. $22.95. Code: VMA

MIND CONTROL AND UFOS:
Casebook on Alternative 3
By Jim Keith

Keith's classic investigation of the Alternative 3 scenario as it first appeared on British television over 20 years ago. Keith delves into the bizarre story of Alternative 3, including mind control programs, underground bases not only on the Earth but also on the Moon and Mars, the real origin of the UFO problem, the mysterious deaths of Marconi Electronics employees in Britain during the 1980s, the Russian-American superpower arms race of the 50s, 60s and 70s as a massive hoax, more.

248 Pages. 6x9 Paperback. Illustrated. $14.95. Code: MCUF

MAN-MADE UFOS
WWII's Secret Legacy
By Renato Vesco & David Hatcher Childress

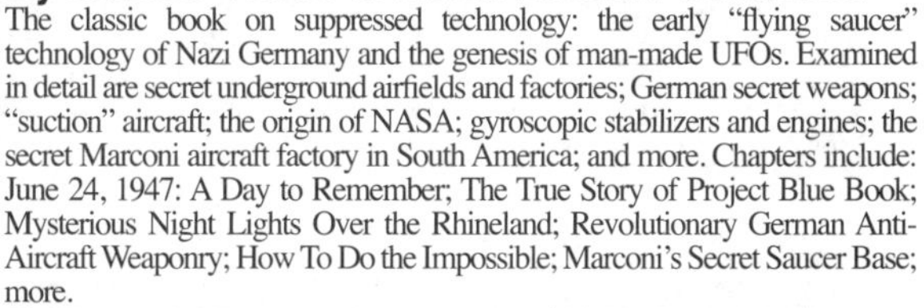

The classic book on suppressed technology: the early "flying saucer" technology of Nazi Germany and the genesis of man-made UFOs. Examined in detail are secret underground airfields and factories; German secret weapons; "suction" aircraft; the origin of NASA; gyroscopic stabilizers and engines; the secret Marconi aircraft factory in South America; and more. Chapters include: June 24, 1947: A Day to Remember; The True Story of Project Blue Book; Mysterious Night Lights Over the Rhineland; Revolutionary German Anti-Aircraft Weaponry; How To Do the Impossible; Marconi's Secret Saucer Base; more.

524 Pages. 6x9 Paperback. Illustrated. $22.95. Code: MMU

ROSWELL AND THE REICH
By Joseph P. Farrell

Farrell comes to a radically different scenario of what happened in Roswell in July 1947, and why the US military has continued to cover it up to this day. Farrell presents a fascinating case that what crashed may have been representative of an independent postwar Nazi power—an extraterritorial Reich monitoring its old enemy, America, and the continuing development of the very technologies confiscated from Germany at the end of the War.

540 pages. 6x9 Paperback. $19.95. Code: RWR

INVISIBLE RESIDENTS
The Reality of Underwater UFOS
by Ivan T. Sanderson

Sanderson puts forward the curious theory that "OINTS"—Other Intelligences—live under the Earth's oceans. This underwater, parallel, civilization may be twice as old as Homo sapiens, he proposes, and may have "developed what we call space flight." Sanderson postulates that the OINTS are behind many UFO sightings as well as the mysterious disappearances of aircraft and ships in the Bermuda Triangle. What better place to have an impenetrable base than deep within the oceans of the planet? Sanderson offers here an exhaustive study of USOs (Unidentified Submarine Objects) observed in nearly every part of the world.

298 PAGES. 6x9 PAPERBACK. ILLUSTRATED. $16.95. CODE: INVS

THE ENERGY GRID
Harmonic 695, The Pulse of the Universe
by Captain Bruce Cathie

This is the breakthrough book that explores the incredible potential of the Energy Grid and the Earth's Unified Field all around us. Bruce Cathie has been the premier investigator into the amazing potential of the infinite energy that surrounds our planet every microsecond. Cathie investigates the Harmonics of Light and how the Energy Grid is created. In this amazing book are chapters on UFO Propulsion, Nikola Tesla, Unified Equations, the Mysterious Aerials, Pythagoras & the Grid, Nuclear Detonation and the Grid, Maps of the Ancients, an Australian Stonehenge examined, more.

255 PAGES. 6x9 TRADEPAPER. ILLUSTRATED. $15.95. CODE: TEG

THE BRIDGE TO INFINITY
Harmonic 371244
by Captain Bruce Cathie

Cathie has popularized the concept that the earth is crisscrossed by an electromagnetic grid system that can be used for anti-gravity, free energy, levitation and more. The book includes a new analysis of the harmonic nature of reality, acoustic levitation, pyramid power, harmonic receiver towers and UFO propulsion. It concludes that today's scientists have at their command a fantastic store of knowledge with which to advance the welfare of the human race.

204 PAGES. 6x9 TRADEPAPER. ILLUSTRATED. $14.95. CODE: BTF

THE HARMONIC CONQUEST OF SPACE
by Captain Bruce Cathie

Chapters include: Mathematics of the World Grid; the Harmonics of Hiroshima and Nagasaki; Harmonic Transmission and Receiving; the Link Between Human Brain Waves; the Cavity Resonance between the Earth; the Ionosphere and Gravity; Edgar Cayce—the Harmonics of the Subconscious; Stonehenge; the Harmonics of the Moon; the Pyramids of Mars; Nikola Tesla's Electric Car; the Robert Adams Pulsed Electric Motor Generator; Harmonic Clues to the Unified Field; and more. Also included are tables showing the harmonic relations between the earth's magnetic field, the speed of light, and anti-gravity/gravity acceleration at different points on the earth's surface.

248 PAGES. 6x9. PAPERBACK. ILLUSTRATED. $16.95. CODE: HCS

TECHNOLOGY OF THE GODS
The Incredible Sciences of the Ancients
by David Hatcher Childress

Popular *Lost Cities* author David Hatcher Childress takes us into the amazing world of ancient technology, from computers in antiquity to the "flying machines of the gods." Childress looks at the technology that was allegedly used in Atlantis and the theory that the Great Pyramid of Egypt was originally a gigantic power station. He examines tales of ancient flight and the technology that it involved; how the ancients used electricity; megalithic building techniques; the use of crystal lenses and the fire from the gods; evidence of various high tech weapons in the past, including atomic weapons; ancient metallurgy and heavy machinery; the role of modern inventors such as Nikola Tesla in bringing ancient technology back into modern use; impossible artifacts; and more.

356 PAGES. 6X9 PAPERBACK. ILLUSTRATED. $16.95. CODE: TGOD

THE COSMIC MATRIX
Piece for a Jig-Saw, Part Two
by Leonard G. Cramp

Cramp examines anti-gravity effects and theorizes that this super-science used by the craft—described in detail in the book—can lift mankind into a new level of technology, transportation and understanding of the universe. The book takes a close look at gravity control, time travel, and the interlocking web of energy between all planets in our solar system with Leonard's unique technical diagrams. A fantastic voyage into the present and future!

364 PAGES. 6X9 PAPERBACK. ILLUSTRATED. $16.00. CODE: CMX

THE TESLA PAPERS
Nikola Tesla on Free Energy & Wireless Transmission of Power
edited by David Hatcher Childress

David Hatcher Childress takes us into the incredible world of Nikola Tesla and his amazing inventions. Tesla's fantastic vision of the future, including wireless power, anti-gravity, free energy and highly advanced solar power. Also included are some of the papers, patents and material collected on Tesla at the Colorado Springs Tesla Symposiums, including papers on: •The Secret History of Wireless Transmission •Tesla and the Magnifying Transmitter •Design and Construction of a Half-Wave Tesla Coil •Electrostatics: A Key to Free Energy •Progress in Zero-Point Energy Research •Electromagnetic Energy from Antennas to Atoms •Tesla's Particle Beam Technology •Fundamental Excitatory Modes of the Earth-Ionosphere Cavity

325 PAGES. 8X10 PAPERBACK. ILLUSTRATED. $16.95. CODE: TTP

PATH OF THE POLE
by Charles Hapgood

Hapgood researched Antarctica, ancient maps and the geological record to conclude that the Earth's crust has slipped in the inner core many times in the past, changing the position of the pole. *Path of the Pole* discusses the various "pole shifts" in Earth's past, giving evidence for each one, and moves on to possible future pole shifts. Packed with illustrations, this is the sourcebook for many other books on cataclysms and pole shifts.

356 PAGES. 6X9 PAPERBACK. ILLUSTRATED. $16.95. CODE: POP

ARKTOS

The Polar Myth in Science, Symbolism & Nazi Survival

by Joscelyn Godwin

Explored are the many tales of an ancient race said to have lived in the Arctic regions, such as Thule and Hyperborea. Progressing onward, he looks at modern polar legends: including the survival of Hitler, German bases in Antarctica, UFOs, the hollow earth, and the hidden kingdoms of Agartha and Shambala. Chapters include: The Golden Age; The Northern Lights; The Arctic Homeland; The Aryan Myth; The Thule Society; The Black Order; The Hidden Lands; Agartha and the Polaires; Shambhala; The Hole at the Pole; Antarctica; more.

220 Pages. 6x9 Paperback. Illustrated. $16.95. Code: ARK

SECRETS OF THE HOLY LANCE

The Spear of Destiny in History & Legend

by Jerry E. Smith

Secrets of the Holy Lance traces the Spear from its possession by Constantine, Rome's first Christian Caesar, to Charlemagne's claim that with it he ruled the Holy Roman Empire by Divine Right, and on through two thousand years of kings and emperors, until it came within Hitler's grasp—and beyond! Did it rest for a while in Antarctic ice? Is it now hidden in Europe, awaiting the next person to claim its awesome power? Neither debunking nor worshiping, *Secrets of the Holy Lance* seeks to pierce the veil of myth and mystery around the Spear.

312 PAGES. 6X9 PAPERBACK. ILLUSTRATED. BIBLIOGRAPHY. $16.95. CODE: SOHL

THE CRYSTAL SKULLS

Astonishing Portals to Man's Past

by David Hatcher Childress and Stephen S. Mehler

Childress introduces the technology and lore of crystals, and then plunges into the turbulent times of the Mexican Revolution form the backdrop for the rollicking adventures of Ambrose Bierce, the renowned journalist who went missing in the jungles in 1913, and F.A. Mitchell-Hedges, the notorious adventurer who emerged from the jungles with the most famous of the crystal skulls. Mehler shares his extensive knowledge of and experience with crystal skulls. Having been involved in the field since the 1980s, he has personally examined many of the most influential skulls, and has worked with the leaders in crystal skull research. Color section.

294 pages. 6x9 Paperback. Illustrated. $18.95. Code: CRSK

THE ANTI-GRAVITY HANDBOOK

edited by David Hatcher Childress

The new expanded compilation of material on Anti-Gravity, Free Energy, Flying Saucer Propulsion, UFOs, Suppressed Technology, NASA Cover-ups and more. Highly illustrated with patents, technical illustrations and photos. This revised and expanded edition has more material, including photos of Area 51, Nevada, the government's secret testing facility. This classic on weird science is back in a new format!

230 PAGES. 7X10 PAPERBACK. ILLUSTRATED. $16.95. CODE: AGH

ARK OF GOD
The Incredible Power of the Ark of the Covenant
By David Hatcher Childress

Childress takes us on an incredible journey in search of the truth about (and science behind) the fantastic biblical artifact known as the Ark of the Covenant. This object made by Moses at Mount Sinai—part wooden-metal box and part golden statue—had the power to create "lightning" to kill people, and also to fly and lead people through the wilderness. The Ark of the Covenant suddenly disappears from the Bible record and what happened to it is not mentioned. Was it hidden in the underground passages of King Solomon's temple and later discovered by the Knights Templar? Was it taken through Egypt to Ethiopia as many Coptic Christians believe? Childress looks into hidden history, astonishing ancient technology, and a 3,000-year-old mystery that continues to fascinate millions of people today. Color section.

420 Pages. 6x9 Paperback. Illustrated. $22.00 Code: AOG

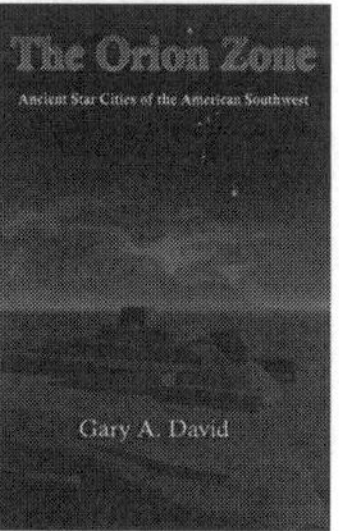

THE ORION ZONE
Ancient Star Cities of the American Southwest
by Gary A. David

This book on ancient star lore explores the mysterious location of Pueblos in the American Southwest, circa 1100 AD, that appear to be a mirror image of the major stars of the Orion constellation. Chapters include: Leaving Many Footprints—The Emergence and Migrations of the Anazazi; The Sky Over the Hopi Villages; Orion Rising in the Dark Crystal; The Cosmo-Magical Cities of the Anazazi; Windows Onto the Cosmos; To Calibrate the March of Time; They Came from Across the Ocean—The Patki (Water) Clan and the Snake Clan of the Hopi; Ancient and Mysterious Monuments; Beyond That Fiery Day; more.

346 pages. 6x9 Paperback. Illustrated. $19.95. Code: OZON

ATLANTIS & THE POWER SYSTEM OF THE GODS
by David Hatcher Childress and Bill Clendenon

Childress' fascinating analysis of Nikola Tesla's broadcast system in light of Edgar Cayce's "Terrible Crystal" and the obelisks of ancient Egypt and Ethiopia. Includes: Atlantis and its crystal power towers that broadcast energy; how these incredible power stations may still exist today; inventor Nikola Tesla's nearly identical system of power transmission; Mercury Proton Gyros and mercury vortex propulsion; more. Richly illustrated, and packed with evidence that Atlantis not only existed—it had a world-wide energy system more sophisticated than ours today.

246 PAGES. 6x9 PAPERBACK. ILLUSTRATED. $15.95. CODE: APSG

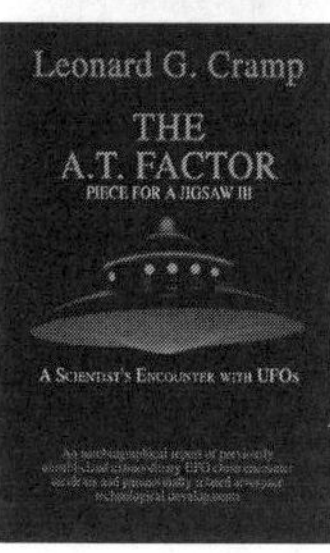

THE A.T. FACTOR
A Scientists Encounter with UFOs
by Leonard Cramp

British aerospace engineer Cramp began much of the scientific anti-gravity and UFO propulsion analysis back in 1955 with his landmark book *Space, Gravity & the Flying Saucer* (out-of-print and rare). In this final book, Cramp brings to a close his detailed and controversial study of UFOs and Anti-Gravity.

324 PAGES. 6X9 PAPERBACK. ILLUSTRATED. $16.95. CODE: ATF

ORDER FORM

10% Discount When You Order 3 or More Items!

One Adventure Place
P.O. Box 74
Kempton, Illinois 60946
United States of America
Tel.: 815-253-6390 • Fax: 815-253-6300
Email: auphq@frontiernet.net
http://www.adventuresunlimitedpress.com

ORDERING INSTRUCTIONS

✓ Remit by USD$ Check, Money Order or Credit Card
✓ Visa, Master Card, Discover & AmEx Accepted
✓ Paypal Payments Can Be Made To:
info@wexclub.com
✓ Prices May Change Without Notice
✓ 10% Discount for 3 or More Items

SHIPPING CHARGES

United States

✓ Postal Book Rate { $4.50 First Item / 50¢ Each Additional Item
✓ POSTAL BOOK RATE Cannot Be Tracked! Not responsible for non-delivery.
✓ Priority Mail { $6.00 First Item / $2.00 Each Additional Item
✓ UPS { $7.00 First Item / $1.50 Each Additional Item
NOTE: UPS Delivery Available to Mainland USA Only

Canada

✓ Postal Air Mail { $15.00 First Item / $2.50 Each Additional Item
✓ Personal Checks or Bank Drafts MUST BE US$ and Drawn on a US Bank
✓ Canadian Postal Money Orders OK
✓ Payment MUST BE US$

All Other Countries

✓ Sorry, No Surface Delivery!
✓ Postal Air Mail { $19.00 First Item / $6.00 Each Additional Item
✓ Checks and Money Orders MUST BE US$ and Drawn on a US Bank or branch.
✓ Paypal Payments Can Be Made in US$ To: info@wexclub.com

SPECIAL NOTES

✓ RETAILERS: Standard Discounts Available
✓ BACKORDERS: We Backorder all Out-of-Stock Items Unless Otherwise Requested
✓ PRO FORMA INVOICES: Available on Request
✓ DVD Return Policy: Replace defective DVDs only

ORDER ONLINE AT: www.adventuresunlimitedpress.com

10% Discount When You Order 3 or More Items!

Please check: ✓

☐ This is my first order ☐ I have ordered before

Name
Address
City
State/Province | Postal Code
Country
Phone: Day | Evening
Fax | Email

Item Code	Item Description	Qty	Total

Please check: ✓

Subtotal ▸
Less Discount-10% for 3 or more items ▸

☐ Postal-Surface — Balance ▸
☐ Postal-Air Mail (Priority in USA) — Illinois Residents 6.25% Sales Tax ▸
Previous Credit ▸
☐ UPS (Mainland USA only) — Shipping ▸
Total (check/MO in USD$ only) ▸

☐ Visa/MasterCard/Discover/American Express

Card Number:
Expiration Date: Security Code:

✓ SEND A CATALOG TO A FRIEND: